E. L. McFarland

PROF. E. L. McFARLAND
DEPARTMENT OF PHYSICS
COLLEGE OF PHYSICAL SCIENCE
UNIVERSITY OF GUELPH
GUELPH, ONTARIO   N1G 2W1

Environmental Radioactivity

From Natural, Industrial,
and Military Sources

THIRD EDITION

Environmental Radioactivity

*From Natural, Industrial,
and Military Sources*

THIRD EDITION

Merril Eisenbud

Professor Emeritus
New York University Medical Center
Institute of Environmental Medicine
New York, New York

1987

ACADEMIC PRESS, INC.

Harcourt Brace Jovanovich, Publishers

Orlando   San Diego   New York   Austin
Boston   London   Sydney   Tokyo   Toronto

ACADEMIC PRESS, INC.
Orlando, Florida 32887

United Kingdom Edition published by
ACADEMIC PRESS INC. (LONDON) LTD.
24–28 Oval Road, London NW1 7DX

Library of Congress Cataloging in Publication Data

Eisenbud, Merril.
    Environmental radioactivity.

    Bibliography: p.
    Includes index.
    1. Radioactive pollution. I. Title.
TD196.R3E57   1987   363.1'79   86-17375
ISBN 0–12–235153–3 (alk. paper)

PRINTED IN THE UNITED STATES OF AMERICA

86 87 88 89   9 8 7 6 5 4 3 2 1

Contents

v

14. Accidents That Resulted in Contamination of the
    Environment

15. Radiation Exposure and Risk: Some Contemporary Social
    Aspects

Appendix. The Properties of Certain Radionuclides

Preface

The first edition of this book was published in 1963, at a time when interest in environmental radioactivity was focused on fallout from testing nuclear weapons. An agreement to ban testing in the atmosphere was signed in that year by the major nuclear powers, and would have resulted in diminished interest in the subject except that 1963 was also when the first commercial order for a privately owned nuclear power plant was placed by a utility company in the United States. Attention thus shifted from the effects of nuclear weapons testing on the environment to the effects of nuclear power plants, an unfortunate coupling that may explain some of the apprehensions of the public about the dangers from the civilian uses of atomic energy.

When the second edition was published in 1973, the nuclear power industry was in a phase of rapid growth that was sustained until the late 1970s, when orders for new nuclear power plants ceased in the United States and many existing orders were canceled. This was due to many factors, including reduced demand for energy as a result of economic recession, greater energy conservation, growing public opposition to nuclear power, inflation of construction costs, and finally the 1979 accident that destroyed Reactor 2 at the nuclear power plant at Three Mile Island, Pennsylvania. Information about the catastrophic accident at Chernobyl in the U.S.S.R. came to the world's attention in the final stages of preparation of this manuscript. This disaster is certain to further complicate the future of nuclear power.

This edition is being written nearly twenty-five years after the first one was published, and it is going to press at a time of great uncertainty with respect to the future of both the military and civilian uses of nuclear energy. Above all is the continuing danger of nuclear war. The stockpiles of nuclear weapons continue to increase, and delivery systems of greater sophistication are being developed. Public anxiety about the products of nuclear technology is to be expected as long as the major powers have the capacity to cover much of the world with life-threatening amounts of radioactivity.

The civilian nuclear energy industry has established an impressive safety record, but it is beset with economic, regulatory, and political problems that will not be resolved easily. This is even true of such benign applications as those in medical research but the problems are most acute in the electric utility industry. However, the wisdom of continued dependence on fossil fuels is also being questioned increasingly because of potential serious long-term environmental effects. Since substitution of renewable resources such as solar energy does not seem feasible for the foreseeable future, it is to be hoped that the present hiatus in the development of nuclear power will be used to design reactors that will be more acceptable to the public. There may be no other choice if the world is to meet its future energy needs.

There is one feature of this text that requires a word of explanation. After considerable thought and many discussions with my colleagues, I made the decision to avoid the use of SI units. I realized that in this interim period when the new units are coming rapidly into general use, it would be preferable to use both systems of units, with one in parentheses. This was in fact done in an early stage of manuscript preparation, but it caused many of the pages, and particularly those with complex tables, to become so cluttered that I decided to use only one system of units.

Many individuals have assisted me in the preparation of this work and its previous editions. In the preparation of this edition, I am especially grateful for the many suggestions I have received from John Auxier, Beverly Cohen, Norman Cohen, Thomas Gesell, Frank Gifford, Edward Hardy, Catherine Klusek, Paul Linsalata, Frank Parker, Norman Rasmussen, Charles Roessler, Keith Schiager, Lauriston Taylor, Arthur Upton, Herbert Volchok, and Robert Watters. As was true in the last edition, Eleanor Clemm labored diligently in my behalf, not only as an excellent typist, but also as editor, bibliographic assistant, and general coordinator.

I wish to express my appreciation to the Office of Energy and Health

Research of the U.S. Department of Energy for a grant awarded me to support the cost of manuscript preparation. I am also indebted to the National Institute of Environmental Health Sciences for the many years of support I received under Center Grant No. ES 00260. Finally, it is with pleasure and pride that I record my indebtedness to the Institute of Environmental Medicine of the New York University Medical Center, where I recently completed a long and pleasant association with stimulating members of the faculty, staff, and student body.

MERRIL EISENBUD

Chapel Hill, North Carolina

Preface to Second Edition

The first edition of this book was published early in 1963 (McGraw-Hill, New York) at a time when worldwide concern existed because of radioactive fallout from testing nuclear weapons. In that year the principal nuclear powers agreed to ban open-air testing, and the environmental levels of radioactivity from that source have accordingly diminished considerably since then despite occasional tests by countries that did not participate in the test-ban agreement. However, interest in the subject of environmental radioactivity has not diminished correspondingly, but has been more than sustained by the rapid growth of the nuclear power industry and the use of radioactive materials in medicine, research, space exploration, and industry.

When the original edition was published, there were only four commercially operated power reactors in the United States, with an installed generating capacity of about 641 megawatts. At this writing, the generating capacity of power plants already in operation in the United States is about 13,400 megawatts and an additional 116,000 megawatts are in the planning stage. Nuclear power will represent an increasing fraction of the electrical generating capacity in many countries, and the potential environmental impact of this new industry is receiving wide attention.

As this second edition goes to press, the United States continues to be

Ever since early in World War II, extensive research has been conducted to understand the physical and chemical properties of radioactive substances, the manner in which they are transported physically through the environment, and the way in which some of them enter into man's food supplies, the water he drinks, and the air he breathes. In 1959, when I accepted an opportunity to develop a graduate teaching program in the general field of radiological hygiene, I was impressed with the need to consolidate the vast amount of information that had been developed on the subject of environmental radioactivity in order that the subject could be presented to students and others in a comprehensive and systematic manner.

The subject of environmental radioactivity has aspects of vast dimensions, and the task of bringing together the pertinent information in so many diverse disciplines proved to be not without its difficulties. There was first the question of what to include, and in this regard I decided that the text should be concerned primarily with the behavior of radioactive substances when they enter the environment. The important and elaborate technology by which passage of radioactive materials to the environment may be prevented and the equally important field of health physics that is concerned with protecting the atomic energy worker were thus placed beyond the bounds of this work, although it has been necessary frequently to deal briefly with both subjects in the present text.

I am greatly indebted to my many associates, past and present, who assisted me in the preparation of this work. It is not possible to acknowledge all the assistance I have received, but certain of my colleagues have been particularly helpful in the review of early drafts of certain chapters. In this regard I am particularly indebted for the help of Norton Nelson, Roy E. Albert, Abraham S. Goldin, Bernard S. Pasternack, Gerard R. Laurer, Harold H. Rossi and Ben Davidson. The index was prepared by Stephen F. Cleary, and every word of every one of the three to five drafts that ultimately evolved into the seventeen chapters of this book was typed with remarkable proficiency and patience by Patricia S. Richtmann. Finally, as is customary for reasons that can only be understood fully by authors, and authors' families, I wish to express my appreciation for the help and encouragement provided by my wife, Irma, to whom this book is dedicated.

MERRIL EISENBUD

Environmental
Radioactivity

From Natural, Industrial,
and Military Sources

THIRD EDITION

Chapter 1

Introduction

The discovery in 1939 that energy contained within the atomic nucleus can be released has led to some of the most far-reaching technical developments in human history. It so happened that the discovery of nuclear fission coincided with the outbreak of World War II, and the first application of nuclear energy was therefore for military purposes. The dramatic announcement of the destruction of Hiroshima by a nuclear weapon created an image of the destructive power of "the atom" that has been imprinted indelibly on the consciousness of the world's citizens. The linkage of nuclear energy and nuclear war is certainly a major factor in the widespread public opposition to even the most benign applications of this new form of energy and its by-products.

The most obvious long-range benefit from the fission process was that it promised to provide society with a source of power that would ensure a higher standard of living in countries that do not have adequate reserves of fossil fuel. In time, as the energy requirements of the world increase and as the reserves of fossil fuel become smaller, nuclear energy is bound to play an increasingly important role in civilian economies. At present, nearly half a century since the feasibility of operating a nuclear reactor was first demonstrated in 1942, nuclear is, in fact, producing a large fraction of the electricity used in many countries of the world. However, a combination of public apprehension about the effects of nuclear reactors, a worldwide recession that started in the mid-1970s, and an unforeseen rise in the cost of building nuclear power plants has resulted in far less dependence on nuclear energy in the United States than was anticipated in 1973, when the last edition of this book was published. A major factor in this change was the accident that occurred at Three Mile Island Unit II in March 1979 (Chapter 14).

A less conspicuous but highly important contribution of atomic energy to mankind has been the copious quantities of radionuclides that have become available. In the fields of medical and biological research, the use of "radioactive isotopes" is now so commonplace that we are no longer conscious of the research that has become possible because these useful substances have been made available. The use of radionuclides as a research tool has entered rapidly and yet with comparative unobtrusiveness into our research laboratories, and hundreds of discoveries in the biomedical sciences would not have been possible if it were not for the ready availability of radionuclides.

Regrettably, we cannot forget the negative side to nuclear energy—the possibility of nuclear war. As with many great technological advances, humans have the wisdom to use their new knowledge constructively if they wish to do so, but they are also capable of great folly. Only time will tell whether, on balance, nuclear energy has been used to bring blessings to mankind or to hasten its social destruction.

The Early History of Radioactivity

Experience with the dangers of radioactive materials preceded by many years the discovery of the phenomenon of radioactivity. As will be seen in Chapter 2, the atmospheres of mines in Central Europe that had been exploited for their heavy metals since medieval times were unknowingly so radioactive that the miners developed a fatal lung disease which was later diagnosed as lung cancer (Lorenz, 1944). It was not until nearly 400 years after the German physician Georgius Agricola wrote his classic book "De Re Metallica", in which he described the high mortality among the miners, that it became known that the air of the mines contained the radioactive emanations of radium (Agricola, 1556).

Some radioactive substances were used even before it was known that they were radioactive. The Welsbach gas mantle, which was developed in 1885, utilized the incandescent properties of thorium–cerium oxide to greatly increase the luminosity of gaslight in many parts of the world, and uranium oxide has long been used to provide a vivid orange color in ceramic glazes. Other oxides of uranium and thorium have also been used as glazes and for tinting glass (see Chapter 10).

Since the turn of the century and continuing up to the present time in some parts of the world, natural radioactivity has been exploited for its supposed benefit to health (see Chapter 7). There is no fully satisfactory explanation for how this custom originated, but it is known that the popularity of mineral waters in spas around the world led to their establishment as health resorts as long ago as Roman times. It is also known that the laxative properties of spring waters having a high mineral content were highly prized, and extensive resorts grew up at places such as Saratoga Springs in New York State and the spas of Europe, Japan, and South America. When the phenomenon of radioactivity was discovered, tests of these mineral waters showed some of them to contain abnormally high concentrations of natural radioelements, and this fact may have given rise to the idea that the newly discovered and mysterious property of matter was the reason why the mineral waters seemed to possess curative powers. The discovery that the radiations from radium could destroy cancerous tissue possibly abetted development of the fad, and the spas of the world soon began to advertise the radioactivity of their waters. To this day the labels on bottled mineral waters in many countries contain measurements of the radioactivity of the spring from which the water was obtained. We will see later that the radioactive sands of Brazilian beaches and the high radon concentrations in the air of old mines in Austria and the United States have attracted tourists who believe that exposure to natural radioactivity can cure arthritis, general debility, and a variety of other diseases (Lewis, 1955; Scheminzky, 1961).

In the early 1920s and continuing up to about 1940, radioactive substances, particularly radium and radon, found a place in the medical faddism of the period. Radium was injected intravenously for a variety of ills but, far from being cured, many of the patients later developed bone cancer. Devices that were sold for home use made it possible to add radon to drinking water, and radioactive poultices were prescribed for arthritic joints (Fig. 1–1).

Finally, a word must be said about the early use of radium in luminous paints. It had been discovered that a slight amount of radium added to a suspension of zinc sulfide caused the material to scintillate, and the dials of thousands of timepieces, compasses, and other devices were painted with such paints during and immediately after World War I, with no precautions to protect the employees. Many cases of aplastic anemia and bone cancer developed among the factory workers engaged in applying these luminous paints prior to about 1940, when hygienic practices were developed which proved practical and prevented further injuries from

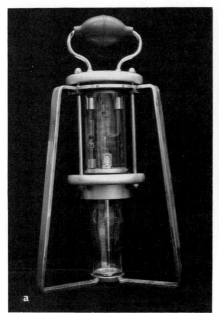

Fig. 1–1. Early uses of radioactivity in health fads. (a) The Radiumator: air was pumped by hand bulb through a small radium source, entraining radon, which was bubbled through the glass of drinking water. (b) The Revigator (patented 1912): the cone is a mildly radioactive "ore" which was placed in the drinking water crock. (c) Radioactive compress used for miscellaneous aches and pains. The compress contained 0.1 mg of ^{226}Ra and was certified by the Radium Institute of the Faculte des Sciences de Paris.

occurring. We will see in Chapter 2 that the information derived from studies of the early radium cases has contributed in a unique and effective way to the excellent safety record of the modern atomic energy program. The use of radioluminescent materials will be discussed more fully in Chapter 10.

The Atomic Energy Industry

This introductory chapter provides a convenient place in which to provide a bird's-eye view of the nuclear industry, the major components of which will be discussed in greater detail in later chapters.

One way to visualize the main parts of the atomic energy industry is by the flow diagram in Fig. 1–2. After being mined and concentrated (Chapter 9), the uranium is shipped to refineries for conversion to a uranium metal or oxide of sufficient purity to be used in reactors (Chapter 8). The refinery products may be shipped directly to fuel-element fabrication plants or the uranium may be converted to UF_4, a green salt, which is then converted to UF_6, a volatile corrosive compound. To produce uranium that is suitable for use in reactors, it is necessary to increase its ^{235}U content. Until the present time in the United States, this has been accomplished by the gaseous diffusion process in enormous plants at Oak Ridge, Tennessee; Paducah, Kentucky; and Portsmouth, Ohio. Depending on the purpose for which the uranium is intended, the degree of enrichment may vary from a few tenths of 1% to more than 90%. It is anticipated that the gaseous diffusion process in the United States will be replaced by a process that employs lasers. Some European countries are now using centrifuges for this purpose (Tait, 1983).

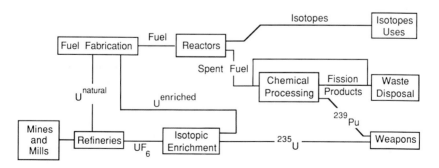

Fig. 1–2. The principal steps in the nuclear fuel cycle. The nuclear power industry in the United States has not processed spent fuel since the mid-1970s (see Chapter 8).

The uranium, either in natural or enriched form or as the metal, oxide, or other compounds or alloys, is then transported to the fuel-element fabrication plants. The exact shape to be taken by the uranium and the manner in which it will be clad with protective cladding of stainless steel, zirconium, or various alloys depend on the needs of the reactor designer.

The fabricated fuel elements are then shipped to reactors (Chapter 9), some of which are used for production of ^{239}Pu for military purposes. Other fuel elements are used in power reactors operated by the civilian power industry and the Navy. In addition, some of the fuel will be used by research reactors located in laboratories and industrial plants.

The products of these reactors may be heat, plutonium, radioisotopes, or radiations for research or industrial purposes. The spent fuel can either be treated as a solid waste and placed in a radioactive waste repository or transported to reprocessing plants, in which the fuel is dissolved and the unused uranium and ^{239}Pu recovered (Chapter 8). The fission products can be processed into a form convenient for waste storage and disposal (Chapter 11), but in some cases radioactive isotopes may be separated from these fission products for research, medical, or industrial applications (Chapter 10). U.S. policy has not encouraged fuel reprocessing since the late 1970s.

In addition to the radioactive by-products of the spent fuel-reprocessing plants, certain radionuclides are produced in research reactors by neutron irradiation. For example, the naturally occurring stable isotope ^{59}Co may be placed in a reactor to produce ^{60}Co. Many of the more common radioactive isotopes such as ^{14}C, ^{131}I, and ^{32}P are produced by neutron bombardment of the appropriate parent nuclide.

Early Studies of Radioactive Contamination of the Environment

A limited amount of information about environmental radioactivity was already available prior to World War II, but there was little diffusion of this knowledge beyond the relatively few highly specialized laboratories that were then equipped to make measurements of radioactivity. The world inventory of radioactive materials was confined to those found in nature, with the insignificant exception of a relatively few millicuries of artificial radioactivity that were produced in cyclotrons during the late 1930s.

During World War II, the construction of large, water-cooled, plutonium-producing reactors at Hanford in the state of Washington and the associated operations for extracting the plutonium from the irradiated uranium resulted in the first major possibilities for major contamination of the environment by radioactivity. This was also true to a lesser extent at two other major nuclear research and production centers at Oak Ridge, Tennessee, and Los Alamos, New Mexico. The early Hanford studies on the behavior of various radionuclides in the environment are classics in the field and have served to demonstrate, on the one hand, the caution one must adopt in discharging radioactive substances to the environment and, on the other hand, the fact that substantial quantities of radioactivity can be discharged safely if the properties of the individual radionuclides and their behavior in the environment are well understood.

The policies laid down by the Manhattan Engineering District (MED) of the Corps of Engineers (Groves, 1962), which was the World War II military organization responsible for the atomic energy program, placed a high priority on the importance of operating in such a way as to keep environmental contamination at a minimum. When the Atomic Energy Commission, a civilian organization, succeeded the MED in 1946 (Hewlett and Anderson, 1962), the cautious policies toward release of radioactive materials to the environment from industrial and research activities were continued, but starting in the late 1940s and continuing at an accelerated rate until 1962, there began a series of nuclear weapons tests in the atmosphere, first by the United States and then by the Soviet Union, the United Kingdom, France, India, and China, that discharged into the environment amounts of radioactivity that were enormous in relation to the prohibitions self-imposed by the AEC in the operation of its research and industrial facilities. The radioisotopes produced in the nuclear explosions in various parts of the world soon permeated the atmosphere, the soils, and the food chains to such an extent that widespread apprehension began to develop, first in certain scientific circles and later among the general public throughout the world. Responding to this concern, the Congress of the United States held a number of hearings on the subject of fallout from weapons testing and also on radioactive waste-disposal practices. At about the same time the National Academy of Sciences in the United States and the Medical Research Council in Great Britain undertook to assess the state of knowledge on the effects of small doses of radioactivity (National Academy of Sciences–National Research Council [NAS–NRC], 1956; Medical Research Council [MRC], 1956), and the United Nations in 1955 appointed a committee, consisting originally of scientific representatives of 15 nations (but later expanded), to investigate the effects of radiation on humans.

The United Nations Scientific Committee on the Effects of Atomic Radiation (UNSCEAR) has published a series of authoritative reports that are classics in international scientific collaboration (UNSCEAR, 1982). The widespread interest in the subject of environmental radioactivity resulted in acceleration of research on the behavior of trace substances in the environment. Many branches of the biological and physical sciences, including genetics, inorganic chemistry, trace-element metabolism, micro-meterology, upper-atmosphere meteorology, and oceanography, have made great forward strides because of the need for a better understanding of environmental radioactivity. Large well-equipped ecological laboratories were established at the major atomic energy production and research centers in the United States and other countries, and funds and equipment for ecological research were supplied to individual investigators at many universities. The concern that began to pervade the scientific community in the mid-1960s about contamination by toxic chemicals was to some extent the result of ecological knowledge that was obtained from studies of fallout from nuclear weapons tests.

Those studies raised many difficult questions, which at first seemed unique to the subject of environmental radioactivity. What are the ecological pathways by which these substances reach humans? Do they accumulate in such a way that they can result in unforeseen ecological injury? Are there synergistic effects with other environmental pollutants? By the late 1960s it was apparent that the same questions could be asked about insecticides, food additives, fossil fuel combustion products, trace metals, and other nonradioactive pollutants of the environment. In many respects, the pioneering studies of the environmental effects of radioactivity provided the tools by which more general problems of environmental pollution could be understood.

Chapter 2

The Biological Basis of Radiation Protection

This chapter is intended to provide sufficient information to permit the general reader to appreciate the relative scale of hazard associated with a given level of exposure and to understand the basis on which levels of maximum permissible exposure are established. More comprehensive reviews of the subject are contained in a number of authoritative reports which are more detailed in regard to the underlying radiobiological basis of radiation injury (UNSCEAR, 1977, 1982; NAS–NRC, 1980).

Although much remains to be learned, more is known about the effects of ionizing radiation exposure than about the effects of any other of the many noxious agents that have been introduced artificially into the environment. To a considerable degree, this is the result of the large amount of research that has been performed in this country and abroad since exploitation of nuclear fission began in 1942. In the United States, the money spent for radiation research by the government has been far greater than the expenditures for studies of the effects of the many human-produced chemical pollutants of air, water, and food.

Early Knowledge of Radiation Effects

Reports of radiation injury began to appear in the literature with astonishing rapidity after the announcement on November 8, 1895, of Roent-

9

gen's discovery of X rays. The first volume of the *American X-Ray Journal*, published in 1897, included a compilation (Scott, 1897) of 69 cases of X-ray injuries reported from laboratories and clinics in many countries of the world. The reason why so many injuries were reported so quickly is related to the way X rays were discovered and the kind of research that had been under way for many years even before the existence of X rays was known.

X-ray effects seem to have been first observed in 1859, nearly 36 years before Roentgen's announcement, when Plucker recorded the fact that an apple green fluorescence was seen on the inner wall of a vacuum tube within which a current was flowing under high voltages (Grubbe, 1933). Plucker's work was followed by a number of experiments in other laboratories, and in 1875 Sir William Crookes made the first high-vacuum tube which thereafter bore his name, and discovered that the apple green fluorescence reported originally by Plucker originated from a discharge at the negative electrode of the tube. Thus was inaugurated the term "cathode rays." Roentgen's contribution was the startling observation that he could see the shadow created by the bones of his hand when it was placed between a Crookes tube and a screen covered with fluorescent chemical. To this penetrating radiation he gave the name X rays.

The relevant point is that research with Crookes tubes was under way in many parts of the world prior to Roentgen's discovery and that, unknown to those investigators, X rays were being generated by many of the tubes they were using. Emil Grubbe, who manufactured Crookes tubes in Chicago and was studying the fluorescence of chemicals at the time of Roentgen's announcement, began immediately to experiment with the newly named X rays and promptly developed an acute dermatitis which was followed by skin desquamation. An early description of his injury was presented at a clinical conference at the Hahnemann Medical College in Chicago on January 27, 1896, at which the interesting observation was made that "any physical agent capable of doing so much damage to normal cells and tissues might offer possibilities, if used as a therapeutic agent, in the treatment of a pathological condition in which pronounced irritative blistering or even destructive effects might be desirable." Two days later, less than 3 months after publication of Roentgen's discovery, a patient was referred to Grubbe for treatment of a breast carcinoma. In his interesting recital of the facts, Grubbe (1933) claimed that he was the first person to be injured by X rays, that he was the first person to apply X rays to pathological lesions for therapeutic purposes, and, incidentally, that he was the first to use sheet lead for protection against X-ray effects. More recently, Hodges (1964) and Brecker and Brecker (1969) have cast doubt on the authenticity of Grubbe's claims because of inconsistencies in the chronology. However, Hodges does conclude that Grubbe was the first to treat cancer with X rays. Although the question of priorities of radiological discoveries may never be fully satisfied in the historical sense, the 1933 Grubbe paper and the subsequent work by others do provide insight into the early history of radiation injury.

An indirect result of Roentgen's announcement was Becquerel's accidental discovery of radioactivity in the same year as the discovery of X rays. By 1900 the radioactive constituents of radium ore had been sufficiently concentrated that burns were produced on the skins of the pioneers in radioactivity research (Becquerel and Curie, 1901).

The 1897 report by Scott of injuries produced by X rays was followed by one by Codman (1902) in which he reported that the number of cases of radiation injury in the literature had increased to at least 147. Interestingly, Codman's study caused him to note that the number of cases had begun to diminish, which he attributed to "the bitter teaching of experience and the fact that the introduction of better apparatus has done away with long exposures and the close approximation of the tube." Codman's observation, though justified at the time, has regrettably not been supported by history. The total number of people who were injured and killed by the use and misuse of X rays and radium prior to the development of proper standards of radiation hygiene will probably never be known, but certainly numbers in the many hundreds.

The early interest of the experimental biologists in the biological effects of ionizing radiation is illustrated by the discovery in 1897 of the ability of ionizing radiation to cause lenticular cataracts in exposed animals (Clapp, 1934). The production of genetic mutations by exposure to X rays was reported by Muller (1927) and opened an era of research on the hereditary effects of radiation.

During and immediately following World War I, the use of radium in luminous paints was attended by hazards arising out of ignorance of the effects of this radioelement when inhaled or ingested (Martland, 1951; Evans *et al.*, 1969). Among a total population of about 3000 luminous-dial workers, a total of 63 cases of bone cancer are known to have occurred (Rowland *et al.*, 1983). These workers, mostly women, ingested radium because of the practice of pointing the paint brushes between their lips. In addition, during the 1920s radium was administered medically as a nostrum for a variety of ailments including arthritis, mental disease, and syphilis; at least 20 deaths are known to have been caused by this practice before it was discontinued. More recently, from 1944 to 1951, a compound containing ^{224}Ra (half-life 3.6 d) was injected intravenously into about 2000 German patients with tuberculosis and other diseases (Spiess and Mays, 1970), and by 1981 bone sarcomas had already appeared in 54 of these patients (Wick and Gossner, 1983; Mays, 1983).

The dial-painting cases have been studied thoroughly by a number of investigators, but the main credit belongs to Evans for having worked out the basic biophysical principles of radium injury in sufficient detail that safe practices could be adopted. These practices proved effective for protection not only against radium but also against many of the later hazards of the atomic energy industry, which utilized the information gained with radium to excellent advantage.

In the first 40 years of this century only about 2 lb of radium was extracted from the earth's crust and at least 100 people died from various misuses of this material. In contrast, since 1942 the atomic energy programs have produced the radioactive equivalent of many tons of radium and, up to the present time, except for uranium mining, there have been no deaths that could be attributed to the internal deposition of the wide variety of artificially produced radionuclides.

The uranium mining experience has been a tragedy that could have been avoided. It had been known for centuries that men who worked in the Eastern European metal mines of Schneeberg and Joachimsthahl (Hartung and Hesse, 1879) were prone to a quickly fatal lung disease, but only toward the latter part of the 19th century was it learned that the disease was bronchiogenic carcinoma. In this century, these mines became a source of pitchblende, and when it was realized that cancer could result from internal irradiation, it was suggested that the high incidence of lung cancer among the miners might be explained by their exposure to radioactive substances in the atmospheres of the mines. Studies of the mine air revealed the presence of high concentrations of radon, and this radioactive gas came to be regarded as the likely etiological agent in the high incidence of lung cancer (Hueper, 1942; Lorenz, 1944). A standard for protection against the effects of radon inhalation already existed by World War II and, and had it been enforced in the uranium mining industry, hundreds of lives would have been saved.

Human experience early in this century also provided the information that ionizing radiations in a sufficient dose could produce sterility, damage to blood-forming tissues, and, in the case of high levels of exposure, a complex of symptoms that came to be known as the acute radiation syndrome. However, not all this information was obtained from the misuse of ionizing radiations. X rays and radium have been used in cancer treatment since early in this century, which created the need to understand the effects of large doses on healthy as well as cancerous tissue to permit the radiologist to limit side effects of radiotherapy.

Summary of Present Knowledge of Radiation Effects on Humans

A great amount of research conducted since 1942 has been directed toward understanding both the mechanisms of radiation injury and the ecological relationships that exist in an environment contaminated with radioactive material. Knowledge about the effects of radiation exposure comes from many sources. The greatest volume of literature comes from work with experimental animals. These studies have taught us much about the mechanisms of radiation injury and the relationships between dose and response in the various species of experimental animals. However, the dose–response relationships are not necessarily transferable to humans. Although the basic mechanisms of radiation injury may be the same in all species, there are major differences among species in susceptibility to radiation injury. Thus, information about the size of the dose required to produce a given effect in a species of laboratory animal cannot be applied directly to humans.

Any general discussion of radiation effects should begin with certain dichotomies according to the following:

1. Whether the source of radiation is external to the body (as in the case of exposure to medical X rays) or is an internally deposited radionuclide (as in the case of radioiodine in the thyroid).

2. Whether the dose was from a relatively massive exposure delivered in a short period of time (less than a few days) or was delivered in small bits over longer periods of time, which may extend to many years.

3. Whether the effects appear soon after exposure ("acute effects" or are delayed for months or years ("delayed effects").

It will prove helpful if these distinctions are kept in mind when the effects of radiation are being reviewed.

THE PROMPT EFFECTS OF EXPOSURE

When a massive dose of whole-body radiation is received instantaneously or when the exposure is received predominantly in the first few days, as in the case of the external radiation from fresh fission products, symptoms of acute radiation injury may be seen as early as a few hours after exposure and will follow a course dependent on the size of the dose received. Table 2–1, in which the expected effects of massive exposure to external radiation are summarized, shows that relatively minor effects would occur at doses less than 100 rem, but that about 50% fatalities would be expected to occur in the range 400–500 rem. As the whole-body dose approached 1000 rem, the fatalities would reach 100%.

Nausea occurs in increasing frequency above 100 rem and will be seen in almost all exposures above about 300 rem. The nausea may occur within hours after exposure and be followed by an asymptomatic period of as much as 2 weeks after a dose of 100–250 rem, but less than 1 day when the dose is greater than 700 rem. The signs and symptoms which then develop

TABLE 2-1

SUMMARY OF CLINICAL EFFECTS OF ACUTE IONIZING RADIATION DOSES[a]

	Dose (rem)					
	0–100	100–200	200–600	600–1000	1000–5000	Over 5000
Incidence of vomiting	None	100 rem: 5% 200 rem: 50%	300 rem: 100%	100%	100%	100%
Time of onset	—	3 hr	2 hr	1 hr	30 min	30 min
Principal affected organs	None	Hematopoietic tissue			Gastrointestinal tract	Central nervous system
Characteristic signs	None	Moderate leukopenia	Severe leukopenia; purpura; hemorrhage; infection; epilation above 300 rem		Diarrhea; fever; disturbance of electrolyte balance	Convulsions; tremor; ataxia; lethargy
Critical period postexposure	—	—	4 to 6 weeks		5 to 14 days	1 to 48 hr
Prognosis	Excellent	Excellent	Good	Guarded	Hopeless	Hopeless
Convalescent period	None	Several weeks	1 to 12 months	Long	—	—
Incidence of death	None	None	0 to 80% (variable)	80 to 100% (variable)	90 to 100%	90 to 100%
Death occurs within	—	—	2 months		2 weeks	2 days
Cause of death	—	—	Hemorrhage; infection		Circulatory collapse	Respiratory failure; brain edema

[a] From Glasstone (1962).

may include epilation, sore throat, hemorrhage, purpura, petechiae, and diarrhea.

Among the most striking changes are those due to injury to the blood-forming organs, which reduces the rate at which the components of blood elements are produced and has a dramatic effect on the composition of circulating blood (Wald *et al.*, 1962).

At whole-body doses in excess of 1000 rem, the predominant symptoms may be due to injury to the gastrointestinal tract and central nervous system. Disorientation within a matter of minutes owing to central nervous system injury was a conspicuous feature of at least one case (Shipman, 1961) after accidental exposure to several thousand rads.

The radioiodines may be the only internal emitters capable of producing prompt effects. These nuclides are short-lived, are readily absorbed into the body, and concentrate in the thyroid, which weighs only a few grams. For these reasons, among the internal emitters, the radioiodines are uniquely capable of delivering a high dose over a short period of time. Exposure to heavy doses of radioactive iodine can reduce thyroid function, and doses of several thousand rem, such as are used in treatment of thyroid cancer, can result in destruction of the gland. It is likely that the external radiation dose from generalized fission-product contamination of the environment would be so high as to be overwhelming relative to the dose from radionuclides other than the radioiodines. It will be seen in Chapter 14 that doses to the thyroid of up to 1800 rem, superimposed on whole-body doses of about 175 rem following exposure due to fallout from weapons testing, resulted in impaired thyroid function (a relatively prompt thyroid effect) in addition to causing thyroid cancer after many years.

Delayed Effects

The delayed effects of radiation may not appear for several decades after exposure, and can result either from massive doses that have caused prompt effects or from relatively small exposures repeated over an extended period of time. The effects which develop in the exposed individual are referred to as *somatic effects* to differentiate them from *genetic effects*, which occur in the progeny of the exposed person and are the results of changes transmitted by hereditary mechanisms. Radiation injury can also occur in the developing fetus, but this would be considered a somatic rather than genetic effect.

Until the early 1960s, the genetic effects of radiation were thought to be the most important effect of penetrating radiation on human health. A major publication prepared for the AEC in 1958 on the biological effects of radiation placed great emphasis on the genetic effects and included hardly any information on cancer (Claus, 1958; Henshaw, 1958).

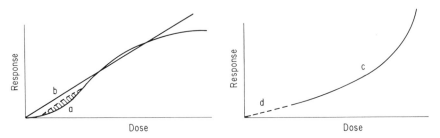

Fig. 2–1. Types of dose–response curves. (a) The classical sigmoid curve, no longer considered applicable in radiobiology. At the low end, the response may be asymptotic with dose, or a true threshold may exist but be obscured by statistical uncertainties inherent in the data. (b) The linear dose–response relationship. Note that (a) and (b) are so close together over such a considerable range that it may not be possible to distinguish (a) from (b) if the experimental or epidemiologic data fall, as in the example shown, between or slightly beyond the two intercepts. The crosshatched portion of (a) is how the dose–response relationship would appear if a fraction of the irradiated population should be exceptionally sensitive to radiation. (c) The linear–quadratic dose–response curve, in which the response is linear at low doses (dashed portion), but takes the form of kD^2 at higher doses (solid portion). The slope of the linear portion may vary, depending on dose rate. Most of the levels to which the public is exposed are assumed to lie within the linear portion (d) of the linear–quadratic curve.

During the 1950s, there was already evidence that the dose–response curve for induction of genetic effects was linear, and it came to be accepted that it was probably without a threshold. This was in contrast to cancer induction, for which a threshold was then believed to exist, and linearity of response was not generally accepted. It was thought that the dose–response relationship for cancer was sigmoidal in form and that a threshold existed below which cancer was not produced. This is the classical dose–response relationship that is widely applicable in toxicology. The sigmoid response is shown in Fig. 2–1a, in which the effects per unit of dose gradually increase until a dose level is reached at which the rate of increase begins to diminish and eventually reaches a plateau when all susceptible members of the population have been affected. As noted in Fig. 2–1a, there may be a threshold below which no effects occur, or the question of whether a threshold exists may be indeterminate because of statistical limitations in the data. Figure 2–1b illustrates the linear–no threshold response, which, because of statistical limitations, cannot always be distinguished from other dose–response relationships.

However, research with animals began to raise doubts about the existence of a threshold, and suggested that the response to gamma radiation might also be linear, as in the case of genetic effects. Moreover, the studies of the Hiroshima and Nagasaki survivors demonstrated during the 1960s that the cancers produced by their radiation exposure were occurring with greater frequency than had been expected and that the kinds of cancers were more varied. Until then, leukemia had been the only cancer seen in excess among the survivors, but other cancers began to appear, and it was soon concluded that there would be at least five "solid" tumors for each case of leukemia (NAS–NRC, 1980). At the same time, experiments with rats indicated that the genetic effects of radiation were less than had been believed (Russell, 1968), and studies in Japan could find no genetic damage among the offspring of survivors of the atomic bombings. For all of these reasons, a shift in thinking took place in the late 1960s and the risk of cancer induction became the main concern of those involved with the health effects of ionizing radiation.

Some Elementary Quantitative Aspects

Further laboratory experimentation began to support the hypothesis that both the genetic and carcinogenetic effects of ionizing radiation were inherently stochastic, which implies that no threshold exists, that the frequency with which effects are seen depends on probabilistic mechanisms, and that the probability of an effect is a continuous function of dose.

A hypothesis that competes with the linear dose–response relationship is the quadratic form (Fig. 2–1c), in which the number of effects produced per unit dose increases progressively with increasing dose. In recent years, the so-called linear–quadratic relationship (Fig. 2–1d) has been favored by radiobiologists and epidemiologists. In this relationship the effects are linear at low doses but become curvilinear at higher doses according to the relationship:

$$I = aD + bD^2 + C$$

in which I is incidence, C is the incidence at zero dose, D is dose, and a and b are constants determined empirically (Upton, 1977; NAS–NRC, 1980).

Levels of exposure to environmental radioactivity that are most frequently encountered result in doses of the order of natural background (about 0.1 rem/year) or less, where the effects are assumed to be governed by the linear portion of the linear–quadratic relationship.

With some exceptions, there is evidence from experiments with laboratory animals that the effects per unit dose are less at low dose rates than at high dose rates. The dose from exposure to low levels of radioactivity in the environment usually results from more or less continuous exposure over an extended period of time. This has the effect of reducing the slope of the linear portion of the linear–quadratic response shown in Fig. 2–1d (Upton, 1977). Based mainly on the results of animal experiments, it has been assumed in recent years that the risk coefficients derived by extrapolation from high dose–high dose rate exposure can be reduced by a factor of 3 when the exposure is protracted in time (NCRP, 1980a). However, a good deal of contrary evidence also exists. The effect per rem on the incidence of lung cancer has been observed to be *inversely* proportional to dose both in experimental animals and in epidemiologic studies of uranium miners (Cuddihy, 1982; NAS–NRC, 1980). There have been similar findings in the production of bone tumors in experimental animals (Schlenker, 1982), and the frequency of radiation-induced breast cancer among women irradiated for postpartum mastitis has been shown to be linear with dose and independent of the manner in which the dose was fractionated in delivery (Boice, 1982).

Another factor that may affect the dose–response relationship is the effect of progressively smaller doses on the time it takes for the cancer to develop. Radiation-induced cancers (other than leukemia) rarely appear sooner than 10 years after exposure and can occur more than 30 years after exposure begins. It has been shown by Evans (1967) that the latency period for the development of bone cancer in luminous-dial workers is inversely correlated with the radium body burden. This caused Evans to postulate that the question of whether or not a threshold exists may be moot because there appears to be a finite bone dose below which the time to tumor appearance is so long as to exceed the life span of the individual. Evans coined the term "practical threshold" to explain this feature of dose–response relationships. An inverse relationship between dose and leukemia latency has been reported both in experimental animals (Upton, 1977) and among the Japanese survivors of the World War II atomic bombings (Ishimaru *et al.*, 1982). Albert and Altshuler (1973) have examined this question both theoretically and epidemiologically and have concluded that there may be a basis for assuming that there is an inverse relationship between dose of chemical carcinogens and tumor latency.

The fact that the dose–response relationship is assumed to be linear and the absence of a threshold have important implications for risk assessment and formulation of public policy. Absence of a threshold implies that there is no such thing as an absolutely safe level of exposure. Every increment of dose above zero, however small, results in an increment of risk as well, but the risk becomes smaller as the dose diminishes. It will be seen in Chapter 3 that this complicates the problem of setting standards of permissible exposure. It was proposed recently that the question of whether a specific cancer is causally related to a given dose of radiation should be answered in probabilistic terms. For this purpose, radioepidemiologic tables have been prepared in the hope that legal decisions as to liability for cancer induction can be based on the probability that the cancer was caused by radiation, taking into consideration such factors as the magnitude of dose, age of exposure, and other relevant factors (NAS–NRC, 1984; U.S. DHHS, 1985).

A dilemma that also arises from the assumption of linearity and the absence of a threshold is that the risk to individuals can be very small, but a finite number of cancers can result if a sufficiently large population is exposed. A lifetime risk of 10^{-6} is negligible to an individual, but if the world's population of 4.5×10^9 persons is exposed, 4500 radiation-induced cancers will result. The question of whether the permissible dose should be defined on the basis of individual or collective risk cannot be answered by existing technical, moral, or political concepts. This will be discussed further in Chapter 15.

Somatic Effects

With the above as background, we can now proceed to a brief discussion of each of the principal cancers and other somatic health effects seen in irradiated populations.

Leukemia. Leukemia is a relatively rare disease which has been observed to occur in increased frequency among Japanese survivors at Hiroshima and Nagasaki (Beebe *et al.*, 1978), among children irradiated in infancy for thymic enlargement (Hemplemann *et al.*, 1975; NCRP, 1985a), among patients irradiated for ankylosing spondylitis (Smith and Doll, 1982), among physicians exposed in the practice of radiology (Cronkite, 1961), and possibly among children who were irradiated *in utero* in the course of pelvic examination during pregnancy (Kneale and Stewart, 1976,; Totter and MacPherson, 1981; Monson and MacMahon, 1984). Of these groups, only the radiologists can be described as having been exposed to repeated small doses, in contrast to the other groups, which were subjected to either single or slightly fractionated exposures.

Among the Japanese atom bomb survivors the incidence of leukemia reached a peak in the early 1950s and returned to near-normal by about 1970. The cases occurred primarily among survivors within 1500 m of the hypocenter. Compared to 1.3 expected cases in Hiroshima, there were 22 observed cases of leukemia among survivors who received doses greater than 200 rad. Among those exposed to more than 10 rad in the two cities, there were 84 cases compared to 2.1 cases expected (UNSCEAR, 1977). Based on the linear–quadratic model, the leukemia risk among Japanese survivors was about 2.3 cases per million persons per rad.

There have been several studies of the incidence of leukemia in children who were irradiated *in utero*, with conflicting evidence that the fetus may be more sensitive to the leukemogenic effects of radiation. MacMahon (1962) and Stewart and Kneale (1970) reported an apparent increase in the incidence of leukemia and other neoplasms in the first 10 years of life in children who were irradiated *in utero* in the course of maternal pelvic X-ray examination. This finding has not always been supported by other studies. Court Brown *et al.* (1960) examined the health records of a large group of children with similar histories but found no increase in leukemia or other cancers. Jablon and Kato(1970) studied the 25-year cancer experience of 1292 children who were irradiated *in utero* in the bombings of Hiroshima and Nagasaki, and who received very much higher doses than the children whose mothers were X-rayed during pregnancy. Although some of the Japanese children were estimated to have received as much as 250 rem prenatally, no increase was found in the incidence of leukemia or other cancers during the first 25 years of life. The dose received by the fetus during the Japanese bombings was about 200 times greater than the doses received from pelvic X ray.

It has been suggested that the excess risk of leukemia (and other cancers) among children irradiated during pelvic examinations while *in utero* may be explained by factors related to the medical indications for which the examination was required, rather than to the radiation itself (NCRP, 1977b; Jablon and Kato, 1970). This hypothesis has not been supported by the data of Monson and MacMahon (1984).

Saenger *et al.* (1968) compared the incidence of leukemia in hyperthyroid patients treated with ^{131}I to the incidence in patients treated surgically. In a total population of 36,170 patients treated between 1946 and 1964, among whom 96.5% follow-up was achieved, there was no difference in the leukemia incidence of the two groups. However, the hyperthyroid population as a whole had a leukemia incidence about 50% higher than that of the general population. The average dose administered to these patients was 10 mCi, which produced a whole-body dose of about 30 rem. Saenger's findings were an interesting example of a procedure in which the medical condition for which the irradiation was required may also predispose the patient to radiomimetic effects. As noted above, this has also been suggested in the case of children irradiated *in utero* in the course of maternal pelvic examination.

From all epidemiologic evidence, it can be concluded that for the purpose of estimating the upper limit of risk to an exposed population, a risk coefficient of 20 cases of leukemia per rad per million persons can be assumed (NCRP, 1984c). These cases would occur during a period of about 15 years.

The Hiroshima and Nagasaki studies have demonstrated an excess of cancers other than leukemia among the survivors of the atom bombing in a ratio of five to one. The major sites at which these tumors occur are the thyroid, lung, digestive tract, and female breast.

Ionizing radiations have also been implicated in the high rates of bronchiogenic lung cancer among uranium miners and of bone cancer among persons exposed to radium.

The various estimates of the risk of cancer suggest that, in round numbers, the lifetime risk of developing cancer from radiation exposure is about 10^{-4} per rad.

Bone Cancer. When radium or other radioelements that are chemically similar to calcium are ingested, they deposit in bone, from which they are removed very slowly. As noted earlier, bone cancer developed among individuals who were exposed to radium both in the course of luminous-dial manufacturing and as patients who received radium medicinally during the early years of this century.

Bone cancers among radium dial painters were first observed and diagnosed as "radium jaw" in 1924 by Theodore Blum, a New York dentist. The cases originated from a luminous-dial plant in northern New Jersey, and by the late 1920s it was already understood that the cases of bone cancer being reported among young women who painted radium dials with radium-containing paint were due to the practice of lip pointing the brushes used to paint the numerals. The cases attracted the attention of a forensic pathologist, Harrison Martland, and a toxicologist, A. O. Getler, who undertook to study the first several cases (Martland, 1951).

Valuable epidemiologic information has been obtained from these cases and also from studies of persons to whom radium was administered either intravenously or orally for medical reasons. Many thousands of patients were treated clinically in thi way. One clinic administered more than 14,000 intravenous injections of radium, usually in doses of about 10 µCi. In addition, a number of popular medicinal waters that contained radium were marketed in the early 1920s. One of the most popular of these nostrums was Radithor, a mixture of 1 µCi of radium and 1 µCi of mesothorium (Evans, 1966).

About 84 cases of cancer due to radium exposure have been reported (Rowland *et al.*, 1983). About one-fourth of these were carcinomas of the nasal sinus and mastoid air spaces, which are believed to have been due to the accumulation of radon and its daughter products within these cavities.

Based on his epidemiologic studies prior to about 1942, Evans (1943) proposed that 0.1 µCi of ^{226}Ra be adopted as a maximum permissible body burden. All the known cases of injury had occurred in individuals whose body burden was greater than 1.2 µCi at the time of observation, which was usually many years after cessation of exposure. Evans has more recently estimated that the maximum permissible body burden has a safety factor of about 15 (Evans, 1967; Evans *et al.*, 1969).

A very much larger number of individuals have shown less severe bone pathology than cancer, ranging from small areas of osteoporosis to extensive necrosis. Advanced osteoporitic or necrotic changes were frequently associated with spontaneous fractures.

The ability of the various bone-seeking radioactive elements to produce osteogenic sarcoma varies because of differences in the kinds of radiations emitted and also because the different elements deposit in different parts of the bone structure. Raabe (1984) concludes, on the basis of laboratory studies with beagles, that the relative biological effectiveness compared to ^{226}Ra is 3 for ^{228}Ra, 9 for ^{239}Pu, 6.4 for ^{241}Am, 10.7 for ^{228}Th, and 15.5 for ^{238}Pu.

The question of whether a threshold exists for production of bone cancer from internal emitters is of considerable importance in relation to standards of permissible public exposure because traces of bone-seeking nuclides such as ^{90}Sr and ^{239}Pu are already present in the general environment and can be found in the tissues of the general population. Although the dose to individuals is very small, the numbers of persons exposed is very large, since everyone in the world is exposed to some extent. If the dose–response relationship is linear, a finite number of bone cancers may result, even when the dose to the individual is exceedingly small. As noted earlier, Evans found that the latency period was inversely related to the Ra body burden. It is possible that for the minute doses present in the general population (see Chapter 13), not only is the theoretical probability of inducing cancer very low, but the latency period exceeds the human life span.

The bone-seeking radionuclides are so important that they continue to be the subject of extensive laboratory and epidemiologic study (Rundo *et al.*, 1983).

Lung Cancer. The high incidence of lung cancer among miners in Eastern Europe has been attributed to the diffusion of radon into the mine atmosphere (Hueper, 1942; Lorenz, 1944; Behounek, 1970). These mines had served for centuries as a source of a number of heavy metals, and the miners suffered from a high incidence of a lung disease, which was not diagnosed as lung cancer until the latter part of the 19th century. Studies in the 1920s led to the conclusion that radon in the mine atmosphere was probably responsible. By the 1950s, based on work initiated by Harley (1952), it was apparent that the principal dose to the lung from inhalation of radon was not due to the radon itself but to the accumulation within the lung of the short-lived daughter products of radon attached to the inert dust normally present in mine atmospheres.

An excess of lung cancers was reported among uranium miners in the

United States beginning in the 1960s (Archer, 1981; Holaday, 1969). The miners in the southwestern states were known to have been exposed to concentrations of radon and radon daughters comparable with those reported earlier in the mines of Eastern Europe.

Lung cancer has also been reported among miners in nonuranium mines in which elevated concentrations of radon are present. This has been true among workers in Newfoundland fluorspar mines, where radon enters the mine via ground water (de Villiers and Windish, 1964), and among miners in Swedish iron, zinc, and lead mines (Axelson and Sundell, 1978). The most severe known occurrence of lung cancer among miners is in a large tin mine in southern Hunnan Province of the People's Republic of China, where more than 1500 cases have been reported. Arsenic is known to be present in the mine air and may act synergistically with the radon daughters (Sun *et al.*, 1981).

These experiences as well as experimental evidence that radioactive substances can cause lung cancer in laboratory animals have resulted in considerable attention being given to the methods of protecting employees and the general public from the effects of radon and its daughters. It was learned recently that the general public may be exposed to relatively high concentrations within private dwellings, and this has led to estimates that radon exposure may be responsible for a significant fraction of the lung cancers that occur among nonsmokers in the general population. This will be discussed further in Chapter 7.

When radon is inhaled, it is usually not in equilibrium with its daughter products, which include a series of short-lived nuclides that decay within hours to ^{210}Pb ($T_{1/2} = 22$ years) (see Chapter 7). Because of the difficulty in describing the mixture in the usual units (picocuries per liter), the term "working level" was devised (Holaday *et al.*, 1957) to express the concentration of radioactivity of the mine atmosphere. The unit is defined as that concentration of short-lived radon daughter products that emit 1.3 × 10^5 MeV of α radiation per liter of air. Conversion to dose of WLM, which is a measurement of concentration in air, is complicated by the fact that the daughter products do not deposit uniformly on the bronchial surfaces but, depending on particle size, deposit differentially according to depth in the bronchial tree. Various estimates of WLM per rad have been made, beginning in 1951, and these estimates vary widely depending on the assumptions made. The values (NCRP, 1984c) have continued to vary widely in recent years, e.g., 0.58 WLM/rad (Harley, 1980) and 8 WLM/rad (ICRP, 1981). Among uranium miners in the United States there is evidence of an excess of lung cancer at cumulative exposures greater than about 100 WLM.

The respiratory tract (Fig. 2–2) is designed to transport air deep into the

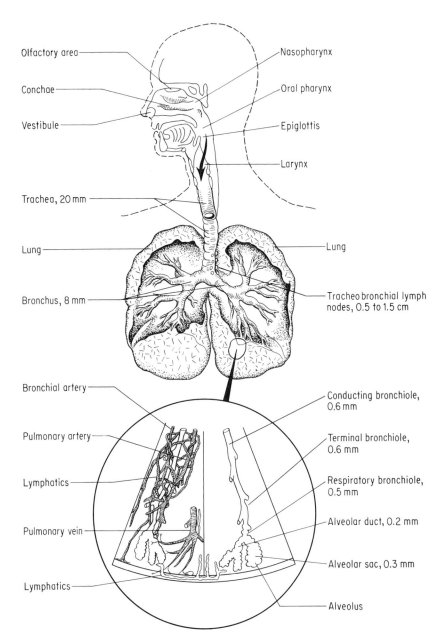

Fig. 2–2. Principal anatomical features of the human respiratory tract. (From National Academy of Sciences–National Research Council, 1961.)

lung to the alveoli, where oxygen and carbon dioxide exchange takes place between the blood and air. Most particles of the dust contained in the inhaled air are removed by deposition within the upper respiratory tract before the alveoli are reached, the exact fraction depending primarily on the particle size and density of the particles (ICRP, 1966a). Deposition within the respiratory tract may result from inertial impaction, settling, or Brownian motion in the case of particles less than about 0.1 μm in diameter. In general, the larger particles tend to be removed between the nasal passageways and the lower bronchi, whereas the smaller particles have a higher probability of penetrating to the alveoli. If the dust is highly soluble, it will be quickly absorbed from the respiratory tract into the blood.

The respiratory tract above the terminal bronchiole is lined with ciliated epithelium that has the ability to move deposits of dust up and out of the respiratory tract, and it has been shown that dust deposited on the bronchial epithelium in humans is removed from the lung in a matter of hours, whereas dust that deposits in the alveolar regions of the lung can remain within the lung for months or years (NAS–NRC, 1961a).

Insoluble dust that deposits on the alveolar walls is removed by phagocytes, which are motile white cells that have the ability to engulf the particles and transport them out of the alveolar spaces. The dust-laden phagocyte will either wander onto the ciliated epithelium, from which it will pass to the mouth by mucociliary action, or pass into the lymphatic vessels and then to the lymph nodes.

The dose received by the different portions of the lungs from inhaled radioactive dust thus depends on the concentration of the radionuclides in the inhaled air, respiration rate, physical properties of the radionuclide, region of the lung in which the dust is deposited, and rate at which the dust is removed. The latter factors depend on particle size and density, as well as on physiological and morphological factors that vary among individuals.

There are several ways of computing the dose from radioactive dust deposited in the lung. The simplest method is based on the assumption that the energy released is absorbed uniformly throughout the total lung mass (taken to be 1000 g in the adult human). However, since most lung cancers originate in the bronchial epithelial tissue, and since most of the radioactive dust is deposited on the bronchial epithelium, the dose from inhaled insoluble dust is calculated by assuming that the energy is absorbed in that tissue. For radon the energy is assumed to be absorbed to a depth of 22 μm, which is the range of the most energetic alpha particle emitted in the radon decay scheme. The dose calculated in this way is orders of magnitude higher than that calculated on the assumption that the energy is absorbed in the whole lung (Altshuler *et al.*, 1964; NCRP, 1984c).

The tissues may not be irradiated uniformly by particles deposited in the lung. This is particularly true when alpha particles are inhaled. A basic question is whether or not the risk of developing cancer is increased because the energy is deposited in a smaller volume of tissue. Both the ICRP and NCRP have taken the position that the hazard of particulate material is probably less than if the same amount of material is distributed uniformly throughout the lung (ICRP, 1977; NCRP, 1975d).

Thyroid Cancer. The thyroid must receive special consideration because of its tendency to concentrate radioactive iodine. When fresh fission products are released to the atmosphere, the dose to the thyroid may be the limiting factor that determines the length of time an individual can be permitted to remain in a given area and the amount of locally grown food that can be consumed (Chapter 6).

The largest number of radiogenic thyroid carcinomas have been seen among individuals who were irradiated with X rays as children for various reasons, including enlarged tonsils, ringworm, and enlarged thymus glands (Shore *et al.*, 1984; NCRP, 1985a). A number of other medical procedures in which the head and neck were X-rayed have also resulted in thyroid cancer.

Many thousands of persons have been administered ^{131}I for diagnostic studies, in which the doses varied considerably but were generally in the range 50–150 rad. There is evidence that the carcinogenicity of ^{131}I is somewhat less than that of external radiation per unit of absorbed dose.

A group of about 4000 women who were treated for hypothyroidism has been followed by Hoffman (1984), who found no difference in the incidence of thyroid or other cancers between those treated with ^{131}I and others treated by surgery. In that series of cases the dose was much greater than 1000 rad. The NCRP has concluded that the carcinogenicity of ^{131}I is no more than one-third that of external radiation (NCRP, 1985a).

A total of 52 cases of thyroid cancer have been reported among Japanese survivors of the atomic bomb (NAS–NRC, 1980). The effect has been found to be proportional to radiation dose, and the increase in frequency has been greater in women than in men (Parker *et al.*, 1974).

In 1954 a group of natives of the Marshall Islands on the atoll of Rongelap were exposed to fallout from a thermonuclear explosion at Bikini (see Chapter 14). Of 67 people exposed on Rongelap, 34 developed thyroid abnormalities, including hypothyroidism, nodules, and seven malignant tumors, at the end of about 27 years. The thyroid doses from absorbed radioiodine were estimated to have ranged from 335 rem in adults to 1800 rem in children less than 10 years old. The whole-body dose was about 175 rem (Conard *et al.*, 1980; Conard, 1984). The thyroid doses are somewhat uncertain because they were reconstructed from urine analysis data several weeks after exposure. It should be noted that the dose received by these cases was received from both radioiodine and external gamma radiation.

Overall, the human experience suggests that thyroid cancers can result from absorbed doses as low as 8–10 rad. Among the Japanese survivors, the induction rate per unit dose in females was about twice that in males, and was higher for children younger than 10 years of age. The age effect was not observed among other cohorts (UNSCEAR, 1977). Among the Marshall Islanders, it is suggested that cancer develops later in persons who received lower doses (NAS–NRC, 1980).

Using data from the various cohorts, the lifetime risk of developing a fatal thyroid cancer is about 250 per 10^6 persons per rad (UNSCEAR, 1977).

For the purpose of calculating the thyroid dose in an individual who inhales or ingests radioiodine, it is assumed that the thyroid absorbs 30% of inhaled or ingested radioiodine (ICRP, 1960), but according to Dolphin (1968, 1971), thyroidal uptake can vary from 10 to 30% and may actually be a function of the thyroid mass, which is usually taken as 20 g, but has been shown by Mochizuki *et al.* (1963) to vary somewhat around a mean weight of 16.7 ± 6.9 g in adults in New York City. The radioiodine uptake can be almost completely blocked by oral administration of potassium iodide in doses in the range 50–200 mg (NCRP, 1977b).

Cataracts. Cataracts are a nonstochastic effect of exposure of the lens of the eye to relatively high doses of X rays, γ rays, β particles, or neutrons. As noted earlier, cataract was reported in experimental animals exposed to X rays within 2 years after Roentgen's discovery. Cataracts in human beings were observed among survivors of the Japanese bombings (Nefzger *et al.*, 1968), among patients whose eyes were treated with X, γ, or β rays for medical purposes (Merriam and Focht, 1957; NAS–NRC, 1961b), and among physicists who were exposed to the radiation from cyclotrons. Although lens changes have been reported from doses as low as 200 rem, the minimum X-ray dose capable of causing clinically significant cataract is thought to lie between 600 and 1000 rad in adults but may be less in children. Neutrons are thought to be 5 to 10 times more effective than X or γ rays in the production of cataracts. Cataract has not been reported as a result of low-level chronic exposure.

Effect of Radiation on Aging. The literature contains many references to the nonspecific acceleration of the aging process. It has been concluded that it is not possible to confirm that such effects occur either in humans or experimental animals (UNSCEAR, 1977; NAS–NRC, 1980). With respect to the survivors of the Japanese bombings, who have received the highest doses of any group studied and have been followed for the longest period of time, "this series has indisputably shown that there is no evidence of life-shortening that could not have been explained by an increased incidence of leukemia and solid tumors" (UNSCEAR, 1977).

Genetic Effects

As in the case of all higher animal forms, the essential characteristics of humans are passed from one generation to the next by means of chromosomes located in the nuclei of reproductive cells. Human cells normally contain 46 chromosomes, of which half are derived from the mother and the other half from the father. The inheritable characteristics are communicated by means of bits known as genes, which are strung together in beadlike fashion to form tiny filaments that are the chromosomes.

The genes from the mother and father are united within the ovum at conception, and thereafter throughout life the chromosomes are almost always faithfully reproduced at each cell division. The 46 human chromosomes are believed to contain on the order of 10^4 genes, and it is this complex of genes and chromosomes that, when passed on to the next generation, will determine many of the physical and psychological characteristics of the individual.

The genes are large molecules which may undergo structural changes (*mutations*) as the result of action by a number of agents including heat, ionizing radiation, and mutagenic chemicals (NAS–NRC, 1980). Mutations can also result from breaks and rearrangements of the chromosomes themselves. It is possible for a mutation to be passed directly to progeny, in whom the mutant characteristic may ultimately manifest itself. This occurs only for those mutations that are *dominant,* i.e., have the ability to express themselves in the child who inherits the mutation. It is estimated that 1% of all persons inherit at least one dominant mutation. Most mutations of the type produced by ionizing radiations are recessive and do not express themselves in the offspring unless a similar mutation is encountered in the chromosomes of the mate. However, the mutated gene will be reproduced and transmitted to successive generations of progeny. As it is highly improbable that two similar mutations would encounter each other in the first postmutation generation, the mutated genes would simply diffuse through the genetic material of subsequent generations, and would accumulate until they became so numerous as to increase the probability of encountering each other in the reproductive process. In the offspring produced from such encounters, the mutations would express themselves as inherited changes in the characteristics of individuals.

The measure of potential damage from radiation exposure is the total number of person–rem delivered to the gonads or, stated another way, the per capita gonadal dose. In any given population, allowance must be made for age distribution and the probability of reproduction by the various members of the population. When the per capita gonadal dose is corrected in this way, it is known as the "genetically significant dose."

In a population subjected to a constant level of ionizing radiation exposure, the total number of mutations produced will continue to increase in the population until the rate at which new mutations are produced is exactly offset by the rate at which new carriers of the mutations die before reproducing. It would take many generations for such an equilibrium to be established.

It is estimated that about 4% of all individuals inherit characteristics that result from recessive mutations due to natural factors in the environment. Many of these characteristics are so minor as to be of little or no concern, such as variation in eye or hair color. Other mutations can result in tragic consequences that include serious deformity, crippling disease, and premature death. The genetically induced effects due to ionizing radiation exposure are generally of the same kind as those that occur spontaneously in the population.

It is estimated that a radiation dose of about 50–250 rad delivered to an entire population would cause the spontaneous mutation rate to double. That is, if the doubling dose were delivered to each generation, a new equilibrium would in time be reached at which genetic mutations would be seen twice as frequently as in the original population. It should be understood that, to be genetically significant, the dose must be delivered prior to the age of procreation, usually taken as 30 years.

The genetic effects of radiation have not been seen in irradiated human populations. Even among the two large populations of atom bomb survivors, no effects have been observed. This does not imply that the effects have not occurred, but rather that if they have occurred, the effects have been too infrequent to be detected (NAS–NRC, 1980).

Radiation-Induced Chromosome Changes

It has been known for many years that aberrations in human chromosome structure occur more frequently in irradiated persons and can be observed in chromosomes from peripheral blood lymphocytes at doses as low as a few rad (UNSCEAR, 1977). The biological significance of such changes, particularly when seen in somatic cells, is not known. Considerable interest in the phenomenon derives from the possibility that the frequency and types of chromosomal aberrations may provide a method of estimating past exposure to ionizing radiation (DuPrain *et al.*, 1980).

Chapter 3

Radiation Protection Standards

Soon after the discovery of X rays and radium, their dangers were so well recognized and their uses were becoming so commonplace that there was a clear need to develop uniform standards of radiation protection. In 1928 the International Society of Radiology sponsored formation of the International Commission on Radiological Protection (ICRP) (Taylor, 1971). This commission still exists, comprising 13 members and 4 committees that have a total of about 40 experts from 14 countries. ICRP has held an important position in the field of radiation protection and in recent years has received financial support from the World Health Organization and other international and national organizations.

The National Council on Radiation Protection and Measurements (NCRP)[1] was founded in the United States in 1929, one year after ICRP was formed. The NCRP was, for many years, housed administratively within the Bureau of Standards, although it was not supported financially by it. In 1964, NCRP was granted a Congressional charter which gave it a broad mandate to continue and extend its activities in the radiation protection field (Taylor, 1971). Since that time it has operated as an independent organization financed by contributions from government, scientific societies, and manufacturing associations. There are 75 members on this

---

[1]The NCRP was originally called the Advisory Committee on X-ray and Radium Protection.

council and about 500 individuals who serve on the scientific committees that develop the NCRP scientific reports and recommendations.

Evolution of the Maximum Permissible Dose

The first efforts to develop safe practices were handicapped because of primitive concepts of dose (Wyckoff, 1980). Until the roentgen was adopted in the early 1930s, radiation dose was related to the length of time it took for a given radiation flux to produce reddening of the skin. The maximum permissible exposure for radiation workers during any month was taken to be 1/1000 of the amount of radiation that would produce erythema in an acute exposure. Later, when the roentgen (R) was defined as the unit of radiation exposure, it was estimated (Failla, 1932) that the erythema dose of X radiation was approximately 600 R, and that the values recommended by various experts between 1925 and 1932 had ranged between 0.04 and 2.0 R/day. The ICRP and NCRP subsequently adopted 0.2 and 0.1 R/day respectively as the *tolerance dose*.

When the first radiation protection standards were proposed, the total population at risk consisted of a few thousand X-ray and radium workers. Until that time the hazard was thought to be due entirely to external radiation, although, unknown to the scientists then concerned, a number of workers in industry had already absorbed radium in amounts that were to prove lethal to some of them. The fact that ionizing radiations could produce genetic mutations was not sufficiently appreciated, and the possibility that there might not be a threshold for the production of radiation-produced neoplasms had not yet been suggested. Thus, radiation protection practices were deemed to be sufficient if no apparent damage was being caused at a given level of radiation exposure. The clinical sign of radiation injury that was considered to be particularly important during that period was a change in the white cell count, and the periodic blood examination was the main procedure by which the health of radiation workers was monitored. A worker was believed to have been protected if he maintained a normal blood count.

The need to develop limits of permissible exposure to internal emitters began to arise in the late 1920s when it was learned that bone cancer (osteogenic sarcoma) had developed among workers applying radium to luminous dials used in timepieces (Martland, 1925). In a group of about 30 ex-dial painters studied by Evans (1967), it was found that the lowest ^{226}Ra body burden of an individual who had developed bone cancer was 1.2 μCi. The NCRP (1941), on the basis of these data, proposed that the maximum permissible body burden of ^{226}Ra be limited to 0.1 μCi. In later years, the number of known radium workers increased manyfold, and X-ray evidence of bone disease other than cancer has been reported for body burdens as low as 0.5μCi. The NCRP recommendation that the permissible body burden be limited to 0.1 μCi remains in effect to the present time, more than 40 years after it was first proposed.

The two benchmarks adopted prior to World War II, 0.1 R/day for whole-body external radiation and 0.1 μCi as the maximum body burden of ^{226}Ra, were the standards on which the World War II atomic energy program was conducted, with an excellent safety record. The fact that these important standards became available just prior to the start of the atomic energy program in 1942 was a remarkable historical coincidence. If the relatively few cases of radium poisoning had gone unnoticed, or if they had not attracted the interest of such perceptive investigators as Martland, Evans, and Failla, the quantitative basis for the safe handling of bone-seeking internal emitters would not have been available by World War II. Similarly, had the uses of X rays developed one or two decades later than they did, ICRP and NCRP would not have had the guidance which the early misadventures with X rays provided, and the maximum permissible dose might not have been available when needed during World War II. The two all-important benchmarks were developed, as though providentially, just in time to serve the needs of the atomic energy program that began in 1942.

Much improvising was needed to fill the gaps that developed during World War II. The recommendations of the NCRP for protection against external radiation from X ray and radium were quickly adapted to the sudden need to protect people against a wider variety of ionizing radiations, many of which, like neutrons, had not been extensively studied by the biologists. Failla and others devised the concept of relative biological effectiveness, and Parker invented two new units, the rep and the rem[2] (Cantril and Parker, 1945). This made it possible to extend to particulate radiation the use of the recommended limit of 0.1 R/day then used in the United States for γ- and X-ray protection. Also, the maximum permissible body burden adopted for ^{226}Ra was used as the basis for computing the permissible body burdens of other alpha-emitting bone seekers, some of which, such as plutonium, did not exist before World War II.

The original "tolerance dose" of 0.1 R/day for radiation workers served

[2]The units of radiation dose are inherently more complex than those used in toxicology or pharmacology, and additional complexity has resulted from frequent changes required by evolving concepts of radiation dosimetry. The rep (roentgen equivalent physical) was the unit of an absorbed dose having the magnitude 93 erg/g of absorbing material. It was replaced by the rad (D), the unit for an absorbed dose of 100 erg/g. The rem has been the unit of dose equivalent (H), which is the product of D, Q, and N at a point of interest in tissue, where D is the absorbed dose, Q is the quality factor, determined by the type of radiation, and N is the product of any other modifying factors (such as the distribution of absorbed dose in space and time) (ICRU, 1971). These units are now in the process of being replaced by the SI (Systeme Internationale) units.

successfully during the period of great expansion in the medical use of X ray, industrial radiography, and the early development of the atomic energy program. There was no evidence that that standard resulted in unsafe conditions for the workers, but development of the concept of stochastic dose–response relationships, and the accompanying statistical problems referred to in Chapter 2, suggested to some that the permissible dose should be reduced. In 1949, in a report that was not published formally until 1954, the NCRP recommended that the dose from occupational exposure be reduced from 0.1 R/day to 0.3 rem/week (NCRP, 1954).

In 1956 another change was made, this time by the ICRP, which limited the dose to the gonads and blood-forming organs to 5 rem per year. The NCRP adopted this recommendation and further stipulated that the accumulated dose to individuals should be limited to $5(N - 18)$ rem, in which N is the age of the individual in years. This formula also reflected the position that an individual should not be exposed to ionizing radiation before the age of 18. A further stipulation by both ICRP and NCRP was that the dose received during a 13-week period should not exceed 3 rem.

The average dose over a working lifetime was thus reduced to about 5 rem/year, one-sixth of the limit used during World War II. The ICRP and NCRP recommendations were quick to find their way into the regulations of the governmental agencies.

Another important development took place in 1956, when reports by the Medical Research Council of the United Kingdom (MRC, 1956) and the U.S. National Academy of Sciences (NAS–NRC, 1956) provided the first quantitative insights into the genetic effects of population exposure. The U.S. government in 1960 established an interagency Federal Radiation Council (FRC) (now inactive as such), which published Radiation Protection Guides that followed the $5(N - 18)$ formula of NCRP and ICRP for occupational exposure and, in addition, recommended that no individual in the general population be exposed to more than 0.5 rem per year and that the "average" dose to members of the general population not exceed 5 rem in 30 years (0.17 rem/year). This limit was based on genetic considerations (FRC, 1960).

The next and, as of this writing, the last major ICRP change took place in 1977 (ICRP, 1977), when it was recommended that exposure be limited to 5 R/year, but that the practice of allowing exposure accumulation by the formula $5(N - 18)$ be discontinued. The age proration formula is still permitted in the United States but is rarely called upon in practice, and is probably destined to be abandoned in the not too distant future. ICRP concluded that the risk to the workers if exposure were limited to 5 R/year would be no greater than the risks taken by employees of other industries that have relatively good safety records.

TABLE 3–1

Dose Equivalent (L) to a Tissue Giving Same Risk
as 0.05 Sv (5 rem) to Whole Body[a]

Organ or tissue	L (rem)
Gonads	20
Breast	33
Red marrow, lung	42
Thyroid, bone (endosteum)	167
Five other tissues receiving the greatest dose in the remainder	83

[a]Vennart (1981). Reproduced from *Health Physics*,
Vol. 40, by permission of the Health Physics Society.

Between 1960 and 1977, the system used by ICRP (ICRP, 1960) to specify the maximum permissible concentration of a radionuclide in air or water was based on the criterion that the dose to the "critical organ," i.e., the organ that receives the highest dose from a radionuclide absorbed into the body, should not exceed a stated annual dose rate after 50 years of exposure. The limit was 5 rem/year for the gonads and 15 rem/year for other organs, except that the limit for bone-seeking radionuclides was based on analogy with 0.1 μCi ^{226}Ra, which, with an RBE of 10, delivers an annual dose of about 30 rem.

The 1977 ICRP report included a system of permissible limits that considered the differences in radiosensitivity of the various organs and tissues based on risk coefficients estimated from epidemiologic data. Under the revised ICRP system, the limits are based on the criterion that the risk of organ irradiation should not be greater than that from 5 rem/year (Vennart, 1981). Based on the information available from epidemiologic sources, ICRP introduced the factor L, which is the dose equivalent to individual organs that results in the same risk as irradiation of the whole body at a rate of 5 rem/year. The L values recommended by ICRP are given in Table 3–1. The major change that has taken place as a result of these ICRP reports is that the dose limits are now being stated on the basis of the risk that will be incurred (Peterson, 1984).

LIMITS FOR EXPOSURE OF THE GENERAL PUBLIC

The question arose during the 1950s as to how the standards should be modified to protect the general public. The subject assumed increased importance at that time because of concerns about worldwide exposure to fallout from testing of nuclear weapons in the atmosphere. The tests coincided with increasing agreement within the scientific community that the assumption of a threshold should be abandoned. The absence of a threshold was first invoked for genetic effects but was soon extended to cancer (Lewis, 1959), and this resulted in an increasingly conservative attitude toward ionizing radiation exposure that has persisted to the present time and has led to great caution in establishing levels of maximum permissible exposure for the population at large.

It is general practice in public health to set lower limits of exposure of the general public to pollutants than are permitted for those exposed occupationally. This practice has developed for at least three reasons:

1.   It seems reasonable to expect an occupationally exposed individual to accept some measure of risk as long as the risk is no greater than other on-the-job risks which are accepted as part of modern living. In most cases the individual accepts occupational risks knowingly. However, these arguments do not hold for members of the community.

2.   Members of the general population may be exposed prenatally or in childhood. Children and the fetus may be more susceptible to injury than adults.

3.   Exposure of the public may be complicated by ecological relationships that influence the passage of radioactive materials to humans. In occupational exposure, one is normally concerned only with external radiation exposure and airborne contamination. However, control over exposure of the general population requires that the regulations also minimize internal exposure that may occur via complex environmental pathways (see Chapters 4–6).

The assumption that there is no threshold for either the genetic or carcinogenic effects of ionizing radiation exposure resulted in the basic precept that there should be no man-made exposure without the expectation of benefit (FRC, 1960). In one way or another, this principle has been stated in many publications of the ICRP, NCRP, and FRC, and it can also be found in statements made by officials of the World War II atomic energy program (Auxier and Dickson, 1983). The concept was first stated by NCRP in a report (NCRP, 1954) which emphasized that radiation criteria involved value judgments that must weigh both the benefits and risks of exposure. The 1954 NCRP report stated further that "exposure to radiation be kept at the lowest practicable level," a principle that is now strongly entrenched in radiation protection practice. It has come to be known by the acronym ALARA (as low as reasonably achievable).

How Radiation Protection Limits Are Established

On the assumption of stochastic dose–response relationships, the process of setting meaningful radiation standards cannot begin until sufficient experimental or epidemiologic information exists to permit estimates of risk coefficients to be made. This was not always so. Prior to about 1950 it was thought that there were threshold doses below which radiation injury would not occur. However, once it was accepted that the genetic and carcinogenic effects of radiation are stochastic as well as nonthreshold, the problem of standards setting became more difficult. The question was no longer "what dose is safe" but "how safe is safe enough." This no longer involved purely scientific judgment; it is a question that could only be answered by making social judgments.

Of primary importance in the process of setting standards is the need to assemble the published scientific literature on the biological effects of radiation so that there can be agreement as to what is and is not known. This role has been performed effectively on an international scale since 1955 by the United Nations Scientific Committee on the Effects of Atomic Radiation, which has published a series of reports that summarize the state of our basic knowledge with respect to biological effects as well as the behavior of radioactive material in the environment. The publications of ICRP and NCRP, in addition to providing guidance for the establishment of radiation standards, include much basic information on the biological effects of radiation. Other important sources of information are the periodic reports of the National Academy of Sciences (NAS) Committee on the Biological Effects of Ionizing Radiation (NAS–NRC, 1980).

UNSCEAR and the NAS have limited their role to assembling the basic scientific information and, in recent years, to estimates of the risk coefficients, i.e., the increased probability of occurrence of genetic effects or cancer per unit of radiation dose per unit of population. Neither UNSCEAR nor NAS recommends limits of acceptable dose.

A number of agencies, including the Nuclear Regulatory Commission (NRC), the Environmental Protection Agency (EPA), the Food and Drug Administration, and the Department of Transportation, have the responsibility of using the available scientific information, including the recommendations of the ICRP and NCRP, to jointly develop the numerical limits of permissible exposure through an interagency committee.

Having decided on a level of permissible dose, there remains the need to specify the design, construction, and operating practices that must be followed to ensure that the limit will not be exceeded. For example, the risks due to transportation of radioactive materials can be more efficiently controlled by specifying packaging and shipping procedures than by simply providing numerical limits of exposure to people along the route. More and more, as experience is accumulated, the radiation protection programs will be implemented by specifying operating practices that preclude the possibility of contamination of the environment above acceptable limits.

BASIC STANDARDS OF MAXIMUM PERMISSIBLE RADIATION EXPOSURE

As noted above, the ICRP and NCRP recommendations do not have the force of law unless the official agencies choose to adopt them, as was the case in the United States during the period 1947–1971, when the now defunct U.S. Atomic Energy Commission incorporated the NCRP recommendations into its regulations.

In more recent years, there has been a general tendency on the part of both EPA and NRC to promulgate standards far lower than those recommended by ICRP and NCRP. Because these regulations are in a constant state of flux, the reader should rely on the official publications of the various agencies for detailed information about the standards in effect at any given time. The basic NCRP and ICRP reports that influence current regulatory policy are listed in Table 3–2. The more important of the Nuclear Regulatory Commission regulations are listed in Table 3–3.

UNDERLYING FACTORS IN STANDARDS SETTING

The promulgation and implementation of standards have become increasingly complex in recent years. This is not because past standards-setting techniques have proved unsatisfactory; there is every reason to believe that the standards that became available a decade or more ago were adequate to provide at least the same margin of safety accepted for other standards applicable to toxic or carcinogenic substances. Rather, the changes that have taken place have been due to growing conservatism in regard to standards setting generally.

The process is also complicated by a number of difficult public policy questions that arise out of the assumption that the dose–response curve is linear at low doses and that there is no threshold. This assumption makes it seemingly possible to quantitate effects at the most minuscule level of exposure. Modern methods of measuring radioactivity can detect quantities of alpha-emitting radionuclides as low as 1×10^{-18} Ci and result in dose estimates that become vanishingly small. Moreover, modern methods of environmental modeling now make it possible to predict levels of exposure far into the future, and at concentrations that greatly exceed the measurement capabilities of even our most sophisticated instrumentation.

TABLE 3–2

SOME BASIC ICRP AND NCRP REPORTS THAT INFLUENCE U.S. STANDARDS

Report no.	Date	Title
		ICRP
9	1966 (b)	Recommendations of the International Commission on Radiological Protection (adopted September 17, 1965)
22	1973 (b)	Implications of Commission Recommendations That Doses Be Kept as Low as Readily Achievable
23	1975	Reference Man: Anatomical, Physiological and Metabolic Characteristics
26	1977	Recommendations of the ICRP
30	1979	Limits for Intakes of Radionuclides by Workers
31	1980 (a)	Biological Effects of Inhaled Radionuclides
32	1981	Limits of Inhalation of Radon Daughters by Workers
		NCRP
17	1954	Permissible Dose from External Sources of Ionizing Radiation (including Maximum Permissible Exposure to Man, Addendum to National Bureau of Standards Handbook (1958)
39	1971	Basic Radiation Protection Criteria
43	1975 (a)	Review of the Current State of Radiation Protection Philosophy
46	1975 (d)	Alpha-Emitting Particles in Lungs
53	1977 (b)	Review of NCRP Radiation Dose Limit for Embryo and Fetus in Occupationally Exposed Women
64	1980 (a)	Influence of Dose and Its Distribution in Time on Dose–Response Relationships for Low-LET Radiations

For example, as a result of the accident that occurred in 1979, the mean per capita dose to persons living within 50 miles of Three Mile Island was estimated to be about 1 mrem. This mean dose was not measured—this is not possible even with the best available instrumentation—but was estimated by using environmental dosimetric models, such as those to be discussed in subsequent chapters.

The assumption of linearity and the use of risk coefficients make it possible to estimate the numbers of cancers that will result theoretically from exceedingly small doses applied to very large populations. The risks to individuals may be minute, but a finite number of cancers can be predicted if a sufficiently large population is exposed. The assumed absence of a threshold has led to the concept of ALARA, but a fundamental problem is that there is no generally accepted definition of the word "reasonable," and the ALARA criteria vary widely from country to country. The

TABLE 3–3

RELEVANT NUCLEAR REGULATORY COMMISSION REGULATIONS CONTAINED
IN TITLE 10 OF THE CODE OF FEDERAL REGULATIONS[a]

Part	Regulations
20	Standards for protection against radiation
30	Rules of general applicability to domestic licensing of by-product material
35	Human uses of by-product material
50	Licensing of production and utilization facilities
60	Disposal of high-level radioactive wastes in geologic repositories
61	Licensing requirements for land disposal of radioactive waste
71	Packing of radioactive material for transport and transportation of radioactive material under certain conditions
100	Reactor site criteria

[a]U.S. Office of the Federal Register (a).

ALARA philosophy of radiation protection has been the subject of considerable controversy (Auxier and Dickson, 1983).

If the number of effects to be expected is a linear function of dose, the same number of effects are to be expected whether 1.0 unit of dose is received by one person, or 0.1 unit is received by 10 persons. This, of course, applies only at low levels of exposure, where the dose–response relationship is assumed to be both linear and stochastic. Thus, in addition to the dose received by individuals, the "collective dose" can be estimated and expressed in person–rem (prem), using the same risk coefficients that are used to calculate the probability of effects in a single individual.

For many types of exposure it becomes possible to estimate the collective dose to the entire population of the world. Radionuclides such as tritium, ^{14}C, ^{129}I, ^{90}Sr, and others are sufficiently long-lived that they can permeate the global biosphere and the collective dose to the world's population can be estimated. Moreover, for very long-lived radionuclides, it is possible to make estimates of collective dose far into the future. Recent risk analyses of systems of high-level radioactive waste management have projected risks for more than 1 million years into the future (NAS–NRC, 1983a). When this is done, very large numbers of people may be involved, so that even though the risk per individual may be very small, the product of risk per individual times the number of persons results in a finite number of genetic or carcinogenic effects. The U.S. Environmental Protection Agency requires (U.S. EPA, 1985) high-level nuclear waste repositories to be designed so that there is no more than 0.1 cancer per year over a period of 10^4 years. Assuming that the world's population stabilizes at 10^{10} persons in the 21st century and that there will be 300 generations of persons during the next 10^4 years, the exposed population during that period will total about 3×10^{12} persons. At a rate of 0.1 cancer per year, there would be 10^3 cancers in 10^4 years, and the risk per individual would be 3×10^{-10}, which is an exceedingly small increment of risk.

It certainly seems absurd to carry such low dose calculations so far into the future, and the sociopolitical problems that, on the one hand, require such calculations and, on the other hand, are created by them will be discussed in Chapter 15. One possible way out of this dilemma would be to define the lower limit of risk worthy of regulatory concern. Following precedent that has long existed in law, this would require definition of the *de minimis* dose, a term that is derived from the legal principle that the law should not be concerned with trivialities. A number of proposals have been made to define the *de minimis* dose in terms of the normal variations that exist in exposure to natural sources of radiation (Adler and Weinberg, 1978; Comar, 1979; Eisenbud, 1981b).

Changing Concepts of Standards Setting

For about 30 years, standards for internal emitters were based on the dose to the "critical organ," which, for any given radionuclide, is the organ that receives the maximum dose when the radionuclide is ingested or inhaled. Thus, the critical organ for radioiodine would be the thyroid and for ^{90}Sr would be the skeleton. The maximum permissible concentration of any radionuclide in air or water was based on the criterion that the permissible dose to the critical organ should not be exceeded after 50 years of exposure to the maximum permissible concentration.

A major change took place in 1977 with the publication of ICRP Report 26 (ICRP, 1977), in which it was proposed that the criterion for maximum permissible exposure should be based on quantitative estimates of risk, and that the acceptability of risks from radiation exposure should be based on comparisons with other risks that society considers to be acceptable. The NCRP has also proposed a system of standards based on risk, and it is likely that the ICRP and NCRP proposals will be adopted eventually by regulatory agencies in the United States and elsewhere.

A basic step in the establishment of standards is the selection of risk coefficients, i.e., number of radiation-induced cancers expected per unit dose, with due allowance for variations in the radiosensitivity of the various organs and tissues of the body. The methods by which this is done were discussed in Chapter 2.

A important tool in dose estimation is the "Reference Man" (ICRP, 1975), which is a detailed compilation of the anatomical and physiological characteristics required for dose calculation. Included in the tabulations are values for the intrapulmonary deposition and clearance of inhaled dusts of varying solubilities, fractional blood uptake of the various chemical elements from the gastrointestinal tract, fraction of the various elements absorbed by blood that are deposited in various parts of the body, and rates of elimination. The characteristics of the Reference Man are used to develop the dosimetric models required to relate the dose that the various parts of the body will receive from a given quantity of inhaled or ingested radionuclides. The use of Reference Man permits the dose calculations to be made in a uniform way, but it is important to remember that the required anatomical and physiological parameters vary with age and sex, as well as among individuals of any given age group of either sex. It has been suggested on the basis of experimental data that, because of individual variability, a few percent of the individuals in a population may receive a dose five times the average (Cuddihy *et al.*, 1979).

USE OF ANNUAL LIMITS OF INTAKE

Once the maximum permissible dose has been established, it becomes possible to calculate the allowable limits of annual intake (ALI) of any nuclide or combination of nuclides either by inhalation or ingestion (Vennart, 1981). The calculations are facilitated by the anatomical and physiological constants given for Reference Man. Having calculated the ALI, it then becomes a simple matter to calculate the maximum permissible concentration in air, food, or water, using the standardized inhalation and ingestion rates given in Reference Man.

The formal mathematics of the procedure for establishing the ALI is complicated by the buildup and elimination of the radionuclides in multiple organs and tissues due to metabolic processes and to the buildup and decay of chains of radionuclides. The reader who wishes to understand the subject more fully is referred to ICRP Reports 26 and 30 (CRP, 1977, 1979) and to a more comprehensive publication of the U.S. NRC (Till and Meyer, 1983). The list of ALIs published by the regulatory agencies should also be consulted as needed.

In many cases, the ALI will be difficult to administer. For example, if low-level radioactive wastes are discharged to an estuary, it is impractical to make estimates of exposure of individual members of specific nearby populations to external radiation (as, for example, to bathers) or to consumers of fish and shellfish. It is more practical to study the exposure pathways and develop models that describe the dose equivalent to hypothetical members of the population who have various assumed recreational and dietary habits. In this way, *the dose equivalent per unit emission* can be calculated, and the *maximum permissible rate of emission to the estuary* can be defined as that which will result in the maximum permissible dose equivalent as determined by the exposure pathway model. Compliance with the standard can then be ascertained by monitoring the rate of discharge from the facility in question. The methods of exposure pathway analyses will be described in the next three chapters.

There have been attempts to assign a lower limit to further dose reduction in terms of the maximum amount of money that should be expended per person–rem averted. When this approach was reviewed by ICRP (ICRP, 1983), it was noted that there has been a wide range of suggested values, ranging from $10 to $1000. Although no official policy has been adopted in the United States, a value of $1000 per person–rem averted has been widely used in informal risk evaluations (U.S. NRC, 1980b). This can involve an expenditure of $10 million per fatal cancer averted if the risk is taken to be 1×10^{-4}/rem.

Some persons find it repugnant to equate death to dollars, but this is done in many ways in the field of public health. The number of ambulances provided for a city, the number of firemen and policemen, and the amount spent for health education are all determined by weighing costs against benefits measured by injuries or deaths averted. Thus, Cohen (1980) has estimated that the cost per fatality averted is about $65,000 for rescue helicopters and $34,000 for improved highway guardrails. It has been estimated that compulsory installation of smoke detectors in bedrooms would save lives at a cost of $40,000 per death averted (Graham and Vaupel, 1981). Society seems prepared to spend far more to prevent death from radiation exposure than from many of the more commonplace hazards that exist. This will be discussed further in Chapter 15.

Organizations Involved in Establishing and Implementing Radiation Protection Standards

Although there is reasonably good agreement among the various scientific groups about many of the quantitative aspects of radiation injury, there is less agreement among the regulating and regulated groups as to how the basic recommendations that issue from the scientific bodies should be administered in practice; in other words, how does one relate the standards based on epidemiologic and experimental evidence to the apparatus of public health regulation? Public health administration is more an art than a science. The health administrator is accustomed to starting with a mixture of scientific information of various grades of quality, which he must evaluate according to existing concepts of permissible risks. The administrator must then fabricate a system of regulation that is understandable, that is practical, and that, above all, protects public health.

Public health officials are accustomed to making compromises. Thus, standards for the biological quality of drinking water are often based on an innocuous group of coliform organisms because one is basically interested in certain pathogenic organisms that originate in fecal pollution for which the coliform organisms are a useful indicator. This system of control does not provide absolute safety, but in most cases it provides adequate safety. For many decades this has been a practical system of control that has lent itself to practical systems of enforcement. Other regulatory mechanisms could be designed which could provide more safety, but they might not be practical to administer or enforce.

NUCLEAR REGULATORY COMMISSION

Primary responsibility for regulating the nuclear energy industry in the United States rests with the Nuclear Regulatory Commission, which was formed in 1974 and was given the responsibilities for health and safety that were originally assigned to the Atomic Energy Commission.

The NRC has responsibility for regulating the civilian nuclear power industry, including the design and operation of power reactors, fuel manufacture, spent fuel processing, and waste management. In addition, the NRC regulates the production and use of radioactive materials ("isotopes") that are produced by reactors and used in research, industry, and medicine (Minogue, 1978). Under the law, NRC has the authority to transfer to the states its responsibilities concerned with regulation of reactor-produced radionuclides. The radionuclides that occur naturally or are produced in a cyclotron or other particle accelerators are not covered by NRC regulation (Moghissi et al., 1978a).

The NRC rules and regulations are published in Title 10 of the Code of Federal Regulations [U.S. Office of the Federal Register (a)], which is divided into many parts, not all of which involve radiation protection. The parts that are concerned with radiation protection are listed in Table 3–3.

ENVIRONMENTAL PROTECTION AGENCY

The environmental functions of the federal government were reorganized in 1970 by creation of the EPA, within which an Office of Radiation Programs was established and given responsibility for the development of radiation protection guidelines and environmental radiation standards. This relieved the NRC of the responsibility for establishing environmental standards and also took over the functions of the Federal Radiation Council, which was an interagency council established during President Eisenhower's administration. Surveillance functions that were formerly the responsibility of the Public Health Service were also transferred to EPA. Thus, the EPA has the responsibility for establishing federal standards of radiation protection and also maintaining surveillance of air, food, and water. The NRC continues to have responsibility (within its area of responsibility) for issuing the regulations by which the EPA standards will be achieved. If EPA specifies that the dose to members of the general population should not exceed a given level, it is the responsibility of the NRC to prepare regulations that will ensure compliance with the EPA limit.

Food and Drug Administration

The Bureau of Radiological Health (BRH) within the FDA has responsibility for developing safety standards for a number of radiation sources not controlled by the NRC. These include naturally occurring radioactive materials used in consumer products, X-ray generators, and radionuclides produced by accelerators (in contrast to reactor-produced radionuclides, which are regulated by NRC). The recommendations of the BRH do not have the force of law, since responsibility for regulation in its area of interest resides in the states. However, BRH fulfills the important function of ensuring uniformity by developing standards which can then be adopted by the states (Neill, 1978).

The States

Under federal law, the NRC is permitted to delegate certain of its health and safety responsibilities to the states, a number of which now control radionuclide use. A basic condition is that the state should have the technical capability to administer its regulations, which must be compatible with those of NRC. By 1983, 26 states had consummated such agreements with the NRC. In some instances, such as the transportation of radioactive materials, local jurisdictions have established regulations without any delegation of authority from the NRC. Dual jurisdiction can thus exist and, although most states have adopted the basic standards proposed by the NRC, there are some differences in interpretation and procedures which result in some confusion. The states are not permitted to issue standards that are less stringent than those of NRC, but the courts have allowed the states to promulgate standards that exceed the NRC requirements.

MUNICIPALITIES

Large cities, such as New York, may have the authority to regulate radiation sources, and, in general, they tend to implement their authority by means of codes similar to those used by the states.

DEPARTMENT OF TRANSPORTATION

The regulations governing shipment of radioactive materials are the responsibility of DOT. Its regulations are published in Title 49 of the Code of Federal Regulations [U.S. Office of the Federal Register (b)]. This is discussed in more detail in Chapter 10.

INTERNATIONAL AGENCIES

A number of international organizations issue recommended safe practices from time to time. Among these are certain specialized agencies of the United Nations, such as the International Labor Organization, the World Health Organization (WHO), the International Atomic Energy Agency, and the Food and Agriculture Organization. These organizations do not have the authority to require adoption of their recommendations by any of the member countries.

Chapter 4

Some General Considerations Concerning Environmental Radioactivity, Including the Use of Models

The next two chapters will be concerned with the mechanisms by which humans are exposed to radioactive materials that enter the general environment. The mechanisms are often complex, and a quantitative understanding of them for the purpose of risk assessment involves information that comes from many scientific disciplines, including biology, physical chemistry, hydrology, meteorology, and oceanography.

Human beings can be exposed to environmental radioactivity in a number of ways. When a γ-emitting substance is deposited on surrounding surfaces, the ambient radiation background will be increased. If the contaminant is initially airborne, the radioactive particles may be inhaled or may deposit on the skin. Particles deposited on surfaces can be resuspended by wind action or human activity and can then be inhaled. The most complex mechanisms are those that involve contamination of food chains.

The ways in which human exposure can occur are illustrated in Figs.

47

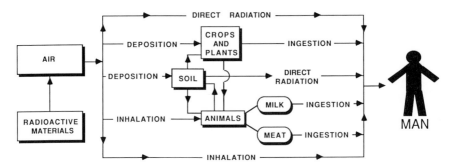

Fig. 4–1. Pathways between radioactive materials released to the atmosphere and man. (From U.S. Environmental Protection Agency, 1972.)

4–1 and 4–2. The pathways are often more complex than shown, and the fine details of some of the transport mechanisms are not always understood. The environment is crisscrossed with physical, chemical, and biological pathways that link the biophysical and biochemical processes on which all organisms depend for their existence. There are subtleties that may increase the opportunities for human exposure, but there are also natural protective barriers.

The importance of ecological pathways was appreciated early in the development of atomic energy when it was found that the hazard to human beings from airborne radioactive iodine was not due entirely to inhalation

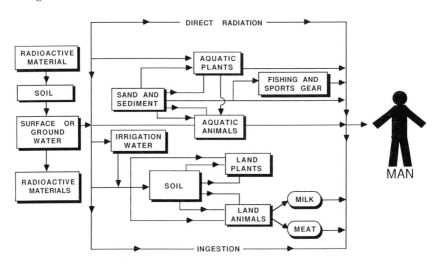

Fig. 4–2. Pathways between radioactive materials released to ground and surface waters (including oceans) and man. (From U.S. Environmental Protection Agency, 1972.)

of the dust or vapor but rather involved a pathway of deposition on leafy surfaces such as forage or grass, followed by grazing of dairy cows and secretion of radioiodine in cow's milk (Chapter 6). Because a dairy cow grazes a large surface area, it absorbs more radioactive iodine by ingestion than by inhalation, and since the radioiodine is secreted with milk, it can in this way be passed directly to humans in a relatively concentrated form.

Comparison of Chemical and Radioactive Contamination

There are both differences and similarities between radioactive and chemical contaminants. For most chemical elements to be toxic, they must be incorporated into specific molecular forms. Thus the aromatic hydrocarbon benzene (C_6H_6) is highly toxic and can cause severe injury to the blood-forming tissues if inhaled. If the benzene is fully oxidized, its toxicity disappears and two new innocuous compounds are formed (CO_2 and H_2O). Far from being toxic, both carbon and hydrogen are essential to all living forms. It is the ring-shaped molecule C_6H_6 that is toxic, not the elements from which the molecule is constructed. However, if the original C_6H_6 was formed from the radioactive isotopes ^{14}C and ^3H, both the CO_2 and H_2O would continue to be radioactive. Changes in the chemical compounds in which radionuclides are incorporated cannot alter their basic radiotoxic properties, except as the particular compounds have different ecologic or metabolic properties.

Many other examples can be cited. Thus, the element zinc may or may not be toxic when inhaled, depending not only on the chemical form in which it exists, but also on its physical state. The effect of inhaling zinc oxide is different from the effect of inhaling zinc carbonate, and the effects are also different depending on whether the zinc oxide is inhaled as a freshly formed fume or as a dust. Moreover, the toxicity disappears when the zinc settles from the atmosphere and becomes part of the pool of inert zinc that already exists in the environment. However, if the zinc is in the form of a long-lived radionuclide, the pool of environmental zinc becomes contaminated, and the isotope will appear as a radioactive contaminant in living plants from the soil or marine organisms taken from the sea.

The fact that the radioactive substances are toxic in such small amounts is often given as an additional example of a basic difference between chemically toxic and radiotoxic materials. For example, industrial hygienists permit workers who are occupationally exposed to lead to breathe air containing up to 50 μg Pb/m^3, whereas the maximum permissible concentration of radioactive ^{210}Pb in air is 4×10^{-4} $\mu Ci/m^3$, equivalent to only 5×10^{-6} $\mu g/m^3$. The limit for occupational exposure to stable lead is thus 10 million times greater than the limit for radioactive lead.

However, while it is true that, atom for atom, the hazard for radioactive substances greatly exceeds that of their stable forms, it is likewise true that the amounts of chemically toxic materials with which we are accustomed to deal are greater by many orders of magnitude than the amounts of radioactive materials. Although ^{210}Pb is more hazardous than stable lead by a factor of 10^7, the quantities of this nuclide available for use are rarely greater than a few micrograms, whereas millions of tons of stable lead are used in industry. The far greater amounts of chemically toxic materials that exist as potential sources of pollution may thus offset the greater toxicity of the radioactive isotopes.

Use of Environmental Transport Models

Assessment of the risk to humans from an existing or contemplated source of radioactive emissions to the environment frequently requires a quantitative understanding of the interrelated pathways by which the radionuclides are eventually ingested or inhaled by humans or result in external radiation exposure. This has led to the development of methods of environmental transport analysis that depend on the use of box models such as those shown in Figs. 4–1 and 4–2. These techniques have developed very rapidly during the past decade. The various models used to calculate dose to humans from radionuclides introduced into the atmospheric, terrestrial, or aquatic environments have been summarized by Miller (1984b).

As a result of the requirements of the Clean Air and Water Acts with respect to chemical pollutants and the regulations of the Nuclear Regulatory Commission that concern nuclear power plants, some of these methods (models) have been given legal status and serve as the basis for demonstrating regulatory compliance. The reliability of the models is of obvious importance to both the regulators and the regulated (EPRI, 1982). For the regulators, the model is essential in judging whether the performance of emission controls will ensure compliance. For those regulated, the models not only ensure compliance but also may help to avoid the unnecessary expenditures that can result from use of overly large safety factors.

Unhappily, in the present state of knowledge, it is often much easier to label the environmental compartments and to assign the nomenclature for the various transfer coefficients than it is to assign credible numerical values to the identified parameters. For example, Table 4–1 lists some of the factors that influence the dose to humans as a result of contamination of a pasture by rainout. The ultimate dose to humans is affected by each of

TABLE 4–1

PARTIAL LIST OF FACTORS THAT INFLUENCE DOSE TO HUMANS FROM DEPOSITION OF A RADIONUCLIDE BY RAINOUT OVER A PASTURE

---

1. Rate of rainfall
2. Size of raindrops
3. Particle size distribution of radionuclide
4. Percent of ground area covered by forage
5. Percent of rainout retained initially by forage
6. Rate of removal of fallout from forage by effects of weathering
7. Soil-to-plant uptake factors via root absorption
8. Quantity of radionuclide ingested by cow
9. Fraction of each of the radionuclides that passes from the gut to blood (for cow and humans)
10. Fraction of the radionuclide in blood that transfers to each organ (for cow and humans)
11. Rates of metabolic elimination from each organ
12. Concentration of the radionuclide in meat or milk
13. Quantity of milk and meat consumed by humans
14. Dose rate conversion factors following intake in humans

---

the 14 parameters listed, as well as others that have not been included. We will see in subsequent chapters that the values of some of these parameters cannot be specified within an order of magnitude. Since the ultimate effect of many of these parameters on the dose assessment will be multiplicative, the uncertainties in the dose estimates may cover several orders of magnitude (Kocher *et al.*, 1983). Thus a word of caution is necessary in regard to the use of mathematical models in their present state of development.

Although many risk assessment models are now in use, and many of them have been incorporated into federal regulations and therefore have the force of law, there have been very few opportunities to verify the model predictions and insufficient information exists about the uncertainties associated with their use. As stated in NCRP (1984a), "Only a few specific quantitative examples of model uncertainty can be given because of the limited extent to which environmental transport models have been validated in the field or evaluated through statistical studies." There are, however, important exceptions. It will be seen in Chapter 13 that the model for the transport of ^{90}Sr from the atmosphere to soil, grass, dietary calcium, and human bone provides excellent estimates of the dose from ^{90}Sr in the environment. However, this model has been under development for 30 years and has the advantage that ^{90}Sr has been present in the environment in sufficient amounts to be readily measurable, thereby permitting model verification and "fine-tuning." Many models, particularly those used for risk assessment in connection with high-level waste management, are used in situations in which verifying measurements cannot be made.

However, there is a pressing need to use predictive models, despite the relatively large uncertainties associated with the model parameters and the paucity of confirmatory studies.

An inherent difficult is that the predictions made by environmental models can be no more accurate or precise than the parameters made available to the modeler. It will take many years for models to evolve to a state where bounds of uncertainty can be assigned with confidence to the predicted dose estimates. This is particularly true for models of transport in aquatic and terrestrial systems. It will be seen in Chapter 5 that atmospheric diffusion models, which have been evolving for about 45 years, have now developed to a state where reliable estimates of concentration can be made downwind from a continuous point source of emission. The fact that this can be done with emissions to the atmosphere is due in part to the fact that atmospheric diffusion, while so complex that it is not yet fully understood, is a relatively predictable process compared to transport through geological media, or convection, diffusion, and sorption processes encountered in the aquatic environment.

Uncertainties in Use of Models

The models used to estimate the dose from stored radioactive wastes or wastes discharged into the ground or surface waters have become so complex that it is virtually impossible for anyone who is not intimately involved with the art of environmental modeling to take the time to form an independent judgment of the uncertainties involved in the dose assessment process. The uncertainties associated with the individual parameters used in the model are usually best understood by specialists from the many disciplines that provide the data used by the model developers. However, if the model developers do not fully appreciate the uncertainties involved in each of the many parameters provided by the specialists, model predictions may result in risk estimates that have substantial but unstated uncertainty.

In his review of atmospheric scavenging processes, Slinn (1984) lists 11 atmospheric diffusion numerical models that include scavenging. The models are of varying complexities; one code uses more than 1500 FORTRAN statements to describe the scavenging process and requires 10^5 words of core memory. This is understandable when one considers the complexity of the physical factors involved, including electric fields, turbulence, chemical compositions, size and flux of raindrops, and size of the particles of contamination, among others.

The air–pasture–cow–milk–child thyroid pathway for ^{131}I is probably the most thoroughly studied of the routes of human exposure and does permit reasonable dose estimates to be made. The dose to the child thyroid can be estimated by using the following linear equation consisting of 10 independent variables (NCRP, 1984a):

$$D/\chi = K{\cdot}V_\mathrm{D}{\cdot}1/\lambda_\mathrm{eff}{\cdot}Q{\cdot}F_\mathrm{S}{\cdot}F_\mathrm{M}{\cdot}F_\mathrm{P}{\cdot}M{\cdot}F_\mathrm{T}{\cdot}1/m_\mathrm{T}{\cdot}1/\lambda_\mathrm{T\ eff} \qquad (4\text{-}1)$$

where D/χ is the thyroid dose in children in the age group 0.5–2 years resulting from a unit concentration (χ) of ^{131}I in air from the pasture–cow–milk pathway, K is a constant (897 mrem g sec/pCi day^2), and the other parameters are described in Table 4–2.

In Table 4–2, Hoffman and Baes (1979) have summarized the uncertainties in each of the 10 variables enumerated, and it is seen that the 84th percentile dose is 2.7 times the median dose and the 99th percentile dose is about 11 times the median. Thus, in a large population, the estimated mean dose would greatly underestimate the dose to many children. This degree of uncertainty, however, is quite acceptable compared to other uncertainties with which one must deal in the field of public health.

In contrast, Garten (1980) has estimated that the most probable dose from [239]Pu for the soil–vegetation–human bone pathway is 4.7×10^{-3} mrad/year pCi g soil, and that the estimated upper 95th percentile dose would be 320×10^{-3} mrad/year pCi g soil, a factor of nearly 100 times greater than the most probable dose.

The terrestrial food chain models, for all their complexity, are among the more simple environmental models and are ones for which there have been a limited number of opportunities for field verification. Models used to predict the dose from radioactive wastes dispersed in the aquatic environment often involve much greater uncertainty, and there have been fewer opportunities for field verification. In one such verification study of the Hudson estuary thaf extended over several years, Linsalata *et al.* (1985a, 1986b) found that the dose to humans from consumption of fish estimated by use of a model recommended by the NRC exceeded the dose estimated from environmental measurements by factors that ranged from 40 to 183.

The Columbia River downstream from the plutonium production reactors at Hanford in the state of Washington is one of the best studied rivers in the world. Effluent transport from the Hanford reactors and chemical processing plants has been studied ever since operation began more than 40 years ago. Models for calculating dispersion of the effluents have been

TABLE 4–2

Results of an Analytical Approach to a Parameter Imprecision Analysis for the $^{131}I_2$ Air–Pasture–Cow–Infant Pathway (after Hoffman and Baes, 1979)[a]

Parameter	Symbol	Geometric mean (units)	μ[b]	σ^{2c}	Contribution to total imprecision (%)[d]
Biomass normalized deposition parameter for I_2	V_D	0.12 (m³/kg sec)	−2.1	0.002	0.18
Effective mean residence time on pasture vegetation	$1/\lambda_{eff}$	6.1 (days)	1.8	0.02	1.8
Total dry matter intake by a dairy cow	Q	15 (kg/day)	2.7	0.014	1.3
Fraction of dairy feed composed of fresh forage	F_s	0.42 (—)	−0.87	0.058	5.3
Milk transfer coefficient	F_M	0.01 (day/liter)	−4.6	0.3	27.6
Fraction of time during a year cow receives fresh forage	F_p	0.37 (—)	−1.0	0.17	15.6
Annual milk consumption of children (ages 0.5–2 years)	M	300 (liter/year)	5.7	0.04	3.7

Parameter	Symbol	Value (units)	μ[b]	σ^2[c]	
Fraction of ingested iodine that deposits in the thyroid gland	F_T	0.30 (—)	-1.2	0.084	7.7
Reciprocal of the thyroid mass	$1/m_T$	0.57 (g^{-1})	-0.56	0.25	23.0
Effective mean residence time of ^{131}I in the thyroid	$1/\lambda_{T_{eff}}$	6.5 (days)	1.86	0.15	13.8
Dose-to-air concentration ratio[d]	D/χ	5100 (mrem m^3/pCi year)	8.53[e]	1.09[d]	100
Dose-to-air concentration ratio[d] D/χ (mrem m^{3}pCi^{-1} yr^{-1})	Mode	1700	X_{84}[f]	14,000	
	Median	5100	X_{95}	28,000	
	Mean	8700	X_{99}	57,000	

[a]National Council on Radiation Protection and Measurements (1984a).

[b]μ, Mean of logarithms of parameter values. All values are accurate to only two significant figures.

[c]σ^2, Variance of logarithms of parameter values.

[d]These values are based on the assumption that all parameters are statistically independent; changes are expected with future quantification of parameter covariance.

[e]Calculated by adding ln(897 mrem g sec/pCi day^2) to 1.77, the sum of the means of logarithms of values for each parameter.

[f]X_{84}, X_{95}, X_{99} are the upper 84th, 95th, and 99th percentiles, respectively, of the predicted distribution of the D/χ ratio.

TABLE 4–3

MONITORED AND PREDICTED CONCENTRATIONS IN THE COLUMBIA
RIVER DOWNSTREAM FROM HANFORD (1984)[a]

Radionuclide	Downstream contribution from Hanford (pCi/liter)		
	Predicted	Measured	Predicted/ measured
^{3}H	1.4	40	29
^{60}Co	0.01	0.009	1.1
^{89}Sr	0.01	0.01	1.0
^{90}Sr	0.07	0.03	2.3
^{131}I	0.04	0.02	2.0
^{137}Cs	0.003	0.006	0.5

[a]Adapted from Price et al. (1985).

developed and improved on, using the abundant site-specific information
that has been gathered. It is therefore not surprising that there is excellent
agreement between the predicted and measured downstream concentrations of the monitored radionuclides. This is seen in Table 4–3, which
shows excellent agreement for all nuclides except ^{3}H, which was found to
be present in much higher concentrations than predicted. The discrepancy
is explained by the fact that the ^{3}H originates not only from the monitored
point sources of discharge, but also from ground water seepage (Price et
al., 1985). Table 4–3 is an excellent example of the value of using site-
specific parameters in the development of models. The same point was
stressed in the Hudson estuary studies, in which it was shown that simple
incorporation of the site-specific bioaccumulation factors for radiocesium
would reduce the overestimates of the NRC model by a factor of 8 (Linsalata et al., 1986b).

Even more complex are the models used to predict the long-term consequences of permanent isolation of radioactive wastes in geologic media
(Chapter 11). These models are intended to be applicable for tens of
thousands of years into the future and cannot possibly be tested in the
field. The models deal with the rates of deterioration of the engineered
safeguards designed to isolate the wastes, the rates of radionuclide dissolution and transport by ground water through geologic media (over
periods greater than 10^{3} years), the eventual movement of the ground
water into surface waters, uptake of the contaminants by biota, and food
chain transport to humans.

The uncertainties that beset our understanding of these processes have been reviewed by Kocher *et al.* (1983). The uncertainties were considered to be quantifiable only after the radionuclides have entered the biosphere, because only then is it possible to incorporate uncertainty estimates based on field data. Yet even within the biosphere, where the greatest amount of experience exists, the uncertainties in dose estimation were shown to encompass several orders of magnitude. The unquantifiable uncertainties in transport within the geosphere include the influence of the heterogeneity of complex geologic systems, the cumulative effects of various geologic processes acting over long periods of time, ground water transport over the required periods of time, and the subtle and complex interacting geochemical factors that influence transport rates.

An important source of bias results from the selection of model parameter values that overestimate the dose because of the understandable tendency of the modeler to select values that provide maximum assurance of safety. Thus, in a model that consists of 10 multiplicative parameters, selection of the 84th percentile for each parameter will result in a predicted dose that exceeds the 99.9th percentile of the predicted dose distribution (NCRP, 1984a; Hoffman and Baes, 1979). The desire on the part of the model user to ensure safety may be an important reason why data obtained from field measurements often yield dose estimates that are much lower than dose estimates made by using mathematical models (NAS–NRC, 1983a; Linsalata *et al.*, 1985a, 1986b).

Despite the uncertainties involved in the use of models, they play an important role in risk assessment and they are frequently required in the course of federal regulatory proceedings. The U.S. Nuclear Regulatory Commission has developed models used to estimate dispersion of emissions to aquatic systems (U.S. NRC, 1976) to the atmosphere (U.S. NRC, 1977a), and to terrestrial systems (U.S. NRC, 1977b). In general, most models tend to be conservative in that the model designer tends to err in the safe direction at each step in which the value of a parameter must be selected.

In the absence of site-specific parameter values, the NRC proposes the use of "default values," which are listed in the regulations. It is often the case in practice that use of the NRC models, including the recommended default values, results in dose estimates that are many orders of magnitude below those of regulatory concern. When this is so, one need go no further, but if the estimates are within one or two orders of magnitude below the level of concern, consultation with specialists and field measurements to provide site-specific data may be in order.

Chapter 5

Transport Mechanisms in the Atmosphere

This chapter will acquaint the nonmeteorologist with the basic principles by which calculations are made of the rates of diffusion and deposition of pollutants that are released to the atmosphere. The treatment will be of necessity superficial, but I hope that it will provide an understanding of some of the methods used in this highly specialized field.

Properties of the Atmosphere

Until World War II, man-made atmospheric contaminants were injected only into the friction layer, which extends to about 1000 m from the ground and which has properties somewhat different from the rest of the atmosphere because of the influences of surface features on atmospheric flow. Transport and mixing within this layer have been studied mainly by meteorologists concerned with dilution of industrial effluents (Randerson, 1984).

The advent of nuclear energy, and more particularly the testing of nuclear weapons, extended the problem of forecasting the fate of atmospheric contaminants considerably above the friction layer to altitudes of

TABLE 5-1

AVERAGE COMPOSITION OF THE ATMOSPHERE[a]

Gas	Composition by volume (ppm)	Composition by weight (ppm)	Total mass ($\times 10^{22}$ g)
N_2	780,900	755,100	38.648
O_2	209,500	231,500	11.841
A	9,300	12,800	0.655
CO_2	300	460	0.0233
Ne	18	12.5	0.000636
He	5.2	0.72	0.000037
CH_4	1.5	0.9	0.000043
Kr	1	2.9	0.000146
N_2O	0.5	0.8	0.000040
H_2	0.5	0.03	0.000002
$O_3{}^b$	0.4	0.6	0.000031
Xe	0.08	0.36	0.000018

[a]Adopted from Mason (1960). Copyright ® 1960, John Wiley and Sons, Inc., reprinted by permission of John Wiley and Sons, Inc.
[b]Variable, increases with height.

100,000 ft or greater. In more recent years the properties of the upper atmosphere, even to the fringes of outer space, have become important to space scientists. An understanding of the physics and chemistry of the upper atmosphere is also required to predict the physical, chemical, and biological effects of emissions from high-flying aircraft.

Although the atmosphere contains many gases, as shown in Table 5-1, more than 99.9% of its weight is due to nitrogen, oxygen, and argon. The relative proportions of these gases remain constant to great heights, but separation due to differences in molecular weight does occur above 60 km. The total mass of the dry atmosphere is estimated to be about 50×10^{17} kg, to which may be added about 1.5×10^{17} kg of water vapor, the most variable constituent of the atmosphere and the one that governs many of its thermodynamic characteristics. Dry air has a density of 1.3 mg/cm^3 at the surface of the earth, where pressure due to the weight of the atmosphere is 760 mm of mercury. At 50 km above sea level, the atmospheric pressure has dropped to 10^{-3} of the sea-level pressure and has attenuated by a factor of 10^{-3} of the sea-level pressure and has attenuated by a factor of 10^{-6} at an altitude of 100 km. The thinness of the atmosphere at these altitudes is illustrated by the length of the mean free path between molecules, about 2.5 cm at 100 km and about 25 m at 300 km (Petterssen, 1968).

Above an altitude of about 600 km, the molecules are thought to behave as satellites in free elliptical orbits about the earth (Mason, 1982).

The atmosphere contains natural and man-made aerosols that originate from many sources. In addition to the air pollutants introduced as the result of human activities, meteorites, volcanic activity, dust storms, forest fires, and ocean spray contribute great quantities of gases and suspended solids.

A gas vapor or aerosol introduced into the atmosphere is diluted by molecular and turbulent diffusion. One can neglect the effect of molecular diffusion, in which the coefficients of diffusivity are many orders of magnitude smaller than those due to turbulence. The total range of values of the diffusivity coefficients that control the rates of atmospheric dilution is enormous, from 0.2 cm^2/sec for molecular diffusion to 10^{11} cm^2/sec for the mixing due to large-scale cyclonic storms in the atmosphere (Gifford, 1968). The atmospheric motions that contribute to the mixing processes thus vary from the microscopic to those measured in hundreds of kilometers.

The motions of turbulent diffusion are so complicated that exact mathematical theories are not available to describe the manner in which a contaminant behaves in space and time when it is introduced into the atmosphere. However, methods have evolved that are basically statistical in nature and which make it possible for one to predict the manner in which a contaminant will diffuse in the atmosphere under a given set of meteorological conditions.

The mixing characteristics of the atmosphere are influenced in a major way by its vertical temperature gradient. A typical temperature profile in the temperate zone is illustrated in Fig. 5–1. As height above the ground increases, the temperature normally decreases at a rate, often referred to as the "lapse rate of temperature," of about 3.5°F per 1000 ft (6.5°C/km). It is seen from Fig. 5–1 that the temperature decreases with height to about 11 km, which is the approximate beginning of an isothermal region of the atmosphere, which, in the example shown, extends to a height of about 32 km. The lower region of the atmosphere, in which the temperature normally decreases with height, is called the troposphere; it contains about 75% of the mass of the atmosphere and almost all its moisture and dust. Above the troposphere, separated by a sharp change in lapse rate at the so-called tropopause, is the isothermal stratosphere. The height of the tropopause varies with latitude and with season of the year. In contrast to the relatively smooth and quiet motions of the stratosphere, the troposphere is a comparatively unstable well-mixed region of the atmosphere. Above the stratosphere, beginning at about 100,000 ft, is the mesosphere, a region characterized by an increasing temperature gradient. The mesophere ex-

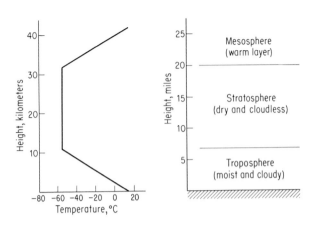

Fig. 5–1. Idealized profile of the atmospheric temperature gradients. (From S. Petterssen; copyright 1958, McGraw-Hill.)

tends upward to the ionosphere, a densely ionized region that begins at a height of about 200,000 ft.

During normal daytime conditions, the earth's surface absorbs solar radiation and becomes warmer than the overlying atmosphere. The temperature gradient then becomes superadiabatic; that is, the temperature decreases at a rate exceeding that which would occur if a parcel of air were permitted to expand adiabatically. The dry adiabatic temperature gradient is normally about $-6.5°C/km$. The influence of the vertical temperature profile on the stability of the atmosphere can be understood from Fig. 5–2. If a parcel of air having a temperature T_1 at altitude H_1 is raised to altitude H_2 hr, it will cool at the adiabatic rate. Since it is assumed in Fig. 5–2 that superadiabatic conditions exist, the parcel of air initially at H_1 will be warmer than the ambient atmosphere when it reaches H_2 hr. It will thus be of lower density than the surrounding air and will continue to rise. Similarly, if a parcel of air is lowered in altitude when the atmosphere is in the superadiabatic condition, the parcel will become more dense than the ambient atmosphere and will continue to descend. Under superadiabatic conditions, all vertical motions tend to be accelerated, and the atmosphere is said to be unstable. Figure 5–3 shows that the reverse situation exists when the lapse rate is less than adiabatic.

In fact, as shown in Fig. 5–3, the temperature gradient may increase with height. This results in the highly stable condition called an inversion. Inversions may be caused by the overrunning of warm air over cold (as along a front between air masses), by advection of cool air at a low level (as during a sea breeze), or by diurnal cooling of the lower layers of the atmosphere.

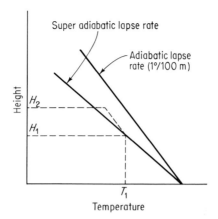

Fig. 5–2. Instability of the superadiabatic atmosphere. A parcel of air raised in height from H_1 to H_2 cools adiabatically, and its rate of rise is accelerated because it becomes warmer and, therefore, less dense than the ambient atmosphere.

This last condition usually develops after sunset, when the surface of the earth, having been warmed during the day, begins to cool more rapidly than its overlying atmosphere. As the night proceeds and the surface of the earth becomes cooler relative to the atmosphere, the temperature gradient may in time become positive.

The above sequence of events is illustrated in Fig. 5–4 (Holland, 1953), which gives an example of how the vertical temperature gradient can

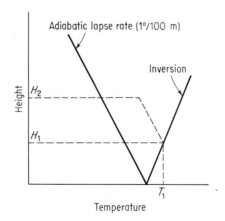

Fig. 5–3. Inherent stability of the inverted temperature gradient. A parcel of air raised in height from H_1 to H_2 cools adiabatically and sinks to its original position because it becomes more dense than the ambient atmosphere.

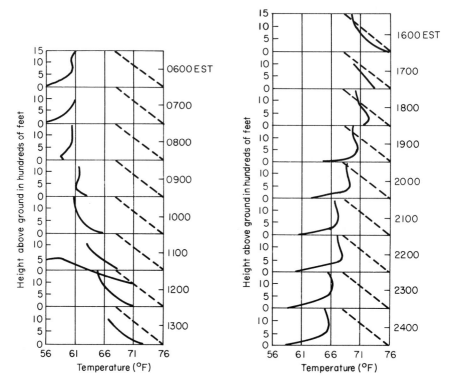

Fig. 5–4. Average diurnal variation of the vertical temperature structure at Oak Ridge National Laboratory during the period September–October 1950 (solid lines). The dashed lines represent the adiabatic lapse rate. (From Holland, 1953.)

change in a 24-hr period. A sharp inversion to a height of about 500 ft is evident at 0600 and has begun to weaken slightly at 0700. By 0800 the morning sun has heated the ground, and the inversion begins to disappear, being replaced by a superadiabatic gradient between 1100 and 1300 hr. By late afternoon, as the ground cools, a surfaced-based inversion develops again and persists throughout the night. If the sun rises on a clear day, the superadiabatic condition will again develop by late morning, and the cycle will be repeated. However, should cloud cover limit solar heating of the earth's surface, the inversion may persist through the day.

The change in temperature profiles greatly affects the characteristics of plumes of stack gases, and it can be appreciated from Fig. 5–5 that the concentration of a contaminant at ground level is influenced by the effect of the vertical temperature gradient on atmosphere stability.

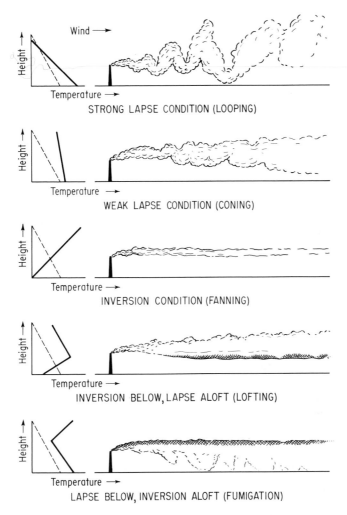

Fig. 5–5. Schematic representation of stack-gas behavior under various conditions of vertical stability. Actual temperature (solid line) and dry adiabatic lapse rate (dashed line) are shown. (From U.S. Weather Bureau, 1955.)

DIFFUSION IN THE FRICTION LAYER FROM A CONTINUOUS POINT SOURCE

There is no exact mathematical description of turbulent diffusion in the atmosphere, but a number of approximations have been developed during the past 50 years. Most of them have in common the assumption that the mean concentration distribution across the vertical and horizontal axes of

the plume cross section follows a Gaussian function. The standard deviations of the mean concentration along the vertical and crosswind axes, σ_y
and σ_z, respectively, depend critically on the degree of turbulence. Both
these diffusion-width parameters increase with downwind distance from a
point source.

An important application of these equations is to describe the rates of
diffusion from a continuous point source of emission. Versions of these
equations have proliferated rapidly in recent years with the development
of computerized methods of handling complex multifactorial processes.

The uncertainties of the models for prediction of downwind concentrations from a continuous point source vary with the type of terrain, period
of time over which observations are made, and distance from the source.
For flat terrain, the models provide predictions that are valid within a
factor of 2 or 3 to distances of 10 km (Crawford, 1978; Little and Miller,
1979; NCRP, 1984a). The models tend to be reasonably accurate for distances up to 140 km for flat areas, but are less accurate for complex
terrain.

In addition to the assumption of a Gaussian distribution, the diffusion
models all assume that the concentration along the downwind axis if the
plume will be directly proportional to the rate of emission of a pollutant
and inversely proportional to the square of the *effective height* of emission,
which is the height of the source above the ground plus the rise due to any
initial buoyancy of the plume.

The most common use for diffusion equations is to estimate the *maximum* concentration of a pollutant along the centerline of a plus at a point
downwind from an elevated point source of emission. If the estimate of the
maximum concentrations proves to be well within the regulatory limits, no
further calculations may be required.

A frequently used equation was developed during early research on
atmospheric diffusion (Sutton, 1953) to describe the ground level concentration downwind from a continuously emitting point source:

$$\chi(x,y) = \frac{Q}{\pi \sigma_y \sigma_z \bar{u}} \exp - \left(\frac{h^2}{2\sigma^2_z} + \frac{y^2}{2\sigma^2_v} \right) \tag{5-1}$$

where χ is the concentration (Ci/m^3) near ground level at downwind point
(x,y); Q is the source strength (Ci/sec); (σ_y, σ_z are the crosswind and
vertical plume standard deviations (m), both functions of x; $\bar{u}$ is the mean
wind speed (m/sec) at the stack elevation, h_s (m); h is the effective stack
height ($h_s + \Delta h$, the plume rise) (m); and x, y are the downwind and
crosswind distances (m).

Numerical values for σ_y and σ_z depend mainly on the degree of atmospheric stability, but also are affected by the roughness and configuration

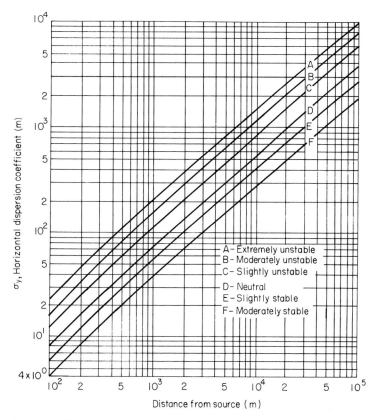

Fig. 5–6. Lateral diffusion (σ_y) vs. downwind distance from source for various turbulence types. (From Gifford, 1968.)

of the terrain, as well as the presence and arrangement of man-made obstacles such as buildings. Therefore, if high accuracy is required, site-specific diffusion coefficients must be determined (Randerson, 1984). However, the values of σ_y and σ_z can be estimated according to stability classification and distance from the source by using Figs. 5–6 and 5–7, as proposed by Gifford (1968). The turbulence classification, based on the degree of atmospheric stability, can be approximated from Table 5–2.

Many authors have developed curves of concentration normalized to wind speed and emission rate ($\bar{u}\chi/Q$) for various stack heights. A widely used set of these nomograms can be found in the handbook by Turner (1970). These curves, which are reproduced in Figs. 5–8 to 5–10, illustrate the principal general features of all equations that describe the diffusion of stack effluents. For a given ambient wind speed, $\bar{u}$ concentration maxima

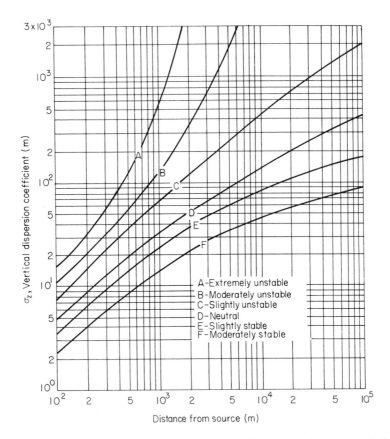

Fig. 5–7. Vertical diffusion (δ_z) vs. downwind distance from source for various turbulence types. (From Gifford, 1968.)

occur at increasing distances within increasing values of h and with increasing degrees of stability. The maximum concentration decreases in inverse proportion to h^2.

The diffusion equation can be used to predict the maximum ground level concentration that will occur downwind from a stack. Solving Eq. (5–1) for $\chi_{\max}$, one obtains

$$\chi_{\max} = \frac{2Q}{e\pi\bar{u}h^2}\frac{\sigma_{\tilde{z}}}{\sigma_v} \tag{5-2}$$

where e is the base of the natural logarithms (2.72). The downwind distance from the stack to $\chi_{\max}$ often is found in the range of 15–20 stack heights, but can occur much closer, as will be discussed.

TABLE 5–2

<small>RELATION OF TURBULENCE TYPES TO WEATHER CONDITIONS[a,b]</small>

Surface wind speed (m/sec)	Daytime insolation			Nighttime conditions	
	Strong	Moderate	Slight	Thin overcast or ≥4/8 cloudiness[c]	≤3/8 cloudiness
<2	A	A–B	B	—	—
2	A–B	B	C	E	F
4	B	B–C	C	D	E
6	C	C–D	D	D	D
>2	C	D	D	C	D

[a]From Gifford (1968).

[b]Conditions: A, extremely unstable; B, moderately unstable; C, slightly unstable; D, neutral (applicable to heavy overcast, day or night); E, slightly stable; F, moderately stable.

[c]The degree of cloudiness is defined as that fraction of the sky above the local apparent horizon which is covered by clouds.

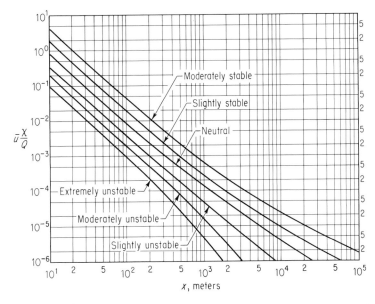

Fig. 5–8. Values of $\bar{u}\chi/Q$ as a function of downwind distance for a source located at the surface. (From Hilsmeier and Gifford, 1962.)

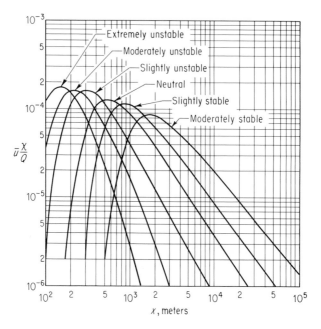

Fig. 5–9. Values of $\bar{u}\chi/Q$ as a function of downwind distance for a source located at a height of 30 m. (From Hilsmeier and Gifford, 1962.)

EFFECTIVE STACK HEIGHT

Pollutants are frequently emitted to the atmosphere at an elevated temperature and with considerable vertical velocity. This is the normal situation for a fossil-fueled power plant as well as for stacks emitting waste gases from many industrial processes. The buoyancy that results from the combined effects of the temperature and velocity of the exhaust gas may result in an *effective* stack height that is considerably higher than the actual height. (Thus, large fossil fuel power plants are designed on the basis of an effective stack height of about 2 h.) The problem of combining the various meteorological parameters together with the effluent temperature and velocity is a complicated one because of the way in which the various parameters interact. For example, while a temperature inversion suppresses vertical motions and might therefore be expected to reduce the effective stack height, the inversion reduces turbulent mixing to an extent that results in a more persistent temperature differential between the plume and the surrounding atmosphere. Consequently, the plume continues to rise at a low rate for a considerable time while slowly spreading downwind. The maximum concentration will be many kilometers from the

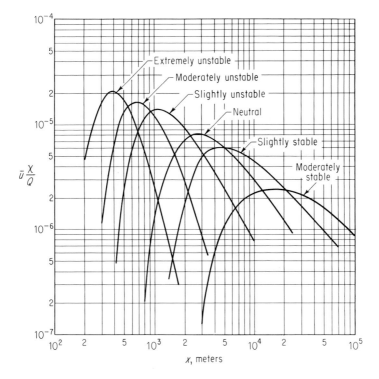

Fig. 5–10. Values of $\bar{u}\chi/Q$ as a function of downwind distance from a source located at a height of 100 m. (From Hilsmeier and Gillford, 1962.)

stack. Conversely, during unstable conditions, with greatly enhanced vertical motions, plume "looping" may occur, and the maximum concentrations may be very close to the stack.

Many investigators have attempted to develop theoretical and empirical methods of computing the effective stack height, taking into consideration the amount of heat in the plume and the ambient vertical temperature gradient (Briggs, 1984; EPRI, 1982). The equations that have been developed are sufficiently complicated and difficult to apply that their use will ordinarily be limited to the micrometeorological specialist. When diffusion equations are used without allowing for the effect of plume boyancy, the estimated concentrations will, as a rule, err on the safe side.

EFFECT OF BUILDINGS AND TERRAIN ON PLUME DISPERSION

One must avoid applying these formulas if the flow patterns are apt to be affected by buildings or local topography. Buildings close to the stack

Fig. 5–11. Perturbation of stack plumes by buildings. Note the effect of increased stack height in eliminating the downwash in the lee of the building. (Courtesy of Professor Gordon Strom.)

may result in the conditions shown in Fig. 5–11, which is a wind tunnel photograph of the effect of local structures on the dispersion of a plume. Micrometeorology in the vicinity of buildings is often too complex for analytical solutions, and in many instances the question of how high a stack should be in order to avoid the down-wash in the lee of a building is best answered by wind tunnel tests.

The general characteristics of the wake downwind of a structure are shown in Fig. 5–12 (Halitsky, 1968), which depicts the flow near a rounded sharp-edged building. Three distinct zones exist: the displacement zone, the wake, and the cavity. The displacement zone is the volume in which the air is deflected around the solid building. Immediately downwind of the building is a region of toroidal circulation, known as the cavity or eddy zone, in which it is possible for high concentrations to accumulate. Beyond the cavity is the true wake, which is a region of high turbulence within which the contaminant spreads throughout. If the area of the wake cross section is taken as the projected area of the building (A), and if it is assumed that the contaminant is mixed uniformly across the wake section, the maximum concentration at the distance of wake formation is

$$\chi = Q/Au \qquad (5\text{-}3)$$

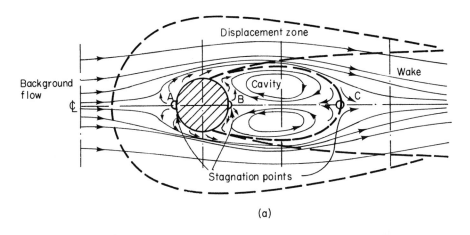

(a)

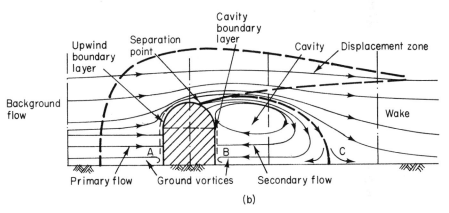

(b)

Fig. 5–12. Flow around a rounded building. (a) Flow in a horizontal plane near the ground; (b) flow in the longitudinal center plane. (From Halitsky, 1968.)

This is a highly oversimplified equation and must be modified to match the complexities of real situations. Hosker (1982) has assembled data on wakes in the lee of arrays of buildings of various shapes in useful handbook form.

Mountains and valleys also tend to distort flow under certain conditions, as shown in Fig. 5–13. There are also occasional situations in which inversions in combination with terrain features result in meteorological isolation of an area in the manner shown in Figs. 5–14 and 5–15. Terrain effects on atmospheric pollutant transport and diffusion are presently under active study and the results of recent research have been summarized by Dickerson *et al.* (1984).

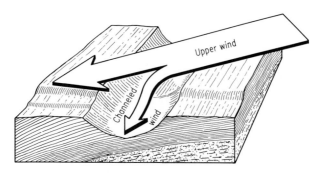

Fig. 5–13. Wind channeling by valley walls. (From U.S. Weather Bureau, 1955.)

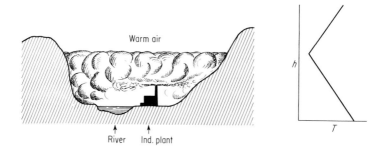

Fig. 5–14. Fumigation of valley floor caused by an inversion layer that restricts diffusion from a stack. (From U.S. Weather Bureau, 1955.)

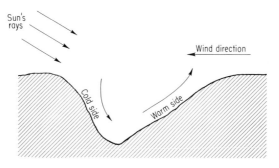

Fig. 5–15. Atmospheric overturn caused by uneven solar heating of valley walls. (From U.S. Weather Bureau, 1955.)

Diffusion of Accidental Releases in the Lower Atmosphere

Accidental releases may involve pulsed discharges of radioactive contaminants to the atmosphere. When this happens, the rate of dispersion depends, as outlined above, on meteorological conditions and the buoyancy of the emission, which in stable air is altered by the relative density of the entrained air. As it is usually not practical to specify the length of time during which the release will take place, one can modify Eq. (4-1) by substituting Qt, the total number of curies released, for Q, curies discharged per unit time. This permits calculation of $\overline{\chi t}$ Ci sec/m^3 in place of χ. This is a convenient form in which to express the calculated consequences of an accident, because χt is a measure of the total exposure, and one may proceed directly to a computation of the integrated dose.

Dispersion of Aerosols

The diffusion equations were derived for gaseous effluents but may be applied to plumes in which the contaminants are in the form of aerosols, provided the particle sizes are such that the settling rates are insignificant compared to the scale of vertical motion due to turbulent mixing. In general, it may be assumed that the diffusion equations will apply to particles small enough to be deposited in the pulmonary tract.

Deposition and Resuspension

The mechanisms by which particles deposit on surfaces vary, depending on whether or not the cloud passage is associated with precipitation. Dry deposition results from gravitational settling and impaction on surfaces exposed to the turbulent atmospheric flow. When precipitation occurs below the rain-forming level, the dust is washed to the surface by falling raindrops (washout). At higher altitudes, dust particles may serve as nuclei for condensing raindrops (rainout), a phenomenon that is responsible for removal of most submicrometer particles from the atmosphere.

The descent of particles through the atmosphere follows certain well-known physical laws (Drinker and Hatch, 1954) which govern the resistance of the motions of particles moving in viscous media. The terminal velocity of a falling particle will be reached when the "drag" on the particle due to the viscosity of the atmosphere is equal to the force of gravity. The terminal velocity of falling particles in still air is plotted for quartz in Fig. 5–16. Because of their irregular shape, the particles will actually fall at slightly lower velocities than spheres. A useful rule of thumb is that a 10-um particle having a density of about 2.5 falls at a rate of about 1 ft/min at sea level.

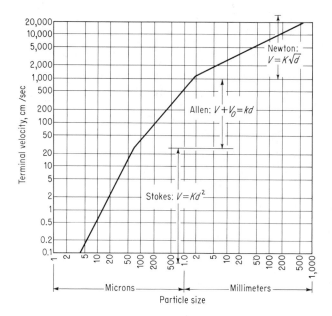

Fig. 5–16. Terminal velocity for quartz particles ($d = 2.6$) in air at sea level. (From P. Drinker and T. Hatch; copyright 1954, McGraw-Hill.)

Suspended particles may be impacted when turbulent air flows across a solid surface. The efficiency of impaction is defined as the ratio of the number of particles deposited on the surface to those originally contained in the volume of air diverted. When air flows around a cylinder, the efficiency of impaction increases as the velocity increases and the diameter of the cylinder decreases (Chamberlain, 1955). Theories have been developed which explain the manner in which impaction occurs as a function of particle size and velocity when regular shapes such as cylinders and planes are involved, but it is not practical to apply these to the irregular surfaces one frequently finds in practice.

Another method of quantifying the rate of air-to-surface transfer is by use of deposition velocity (v_g), which was originally designed to predict the transfer of iodine from the atmosphere to the ground (Chamberlain and Chadwick, 1966). The deposition velocity can be useful when settling velocities are inappreciable (i.e., for particle diameters less than about 10μm), in which case

$$v_g = \frac{\text{amount deposited per square centimeter per second}}{\text{volumetric concentration per cubic centimeter above surface}}$$

where v_g is in centimeters per second. The deposition velocity incorporates all relevant physical and chemical processes into a single observable parameter.

In the absence of a theoretical basis for estimating v_g, a great many investigators have undertaken field studies designed to provide measurements of the velocity of deposition for a variety of surfaces. These data have been summarized by Sehmel (1984) for both dusts and gases and for a variety of meteorological conditions and types of ground cover. Unfortunately, the reported values of v_g are so variable that it is difficult to draw simple conclusions. It has been noted that the variation in v_g is influenced by the type of ground cover, and a "leaf area index" has been used by modelers to correct for this factor. The index varies from 0.5 for grassland to greater than 5 for raw crops. Other factors apparently introduce even greater uncertainty, because the reported values of v_g for dusts range over five orders of magnitude from 10^{-3} to 180 cm/sec. The range of variability reported for the deposition velocities of gases is comparable (2×10^{-3} to 26 cm/sec). Miller (1984a) quotes Heinemann and Vogt (1979) as having determined that v_g for forage grasses is 2 cm/sec for reactive gases such as molecular iodine, 0.1 cm/sec for particles less than 4 μm, and 0.018 cm/sec for unreactive gases. In view of the enormous variability in the reported results, it is not possible to conclude that the published values of deposition velocity are generally useful. However, most studies conclude that v_g for molecular iodine is about 2 cm/sec, and since this is often the most important radioelement released to the atmosphere, the v_g can serve a useful purpose.

Washout (scavenging) of pollutants from a plume by precipitation can be important because in this way the concentration in the atmosphere is reduced downwind from the point of washout, where ground deposition is increased. The mechanisms involved in precipitation scavenging have been treated mathematically by many investigators and more than a dozen numerical codes have been published that describe the phenomenon (Slinn, 1984). The mathematical treatments have increased in complexity since the classic work of Chamberlain (1955), and the subject must remain, at least for the present, outside the scope of this text. Useful guidelines for predicting washout as a function of the type and rate of rainfall are not yet available.

After particles have been deposited on the ground, they may become resuspended as the result of wind action. This is another aspect of the general problem of airborne radioactivity that has thus far defied analytical solution. Sehmel (1984) has listed several dozen factors that influence resuspension, including the specific characteristics of the soil, surfaces, topography, and prevailing meteorology. The resuspension factors (airborne concentration/surface contamination) reported from all field experiments vary by seven orders of magnitude. Sehmel notes that K, the resuspension rate for plutonium (the fraction of a deposit resuspended per second) at the Nevada Test Site, was shown to vary by about four orders of magnitude. However, the spread in the data from Nevada has been shown to be reduced when K is divided by u^3 (the third power of wind speed) (Anspaugh *et al.*, 1975).

It is difficult to see how generalized information about either deposition or resuspension rates can be used with confidence, because of the variety and complexity of the physical processes in any specific case and the extreme difficulty involved in determining the host of relevant parameters required to model the process. However, in some cases it may be possible to develop useful information for a specific site and for a specific form of contamination.

Dose Calculations from Radioactivity in the Atmosphere

In the event of an emergency involving the release of radioactive materials to the atmosphere, it may be necessary to estimate the β or γ dose from the passing cloud or from the radioactivity that it deposits on the ground and, in addition, to estimate the dose to specific organs from inhalation or ingestion of contaminated food or water. Methods of making such estimates have been developed by Healy (1984), to whom the reader is referred for a more detailed presentation of the following discussion.

External Dose from a Passing Cloud of Beta Emitters

The dose rate from β radiation to an individual located on the ground over which a cloud of radioactive gas is passing that is greater in size than the range of the highest-energy β particles can be estimated from the following:

$$D_\beta = 0.23\bar{E}_\beta\chi \tag{5-4}$$

where D_β is the instantaneous β dose rate to the skin (rad/sec), $\bar{E}_\beta$ the mean β energy per disintegration (MeV/dis), and χ the concentration of β-emitting nuclide at a given point downwind of the source (Ci/m^3). The integrated dose delivered by the passing cloud will be

$$D_{\beta\infty} = 0.23\bar{E}_\beta\bar{\chi}t \tag{5-5}$$

where $D_{\beta\infty}$ is the infinite β dose to the skin (rads), and $\bar{\chi}t$ is the integral of concentration $\times$ time for the entire emission, obtained by substituting Q_Σ

(total curies released) for Q (Ci/sec) in Eq. (5-1), and has units curies per second per cubic meter. If more than one β-emitting radionuclide is involved, the dose must be summed from the total of the dose from the individual nuclides.

The range of the β particles in tissue is only a few millimeters, and the dose from external β radiation is thus limited primarily to skin and varies with depth, depending on the energy of the incident radiation. The β-emitting gases are mainly isotopes of the fission products krypton and xenon, for which the dose from inhalation is much less than the dose to the skin due to immersion in the cloud.

EXTERNAL GAMMA DOSE FROM PASSING CLOUD

Although more rigorous solutions to the problem may be found (Kocher, 1983; Healy, 1984), a simplified approach to estimation of the dose from a passing cloud is based on the assumption that the individual is standing on the ground immersed in a cloud that is infinite in size, through which radioactivity is uniformly dispersed for a period during which the total exposure is given as χ^t. The dose estimate may be simplified by neglecting backscatter from the ground. This tends to reduce the dose estimate, but is somewhat offset by the error in the opposite direction that is introduced by the assumption of an infinite cloud.

The dose from a cloud containing 1 Ci sec/m³ can be estimated in this way to be:

$$D\gamma = \frac{1/2\bar{\chi}t(3.7 \times 10^4 \text{ d/sec uCi}) \, (E) \, (1.6 \times 10^{-6} \text{ erg/MeV})}{(100 \text{ erg/g rad}) \, (0.0012 \text{ g/cm}^3)} \quad (5\text{-}6)$$

$$= 0.25\chi tE$$

where D_γ is the γ dose (rads), ($\bar{E}$ the average γ energy (MeV), and $\bar{\chi}t$ the product of concentration × time (Ci sec/m³) calculated by substituting Q_Σ (total curies released) for Q (Ci/sec) in Eq. (5-1).

The dose from the passing cloud of fission products, as from a nuclear explosion or reactor release, will usually be much lower than either the dose received by inhalation (particularly inhalation of radioiodine) or the γ dose that results from the large-scale deposition of the cloud on surfaces.

Tropospheric and Stratospheric Behavior

The emissions discussed thus far are from near-surface sources that are diffused over distances measured in tens of kilometers. The concentrations beyond about 50 km are usually of no interest for releases from industrial or research sources of emission.[1] In contrast, when the contamination originates from nuclear or thermonuclear explosions, it is spread throughout the atmosphere on a global scale.

Our knowledge of stratospheric dispersion has been obtained from gas and dust samples obtained by aircraft or balloons penetrating to an altitude of about 115,000 ft (Holland, 1959a). Studies have been made of the distribution of ozone and water vapor and of a wide variety of radioactive substances, including debris from high-yield nuclear explosions. The transport of naturally occurring radionuclides such as ^{7}Be, ^{32}P, and ^{14}C, all of which are induced by cosmic-ray bombardment of the upper atmosphere, has also been studied. On two occasions, tracers have been incorporated in nuclear weapons exploded at unusually high altitudes. Rhodium-102 was injected into the stratosphere by an explosion 43 km above Johnston Island in the Pacific in August 1958, and the cloud from this nuclear explosion was believed to have risen to about 100 km. In July 1962 an explosion conducted about 400 km above Johnston Island contained a known amount of ^{109}Cd, which has been measured by investigators in many parts of the world (Krey and Krajewski, 1970).

The first significant stratospheric measurements were of the movement of stratospheric water and ozone by Brewer (1949) and Dobson (1956). Their observations were studied by Stewart et al. (1957), who developed a model of stratospheric–tropospheric exchange that is consistent with the observed pattern of fallout from nuclear weapons that will be discussed in Chapter 13. According to this model (Newell, 1971; UNSCEAR, 1982), air enters the stratosphere in the tropical regions, where it is heated, and rises to an altitude of about 30 km, at which level it begins to move toward the poles. As shown in Fig. 5–17, the tropopause is lower in the polar regions than at the equator, and discontinuities in the tropopause in the temperate regions facilitate transfer from the stratosphere to the troposphere. The westerly jet streams occur at these discontinuities with transport velocities of 100–300 km/hr, and they are accompanied by vigorous vertical mixing. The rate of transfer from the lower stratosphere is most rapid in the winter and early spring.

The mean residence time of a stratospheric aerosol consequently depends on the altitude at which it is introduced, the time of year, and the latitude. Dust injected into the lower polar stratosphere by Russian thermonuclear explosions was found to have a mean residence time of less than 6 months, whereas in tropical latitudes the residence time has been

---

[1]This generalization is valid only for radioactive contaminants injected into the atmosphere. Nonradioactive pollutants such as SO_2 and CO_2 that originate from power plants which consume fossil fuels are of interest for much larger distances.

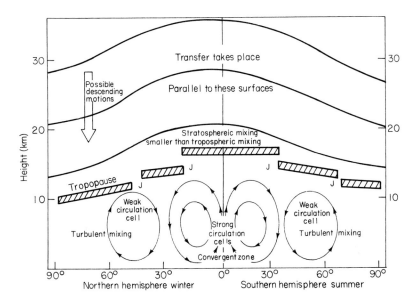

Fig. 5–17. Schematic cross section displaying characteristics of meridional transport ("J" locates typical jet stream positions). (From UNSCEAR, 1964.)

found to be from 2 to 3 years for radionuclides introduced into the middle stratosphere and 5 to 10 years for those injected at 100 km (see Chapter 13).

Sedimentation is evidently a significant factor in the thin stratospheric atmosphere, since Feely *et al.* (1965) have shown that ^{14}C injected as carbon dioxide has a longer residence time than ^{90}Sr injected in particulate form (see Chapter 13). It has also been suggested that the long stratospheric residence times of some nuclides may be due to their being reintroduced into the stratosphere from the troposphere (UNSCEAR, 1977; Reiter, 1974).

Aerosols introduced into the troposphere are distributed by the planetary winds and deposit on the surface of the earth mainly by the scavenging action of rain. A remarkable correlation of ^{90}Sr deposition with rainfall was demonstrated on the Olympic Peninsula by Hardy and Alexander (1962) and will be discussed in Chapter 13. The mean residence time of dust injected into the troposphere is about 30 days on the average, but can vary from 5 days for dust in the lower rain-bearing region of the troposphere to 40 days in the higher initial altitudes (UNSCEAR, 1977). Rainfall removes aerosols from the troposphere primarily by droplet formation around the particle (rainout) and also by a scrubbing action (washout).

There have been suggestions that ocean spray is effective in scavenging dust near the ocean–atmosphere interface, and that this phenomenon might explain reports that fallout of ^{90}Sr into the oceans is higher than on land. However, field studies conducted by Freudenthal (1970a,b) concluded that ocean spray is not significant in this respect.

Chapter 6

Terrestrial and Aquatic Pathways

Contamination of land and water can occur either from deposition of material originally introduced into the atmosphere, or from waste products discharged directly into surface or subsurface waters or placed in or on the ground, from which they are eventually mobilized by ground water or erosion.

The primary reason for being concerned about radioactive contamination of the environment is that it results in exposure of humans. The dose to plants and lower animals is not an important consideration because they are orders of magnitude more resistant to the effects of radiation than humans (Blaylock and Trabalka, 1978; Auerbach *et al.*, 1971; Templeton *et al.*, 1971). As an example, Auerbach cites the work of Donaldson *et al.* (1969), who found no effects following irradiation of chinook salmon at dose rates up to 5.0 rem/day beginning immediately after fertilization of the egg and continuing for 80–100 days until the fish were completely formed.

One contrary finding was reported by Strand *et al.* (1977), who found a 50% suppression of immune response in rainbow trout who received 4 rem from tritiated water during embryogenesis. The significance of this finding has been discounted (Blaylock and Trabalka, 1978) because of the high reproduction rate of aquatic species. Moreover, the doses administered were much higher than would be experienced in practice.

Whicker (1980) found no effects of Pu on ecosystems despite concentrations of 10–1000 $\mu Ci/m^2$ in the upper 3 cm of soil.

The Food Chain from Soil to Humans

Most of the food consumed by human beings is grown on land and, except for elements like carbon and oxygen, which may be obtained from the atmosphere, it is the soil that nourishes the terrestrial ecosystem that supplies human food.

Radionuclides such as ^{210}Po and ^{226}Ra that occur naturally in soil are incorporated metabolically into plants and ultimately find their way into food and water. Artificial radionuclides behave in a similar manner, and worldwide contamination of the food chains by radionuclides produced during tests of nuclear weapons in the atmosphere has taken place during the past 40 years.

In addition to root uptake, direct deposition on foliar surfaces can occur, in which case contaminants can be absorbed metabolically or, more likely, can be transferred directly to animals that consume the contaminated foliage. Foliar deposition is potentially a major source of food chain contamination by both radioactive and nonradioactive substances (Russell, 1965; Russell and Bruce, 1969).

SOME PROPERTIES OF SOILS

Soils (USDA, 1957; Hillel, 1971) consist of mineral and organic matter, water, and air arranged in a complicated physicochemical system that provides the mechanical foothold for plants in addition to supplying their nutritive requirements.

Vertical profiles through soils reveal horizontal layers (horizons) which differ in their physical characteristics and which, in part, determine the kinds and amounts of vegetation that a soil will support. Broadly speaking, three major horizons may be identified. The uppermost, which may be from 30 to almost 60 cm thick, is the surface soil, in which most of the life processes take place. The second horizon is the subsoil, extending to about 1 m below the surface. Still farther below the surface, to a depth of about 1.5 m, is a layer of loose and partly decayed rock, which is the parent material of the soils. These layers are conventionally designated as the A, B, and C horizons and can be differentiated further, as in Fig. 6–1.

The inorganic portion of surface soils may fall into any one of a number of textural classes, depending on the percentages of sand, silt, and clay. Sand consists largely of primary minerals such as quartz and has a particle

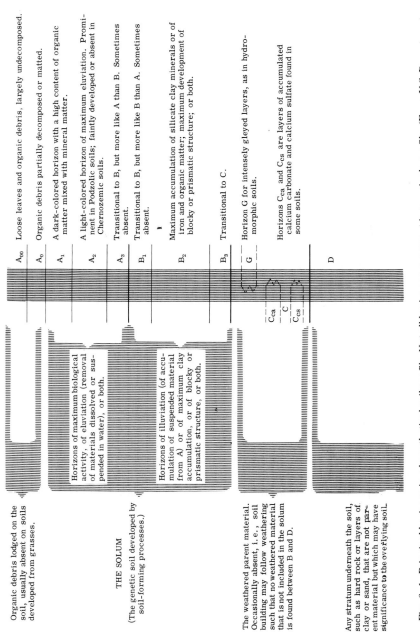

Fig. 6–1. Principal horizons in a hypothetical soil profile. Not all horizons are present in any single profile. (From U.S. Department of Agriculture, 1957.)

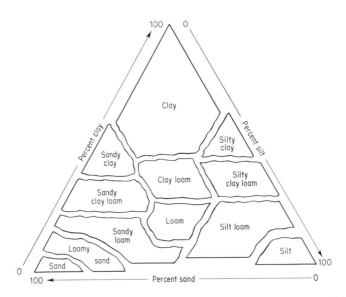

Fig. 6–2. Textural classes of soils according to the percentages of clay, sand, and silt. (From U.S. Department of Agriculture, 1957.)

size ranging from 60 μm to about 2 mm. Silt consists of particles in the range 2–60 μm, and clay particles are smaller than 2 μm in diameter. Figure 6–2 illustrates the textural classes into which soils may be differentiated, depending on the percentages of the three principal constituents.

The important physiocohemical processes by which soils provide the nourishment for plants are controlled largely within the clay fraction of the soil. As essential characteristic of the platelike particles of secondary aluminum silicates that comprise clay is the abundance of negative surface charges. The resultant ability of the clay particles to attract ions, especially positive ions, to their surfaces is one of the most important properties of soils.

Most of the nutrient ions are not dissolved in the soil water, but are sorbed on the surfaces of the soil particles. A much greater reservoir of nutrient elements can be held in this way than in solution. Dissolved nutrients would not remain long in soil but would be leached from it were it not for the extraordinary ability of the clays to bind elements in ionic form. Since it is estimated that 1 ft^3 of loam has a surface area of about 50,000 m^2, the opportunities for ion exchange are very great.

Cations in water solution are exchanged with cations sorbed on the surface of the clays. Most soils tend to become acidic after a period of time

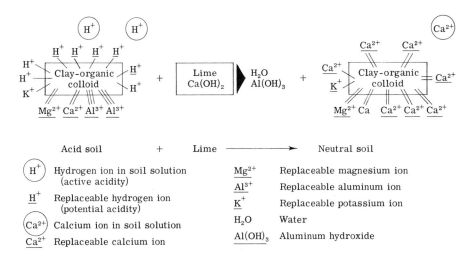

Fig. 6–3. Cation-exchange reactions when an acid soil is limed. (From U.S. Department of Agriculture, 1957.)

because of the replacement of adsorbed cations by the excess of hydrogen ions in rainwater, and for this reason such soils must be limed from time to time. This procedure replaces the hydrogen ions with ions of calcium and magnesium, as illustrated schematically in Fig. 6–3.

The ability of a given soil to exchange cations is quantified by its exchange capacity, customarily expressed as the milliequivalents of cations required to neutralize the negative charge of 100 g of dry soil at pH 7. Montmorillonite clay has such excellent cation-exchange properties that soils rich in this mineral may have cation-exchange capacities of about 100 mEq/100 g, compared to less than 10 mEq/100 g for predominantly kaolin-type soils. Organic matter derived from the decay of plant material can also furnish a major part of the exchange capacity.

According to Jenne (1968), neither the presence of clays or organic matter nor precipitation as oxides or hydroxides is sufficient to explain the observed binding of transition elements such as Co, Ni, Cu, and Zn, and it is proposed that fixation is due to complexation with hydrous oxides of Fe and Mn that occur as coatings on particles of soils and sediments. Means *et al.* (1978) subsequently found the role of Mn to be predominant not only for fixation of the transition elements, but for the actinides as well.

As would be expected, sandy soils tend to have a low exchange capacity, which can be increased by adding organic matter.

BEHAVIOR OF RADIONUCLIDES IN SOILS

Uptake of a long-lived radionuclide by plants depends to a considerable degree on whether it remains within the root zone and on the extent to which it is chemically available for transport to root endings and translocation to edible portions.

When a radionuclide (in soluble form) comes in contact with soil, it can adsorb to reactive coatings on particles, undergo ion exchange, precipitate as an oxide–hydroxide or sulfide, be complexed with organic compounds, or remain in ionic form (Schulz, 1965). The manner in which the radionuclide partitions among these fractions determines to a large extent the length of time it will remain at the site of deposition and the extent to which it will be available for root uptake.

The manner in which a trace element is partitioned between the soil and interstitial water is usually described by the distribution coefficient K_d, which is expressed as the quantity of the radionuclide sorbed per unit weight of solids divided by the quantity of the radionuclide dissolved per unit volume of water, i.e., K_d = milliliters per gram. When radionuclides are transported by water moving through porous or fractured geologic media, the phenomenon of sorption causes the rate of movement of the radionuclide to be reduced relative to the rate of movement of the water. This "retardation factor" is determined by the rates of sorption and desorption on the media surfaces and, at equilibrium (i.e., rate of sorption = rate of desorption), is related directly to K_d. The slowing of radionuclide transport in this way has important implications for radioactive waste management, as will be discussed in Chapter 11. Measured values of K_d for various elements in selected media are given in Table 6–1.

The ions of some elements are bound so tightly to soil particles that they are nearly immobile. It has been shown that thorium and certain of the light rare earth elements (REE) sorbed on clays and iron oxides in a massive ore body near the surface of a hill in the state of Minas Gerais, Brazil, are transported from the deposit by ground water solubilization at a rate of only 10^{-9} per year. This, despite an annual rainfall that averages 170 cm/year (Eisenbud et al., 1984; Lei et al., 1986).

When water moves through geologic media, sorption of ions occurs at different rates depending on the physiochemical properties of the ions and the media. Different ionic species sorb at different rates, and in this way form a chromatographic sequence with the most weakly bound species at the leading edge of the pattern. Cations are generally more strongly sorbed than anions because of the preponderance of negatively charged particles on soil surfaces. Smaller multivalent ions usually exhibit larger K_d values

TABLE 6–1

SOME REPORTED VALUES OF $K_d{}^a$

Locale	Element	Reported values
Lake Michigan	Cs	$<10^5$
Clinch River	Cs	1360
Freshwater Pond	Cs	4600
Different soils	I	0.007–52.6
Hudson River	Pu	7×10^4
Lake Michigan	Pu	3.3×10^5
Savannah River	Pu	$0.4–4.1 \times 10^5$
Nevada Tuff	Ra	6700
White Oak Lake, Tennessee	Ru	≥ 1
Great Lakes	Sr	85
White Oak Creek	Sr	100–150

^{a}Condensed from National Council on Radiation Protection and Measurements (1984a).

(are more strongly sorbed) than larger univalent ions, and for unconsolidated media, such as soil, the fraction sorbed is usually inversely proportional to particle size. This is one of the reasons why clays are such effective sorbents (NAS–NRC, 1978).

Large volumes of radioactive waste solutions have been stored at Hanford, Washington, on and under the ground in trenches and cribs. Studies of the migration of the individual fission products toward the water table below the stored wastes have shown that the individual radionuclides form a sorption cone in which the elements are sequentially removed in an order dependent on their relative affinities for the adsorption media. The soils at Hanford are sandy with relatively low cation-exchange capacity (5–10 mEq/100 g) (Pearce *et al.*, 1960).

Field studies in the Soviet Union (Spitsyn *et al.*, 1960) showed that ^{90}Sr migrated at a rate of 1.1–1.3 cm/day (less than 5 m/year) through soils that had moderately high exchange capacities and that were permeated with ground water. Since the average life of an ^{90}Sr atom is about 40 years, the mean distance that would be traversed by an atom of ^{90}Sr before its decay would be less than 200 m under the given conditions. The total amount of ^{90}Sr would diminish to 0.1% of the original quantity in 10 half-lives (280 years), by which time the mean distance traveled would be 1400 m. The capacity of the soils to store some fission products in ionic form is seen to be substantial.

Field and laboratory studies by Spitsyn et al. (1958, 1960). Alexander et al. (1960), and the United Kingdom Agricultural Research Council (1961) all found that the ^{90}Sr fallout from tests of nuclear weapons is held tightly in the upper 10 cm of soil. Alexander (1967) has studied the downward movement of ^{90}Sr and ^{137}Cs in fields having a variety of soil types and vegetation characteristics. The studies were conducted shortly after the peak fallout of weapons debris occurred in the early 1960s. At all sites studied, the two nuclides were contained mainly in the upper 10 cm of soil. At one site that was examined in more detail than the others, both nuclides were present in appreciable amounts in the leaf litter, but this may have been due to the fact that the sampling was done at a time when weapons testing in the atmosphere had recently occurred. From estimates of the cumulative fallout at the time of sampling, it was concluded that only 50% of the deposit could be accounted for in leaf litter and soil and that the remaining half was presumably contained in the vegetative stand. The greater part of the fallout, after 2 or 3 years, is thus contained in the soil cycle: soil–plant–litter–microbial utilization–soil (Ritchie et al., 1970). This is in contrast to findings in sandy loam at another site, years following the maximum fallout pulse in 1963. Only a very small percentage of the ^{90}Sr, ^{137}Cs, and ^{239,240}Pu present in the soil column was found to be associated with leaf litter. The integrated accumulations in the top 30 cm of undisturbed soil were nearly equivalent to the quantities estimated to have been deposited at this latitude, indicating little removal from the site (Hardy, 1974a). Similar findings have been reported for fallout-derived ^{137}Cs and ^{239,240}Pu in a variety of soils sampled in the late 1970s and early 1980s (Alberts et al., 1980; Linsalata, 1984).

In a study of ^{129}I accumulations in soil near the Savannah River site, it was estimated that the residence half-time in the top 30 cm of soil during a 25-year period of observation was 30 ± 6 years (Boone et al., 1985).

Alexander's measurements confirmed that ^{137}Cs is more tightly bound by soil than ^{90}Sr. For the purpose of forecasting the dose commitment from ^{90}Sr deposited on soil, the United Nations Scientific Committee on the Effects of Atomic Radiation has assumed that ^{90}Sr is leached away from the root zone of cattle fodder at a rate of 2% per year (UNSCEAR, 1969).

Although some generalizations concerning the relative degree of fixation of many radionuclides in soil are possible, the behavior of the radionuclides is so dependent on site-specific factors such as rates and amounts of rainfall, drainage, and extent of tillage that more general quantitative forecasts are not practical at this time. The behavior of radionuclides deposited on soils by fallout from nuclear weapons testing is further discussed in Chapter 13.

UPTAKE FROM SOILS

Although almost every element can be identified in soils, only 16 of them are considered necessary for the growth and reproduction of vegetation. These are carbon, hydrogen, oxygen, nitrogen, phosphorus, sulfur, potassium, calcium, magnesium, iron, manganese, zinc, copper, molybdenum, boron, and chlorine. All are obtained by plants from soil except carbon, hydrogen, and oxygen, which can be supplied by the atmosphere.

It may be assumed that if an ion is present in the soil it will probably be present in plants grown on the soil. Many elements are required for normal metabolism, but some, e.g., iodine, cobalt, uranium, and radium, are known to be present in plants although they serve no known metabolic function.

The extent to which plants absorb radionuclides from soil depends on the chemical form of the nuclide, its distribution coefficient, the metabolic requirements of the plant, and physicochemical factors in the soil.

There are many compilations of the reported transfer factors from soil to plants (Ng, 1982). According to Nishita *et al.* (1961), the relative uptake of various radioelements from soils is $Sr \gg I > Ba > Cs, Ru > Ce > Y, Pm, Zr, Nb > Pu$. To provide assistance in estimating plant uptake of various elements in the absence of field data, the NCRP (1984a) has published the data in Table 6–2, which gives the recommended "default values" that can be used in dose assessment studies when no field data are available. The range of recommended values is huge for some elements, and these values must therefore be used with great caution.

TABLE 6–2

RANGE OF RECOMMENDED DEFAULT VALUES FOR
PLANT UPTAKE FACTORS[a]

Element	Concentration in wet vegetation/dry soil	$\dfrac{Max}{Min}$
Co	$1 \times 10^{-3} - 9.4 \times 10^{-3}$	9
Sr	$1.7 \times 10^{-2} - 1.0$	59
Ru	$3.8 \times 10^{-3} - 6.0 \times 10^{-2}$	16
I	$2.0 \times 10^{-2} - 5.5 \times 10^{-2}$	3
Cs	$6.4 \times 10^{-4} - 7.8 \times 10^{-2}$	121
Ra	$3.1 \times 10^{-4} - 6.2 \times 10^{-2}$	200
U	$2.9 \times 10^{-4} - 2.5 \times 10^{-3}$	8.6
Pu	$1 \times 10^{-6} - 2.5 \times 10^{-4}$	250

[a]Adapted from National Council on Radiation Protection and Measurements (1984a).

Radioisotopes of elements that are ordinarily present in soil and that are utilized in plant metabolism are absorbed in a manner independent of the radioactive properties of the element. Thus, ^{45}Ca in soil becomes part of the pool of available calcium in the soil, and the plant will not differentiate significantly between ^{45}Ca and the stable isotopes of calcium. By available calcium is meant that portion of the total soil calcium which exists in exchangeable form and is available for transport to the root system. The roots are sometimes unable to distinguish between chemical congeners. Thus, plants grown in inorganic solutions containing calcium and strontium are unable to discriminate between the two (FAO, 1960).

The elements sorbed in exchangeable form on the surfaces of the soil particles constitute a reservoir that supplies many of the nutritive requirements of the plant. The elements pass to the plant root tips with the soil water, in which the dissolved elements are in equilibrium with the sorbed solid phase.

The roots of plants are located at soil depths characteristic of the species: less than 30 cm in the case of spinach, in contrast to alfalfa and asparagus roots, which penetrate to 3 m or more. The effectiveness of root structures as an absorbing surface is illustrated by winter rye, single plants of which were shown to have a surface area of 638 m^2 (Wadleigh, 1957).

FOLIAR DEPOSITION OF RADIONUCLIDES

Radioactive substances can contaminate plants by direct foliar deposition (Russell, 1965, 1966). The radionuclides may then pass directly to grazing animals or humans in the form of superficial contamination or may be absorbed metabolically from the plant surface.

It has also been noted that trace substances present in soil can, in addition to root uptake, contaminate plants by the mechanism of rain splash (Dreicer et al., 1984). Particles less than about 100 μm can be deposited on plant surfaces to a height of 40 cm by impacting raindrops. The importance of soil ingestion by grazing animals has been discussed by Zach and Mayoh (1984), who concluded that the daily intake of radionuclides for cattle predicted by models of the U.S. Nuclear Regulatory Commission (U.S. NRC, 1977b) should be increased by the radionuclides contained in 0.5 kg of ingested soil per day. For certain radionuclides which exhibit high K_d values and are highly discriminated against by plants (i.e., Pu, Th, Am), soil ingestion may be the predominant route of intake.

How to apportion the ^{90}Sr content of cow's milk between that which originated from foliar deposition and that which was taken by the plant from the soil attracted a good deal of interest when nuclear weapons tests were in progress (see Chapter 13). The importance of this arose from the fact that in order to make forecasts of future human exposure from expected deposition of ^{90}Sr, it was necessary to know how much of the dose from consumption of milk was due to fresh fallout that would disappear after weapons tests stopped, and how much was due to ^{90}Sr that remained in the soil. A number of investigators (Tajima and Doke, 1956; Russell, 1965) attempted to fit proportionality factors to equations of the following type:

$$C = p_r F_r + p_d F_d$$

where C is the 12-month mean ratio of ^{90}Sr to calcium in milk (pCi/g Ca), F_r the the annual deposit of ^{90}Sr (mCi/km^2 year), F_d the cumulative deposit of ^{90}Sr (mCi/km^2), p_r the proportionality factor for the rate-dependent component of C, and p_d the the proportionality factor for the deposit-dependent component of C.

In order to account for the effect of fallout during the previous growing year on the ^{90}Sr content of milk from cows fed stored feed for part of the year, Bartlett and Russell (1966) proposed inclusion of a third term, the "lag factor," thus

$$C = p_r F_r + p_d F_d + p_l F_l$$

in which F_l is the fallout deposition during the last 6 months of the previous year (mCi/km^2) and p_l is the "lag" proportionality factor.

This refinement could be important during periods when the rate of fallout is varying significantly, but Aarkrog (1971b) did not find it necessary to utilize this additional term to relate the observed rates of fallout in Denmark during the period 1962–1968.

During the period of relatively heavy fallout in the early 1960s, direct uptake from soil apparently accounted for as little as 10% of the ^{90}Sr present in milk. The average of the worldwide soil proportionality factor for ^{90}Sr contamination of milk was 0.3 (UNSCEAR, 1964, 1966), and the proportionality factor for the rate-dependent component was 0.8.

The soil pathway is usually unimportant for a short-lived radionuclide such as ^{131}I, which has a half-life of only 8 days and would decay before being assimilated by the plant.

The significance of surface contamination varies with the growing season, since the possible risk from direct contamination of crops is obviously greater just before or during a harvest, or when active grazing by stock animals (particularly milk-producing animals) is in progress. Conversely, the danger may be lowest in winter months, when there are no standing crops, although it is possible that even during these months direct fallout on the basal structure of grasses in permanent pastures may be stored until the following spring, when contaminants may be absorbed by the growing plants. Retention of this type will be greatest for plants that develop a "mat" of basal parts, old stems, and surface roots (Russell, 1965).

It has been shown that the major mechanism for contamination of soybeans and wheat grown in soil that is superficially contaminated with plutonium is resuspension due to mechanical harvesting (McLeod et al., 1980; Adriano et al., 1982).

The relative importance of foliar contamination also depends on the structure of the plant and the role of the various parts of the plant in relation to the dietary habits of humans. The inflorescences of wheat have a shape that tends to maximize entrapment of fallout particles. Possibly for this reason, wheat was found to be a major source of ^{90}Sr from weapons testing fallout in Western diets (U.S. AEC, 1960b). It has been reported that cereals generally are subject to relatively high foliar retention. The influence of dietary practices is illustrated by the fact that white bread has been shown to contain less ^{90}Sr than whole wheat bread, which is made from unsifted flour that contains the brown outercoat as well as the inner white portion of the wheat grain.

Foliar contamination can be removed by radioactive decay, volatilization, leaching by rain, other weathering effects, death and loss of plant parts, and, of course, by washing prior to human consumption. Chamberlain (1970) examined the ^{90}Sr data from a number of investigators and found the half-life due to field loss during the growing season to be about 14 days, not considering radioactive decay. This appears to be a useful "average" value, but this half-time is influenced by differences in the physical and chemical properties of the radionuclides as well as by the type of plant (Miller and Hoffman, 1982). The half-time of removal of strontium sprayed on crops varied from 19 days in the summer to 49 days in winter. New growth contained less than 1% of the initial deposit. Krieger and Burmann (1969) found that a fractional loss of 0.05/day was satisfactory for the first few weeks after fallout, but that field loss was slower thereafter.

Hansen et al. (1964), in a 3-year study of the ^{131}I and ^{90}Sr content of milk from dairy herds, found the radionuclide content of the milk from cattle pastured in well-fertilized fields to be 50% lower than that from cattle grazing in badly fertilized fields, because the faster-growing grass diluted the contamination present as foliar deposition.

TRANSPORT OF SOIL PARTICLES BY EROSION

Erosion by rainfall runoff is one mechanism for transport of radionuclides incorporated in surface soil. An understanding of soil erosion is important to soil conservation and agricultural scientists, and formulas for predicting soil loss have evolved over the years under the sponsorship of the U.S. Soil Conservation Service of the Department of Agriculture (Musgrave, 1947; Wischmeier and Smith, 1978). In addition, a modest amount of recent research has specifically addressed the mobilization of long-lived nuclides by erosive processes in large watersheds (Foster and Hakonson, 1986; Simpson *et al.*, 1986; Muller *et al.*, 1978; Linsalata, 1984).

Erosion by wind action causes contaminants that have settled to the surface to become resuspended (Chepil, 1957). This is particularly important for alpha-emitters such as plutonium because plutonium because of the increased likelihood of inhalation of the particles (Chapter 5).

METABOLIC TRANSPORT THROUGH FOOD CHAINS

From the foregoing it is seen that root uptake and foliar deposition are the two ways by which fallout has contaminated crops that are eaten by humans or that serve as food for stock animals. Much remains to be learned about the mechanisms by which individual radionuclides pass from soil to the root, from the root to the edible portion of the plant, through the body of the stock animal, and into the milk, flesh, internal organs, and eggs.

It is sometimes possible to predict the behavior of a radioelement from knowledge of its chemical congeners. Comar *et al.* (1956) first noted that in studies of the transfer of radionuclides and their congeners in food chains, the extent to which discrimination takes place at any step can be described by the observed ratio, OR, as follows:

$$\mathrm{OR}_{\text{sample-precursor}} = \frac{C_e/C_c \text{ sample}}{C_e/C_e \text{ precursor}}$$

where C_e and C_c are the concentrations of the element and its congener, respectively.

The use of the observed ratio is illustrated by the case of a cow that is fed herbage containing a known ratio of Ca to ^{90}Sr. The concentration of ^{90}Sr that will appear in the milk can be predicted by use of the $\mathrm{OR}_{\text{milk-herbage}}$, which describes the ability of the cow to discriminate metabolically between Ca and ^{90}Sr in the production of milk. Since this particular OR is known to be about 0.1, we would expect cow's milk to contain 10 pCi ^{90}Sr/g Ca, if the herbage on which the cow feeds contains 100 pCi ^{90}Sr/g Ca.

Of the many artificial radionuclides that have contaminated soils and plants, ^{90}Sr, ^{137}Cs, and ^{131}I have been studied most thoroughly.

Strontium-90 (U.S. AEC, 1972a)

Most discussions of ^{90}Sr in food chains use the ratio picocuries of ^{90}Sr per gram of Ca, in which 1 pCi ^{90}Sr/g Ca is defined as the strontium unit (SU). This practice continues despite the fact that the calcium content of plants has been found to vary considerably, and contamination reported as picocuries per gram of food is sometimes less variable than when related to the unit mass of calcium (UNSCEAR, 1969). However, all things considered, the ratio is useful in following the ^{90}Sr from one biological level to the next.

However, the ratio may be meaningless when applied to soil under practical conditions because the ^{90}Sr is not normally homogeneously mixed throughout the soil, and there is no way to express the strontium–calcium ratio of the nutrients to which the roots are exposed. Under laboratory conditions where the ^{90}Sr was well mixed with the soil, Frederickson *et al.* (1958) showed that the observed ratio from plant to soil varied only within 0.7 to 0.8 over a wide range in the amount of soil calcium. The discrimination against strontium was apparently unaffected as long as the soil was not oversaturated with calcium ions. This would indicate that only a small degree of differentiation takes place at the soil–root interface, but it is difficult to determine in practice exactly how the ^{90}Sr is distributed in the soil, and use of the OR may not always be practical.

Roberts and Menzel (1961) found that a portion of the ^{90}Sr may become unavailable to plants as a result of reactions in the soil. Studies over a 3-year period showed this fraction to be variable from about 5 to 50%.

Measurements of the $OR_{\text{human bone–diet}}$ have been reported from seven countries (UNSCEAR, 1969) and have been shown to average about 0.15, with little variation from country to country. In the United States, the $OR_{\text{human bone–diet}}$ is 0.18 and ranges from 0.15 (Chicago) to 0.22 (San Francisco). There appears to be no systematic difference between dairy and nondairy foods.

The observed ratio in passing from plants to milk by way of the cow has been shown (Comar and Wasserman, 1960) to be about 0.13. Thus, the overall observed ratio ($OR_{\text{human bone–plant}}$) would be about 0.15 if the plants are consumed directly and about 0.020 if the calcium is consumed from milk. In short, the lactating animal has been shown to be a strontium decontaminator, with an efficiency of about 85%. Lough *et al.* (1960) showed that the observed ratio in passing from diet to human milk is also about 0.1.

The overall effect of metabolic differentiation between strontium and calcium in passing from soil to human bone can be summarized for milk and vegetable diets as follows, starting with soil containing 1 pCi ^{90}Sr/g Ca(1 SU):

1 SU		1 SU[1]		0.13 SU		0.020 SU
in	→	in	→	in milk	→	in
soil		plant				human bone

1 SU		1 SU[1]		0.15 SU
in	→	in	→	in
soil		plant		human bone

The net ^{90}Sr/calcium ratio in human bone will thus vary, depending on whether the dietary calcium is derived primarily from dairy foods or from other sources. The U.S. population derives almost all its calcium from milk or milk products, but this is not true of all other countries. In Chile, the population receives only 8% from dairy products, compared to 77% in the United States and 87% in Finland. Cereals, legumes, and vegetables are the dominant sources of calcium in the Far Eastern countries (FAO, 1960).

In some parts of the world the calcium naturally present in food is supplemented with mineral calcium, a practice which would tend to lower the ratio ^{90}Sr/Ca. Calcium carbonate is added to the maize used in the preparation of tortillas in Mexico and to the flour used for making bread in the United Kingdom (FAO, 1960).

Several methods have been proposed for reducing the plant uptake of ^{90}Sr from soils (Menzel, 1960). They include the application of lime, gypsum, fertilizer, and organic matter. According to Menzel, these techniques are only moderately effective and could not be expected to reduce the plant uptake of ^{90}Sr by more than 50% in productive soils. This is a modest diminution in uptake considering the rather large quantities of required soil amendments. For example, at the levels of exchangeable calcium ordinarily found in productive soils, it would require several tons of lime per acre to effect a measurable change in strontium uptake. However, unproductive soils that have low cation-exchange capacity and low exchangeable calcium have relatively large uptakes of ^{90}Sr which can be reduced appreciably by the addition of calcium in available form. In the United Kingdom, the ^{90}Sr uptake in herbage was shown to increase manyfold as the amount of exchangeable calcium diminished from 3–4 g/kg of soil to less than 2 g/kg (UK Agricultural Research Council, 1961).

[1]This holds only when the strontium and calcium are uniformly mixed throughout the root zone of the plant.

Cesium-137

It has been well established by several investigators that cesium is so tightly bound by the clay minerals of the soil that root uptake is slight, and foliar absorption is, therefore, the main portal of entry of ^{137}Cs to the food chains during periods of active fallout. As late as 18 years after the cessation of tests in 1963, ^{137}Cs was still confined to the upper 24 cm of silty clay and exhibited a half-value depth (that depth at which the concentration is 50% of the surface contamination) of about 6 cm in undisturbed soil (Linsalata, 1984). In a study of tilled and untilled soils of five midwestern watersheds 10 years after cessation of atmospheric weapons tests, it was found that ^{137}Cs remained largely in the upper 5 cm of untilled soil but was evenly mixed in the upper 20 cm of tilled soil (Ritchie and McHenry, 1973).

Cesium is a congener of potassium, but the Cs/K ratio is not as constant in biological systems as is the ratio Sr/Ca. The uptake of cesium from soil has been shown to be inversely proportional to the potassium content of soils in which there is a potassium deficiency (Nishita *et al.*, 1961; Menzel, 1964), and Broseus (1970) has shown that this inverse dependence on the potassium content of soil explains the high cesium content of milk from cows grazing in certain parts of the island of Jamaica. A similar observation was made in the Tampa, Florida, milkshed, where the high concentration of ^{137}Cs in milk was traced to the practice of using pangola grass for cattle fodder. This grass has a lower potassium content than other feeds (Porter *et al.*, 1967).

Although cow's milk is the largest single contributor of ^{137}Cs to the U.S. adult diet, other foods containing grain products, meat, fruit, and vegetables contribute about two-thirds of the dietary cesium intake (Gustafson, 1969). Measurements by Gustafson on representative U.S. diets from 1961 through 1968 are given in Table 6–3.

Wilson *et al.* (1969) studied the transport of ^{137}Cs from the atmosphere to milk and concluded that root uptake was so slight that it could be neglected in any model designed to forecast the dose to humans from ^{137}Cs in fallout. For reasons not known, the crude fiber content of the forage was found to influence uptake by the cow, with transfer coefficients varying from 0.0025 for alfalfa and corn silage to 0.01 for mixed grain.

Wilson and associates developed a model for predicting the ^{137}Cs content of milk from cattle fed on stored feed. It was assumed that for the first 6 months in a given year the cows are fed stored feed that had been exposed to the previous year's fallout and are fed on feed contaminated by the current year's fallout during the second 6 months. The mean con-

TABLE 6–3

SOURCES OF ^{137}Cs IN THE U.S. ADULT DIET[a]

Year	Milk (%)	Grain products (%)	Meat (%)	Fruit (%)	Vegetables (%)
1961	31	17	12	20	15
1962	38	17	13	15	14
1963	39	21	22	8	6
1964	34	26	21	8	5
1965	28	23	26	9	4
1966	25	30	23	11	5
1967	24	28	17	8	7
1968	31	19	19	10	9

[a]From Gustafson (1969).

centration of milk in picocuries per liter was then calculated for each of the half-years, using the equation

$$C_m = BC_a$$

where C_m is the average concentration in cow's milk (pCi/liter), C_a the average concentration in air (pCi/m^2), and B a coefficient obtained by multiplying the fallout contamination factor and transfer coefficient shown in Table 6–4.

This technique was shown to be useful for forecasting the ^{137}Cs content of milk from seven milksheds across the nation from 1962 to 1967, although for reasons that were not understood by the investigators, the model did not work well for the milksheds that supply Tampa, Florida, and

TABLE 6–4

SUMMARY OF INFORMATION FOR MODEL OF ^{137}Cs IN MILK FOR DRY LOT HERD[a]

	Dry weight intake (kg/day)	Fallout contamination factor (m^3/kg)	Feed to milk transfer coefficient (day/liter)
Hay	14	9100	0.0025
Grain	7	1470	0.010
Silage	5	6740	0.0025

[a]From Wilson *et al.* (1969).

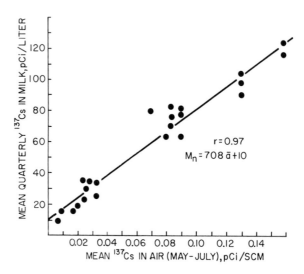

Fig. 6–4. Correlation between surface air activities of ^{137}Cs and quarterly averaged ^{137}Cs in milk from seven milksheds across the nation from 1962 to 1967. (From Wilson *et al.*, 1969.)

Seattle, Washington. The excellent correlation between the observed ^{137}Cs concentration in air during the growing season (May–July) and the mean quarterly ^{137}Cs content of milk is shown in Fig. 6–4, from which the Tampa and Seattle data are excluded. The ^{137}Cs levels in Tampa were anomalously high, a fact that has been attributed to the pangola grass prevalent in the area, as noted earlier.

When ^{137}Cs is ingested by humans, about 80% is deposited in muscle and about 8% in bone (UNSCEAR, 1969; Spiers, 1968). The half-life in adults depends on body weight, sex, and dietary habits and varies between 19 ± 8 days for infants and 105 ± 25 days for men. Women have a more rapid Cs turnover than men (NCRP, 1977a). A burden of 1 pCi/g K produces an absorbed dose of 0.018 mrad/year.

One must be careful that any generalization about pathways to humans is not negated by some special dietary consideration. In the case of ^{137}Cs, Laplanders and other residents of the far north are subject to relatively high ^{137}Cs intake owing to their dependence on reindeer, which feed on lichens that have a tendency to concentrate a variety of trace substances present in the atmosphere (Miettinen, 1969). About 25% of the ^{137}Cs contained in lichens is absorbed by reindeer (Holleman *et al.*, 1971). As a result, the Lapps contain about 50 times the ^{137}Cs body burden of Finns in the southern part of the country. This phenomenon has been said to be characteristic of all arctic and subarctic regions in the northern hemisphere (Rahola and Miettinen, 1973).

Radioiodine

Iodine is an essential nutrient that is required for the functioning of the thyroid hormones. It is generally believed that sea spray serves as the source of iodine in the earth's atmosphere and soils, but it has also been proposed that iodine may be released by the action of atmospheric ozone in reducing the iodide in rocks and soils (Whitehead, 1984).

Because of its short half-life (8 days), ^{131}I is not a significant environmental contaminant insofar as uptake from the soils is concerned. The decay is relatively rapid in relation to the growing time of crops, and significant contamination by means of root uptake would be improbable. On the other hand, radioiodine deposited on the surfaces of plants can be ingested directly by cattle and pass in this way to milk or other dairy products. Since the time is usually short from the collection of milk to its consumption, the possibility of iodine contamination of fresh milk must be considered. If the milk is processed into powdered form, radioiodine contamination will be less of a problem because longer storage time will permit decay of the isotope. However, cottage cheese reaches the consumer almost as quickly as fresh milk and may be contaminated with short-lived nuclides.

Mathematical models for the transfer of radioiodine to food are handicapped by the fact that since radioiodine has a short half-life, contamination must take place over a relatively short period of time. Thus, transfer from air to forage is apt to be variable because of meteorological factors that tend to average out in the case of longer-lived nuclides such as ^{137}Cs or ^{90}Sr. Field observations of the various parameters needed for mathematical modeling have been variable. Chamberlain found the values of deposition velocity (v_g) to range from 0.1 to 0.4 cm/sec following the Windscale accident (Chamberlain, 1960; Chamberlain and Chadwick, 1966). Hawley *et al.* (1964) found the deposition velocity to be 0.6 cm/sec on the average for radioiodine released experimentally at the National Reactor Test Station in Idaho, but the individual observations varied from 0.17 to 1.1 cm/sec.

Soldat (1963), reporting on studies at Hanford Laboratories, found that over a 2-year growing season the ratio of picocuries per kilogram of grass to picocuries per cubic meter of air was about 4200, and the ratio of picocuries per liter of milk to picocuries per kilogram of grass was about 0.15. Various estimates have been made of the extent to which the dose to the human thyroid is increased by the grass–cow–milk pathway compared to inhalation. A widely used multiple of 700 has been justified by Burnett (1970) and is intermediate among various proposed factors.

Chamberlain (1970) concluded from an analysis of data from various investigators that the ^{131}I foliar deposition data conformed to the concept that vegetation behaves as a filter, with deposition dependent on the density of the foliage.

Once the radioiodine deposits on foliage, it is removed by weathering and death of plant parts at a rate of about 5%/day, which, with the 8-day radioactive half-life of ^{131}I, gives an effective half-life of removal from grass ranging from 3.5 days in the Idaho experiments to about 6 days at Windscale. An effective half-life of about 5 days is usually used for risk estimation. Soldat (1965) found that radioiodine in milk reached a peak 3 days after an accidental release.

According to Garner (1960), a cow consumes about 20 kg of grass per day on the average, and an average vegetative stand per square meter of pasture is about 0.125 kg. It follows that a cow grazes an area of about 160 m^2/day. It can be assumed that about 5% of the ingested radioiodine will be secreted in cow's milk (Lengemann, 1966).

The fraction of ingested iodine that reaches the human thyroid following ingestion is usually taken as 0.3 and the effective half-life for elimination from the thyroid as 7.6 days (ICRP, 1960). However, thyroidal radioiodine uptake can be blocked almost completely by 50–200 mg of stable iodine as the iodide or iodate (Blum and Eisenbud, 1967; NCRP, 1977d). Substantial dose reduction will result from administration of KI as long as 2 hr after exposure.

Uranium, Thorium, and the Transuranic Actinide Elements

An understanding of the food chain behavior of uranium, thorium, and the artificially produced transuranic actinide elements is important because of their long half-lives, the fact that they are alpha-emitters, and their persistence in the environment. Although some of these elements, such as plutonium, were unknown until World War II, they are now produced in great quantities. Because of their radiotoxicity, the physical and chemical properties of the transuranic elements have been under intensive investigation for many years, and much has been learned about their behavior in the environment (Hanson, 1980; Watters *et al.*, 1980).

The usefulness of the available information on transfer factors from soil to plant and from plant to animals is limited by the great variability in the reported data. For example, it has been noted by Pimpl and Schuttelkopf (1981) that the concentration ratios (CRs) for plutonium, measured by different groups, vary from 10^{-9} to 10^{-3} for the same plant! The range is from 10^{-6} to 10^{-1} for americium and 10^{-4} to 10^{-1} for curium. Thus, a fundamental weakness in the risk assessment models now in use to predict the effects of the actinide elements is the inadequacy of reliable information on transfer factors, particularly under field conditions.

The range of reported values can be narrowed considerably by selecting carefully designed experiments in which surface contamination was not a factor. Thus, the concentration ratios (micrograms per gram of dry plant divided by micrograms per gram of dry soil) for actinide element analogues in 43 samples of various vegetables grown in Brazilian lateritic soils have been reported by Linsalata *et al.* (1986c). Unweighted mean CRs decreased in the order: ^{228}Ra (6×10^{-2}) $\geq$ ^{226}Ra (3×10^{-2}) $>$ La (2×10^{-3}) $=$ Nd(2×10^{-3}) $\geq$ Ce (8×10^{-4}) $>$ Th (1×10^{-4}). Vegetable uptake of these elements from soil can be simplified as $Ra^{2+} >$ $REE^{3+} >$ Th^{4+}. Similar concentration ratios have been published, based on field data by others (Bondietti *et al.*, 1979; Schreckhise and Cline, 1980a; Romney *et al.*, 1982; Pimpl and Schuttelkopf, 1981; Trabalka and Garten, 1983).

Because the transuranic elements do not exist in nature, their environmental behavior has also been investigated by the use of naturally occurring chemical analogues. Thorium has been shown to be a suitable analogue for plutonium in the quadrivalent state, and the light rare earth elements lanthanum and neodymium can be used as analogues for americium and curium. It has also been suggested (Krauskopf, 1986) that, under certain reducing conditions, thorium may be a suitable analogue for neptunium. It was noted earlier in this chapter that, based on analogue studies, the mobilization rates of Th and La from a weathered ore body being studied in Brazil is about 10^{-9}/year, which means that under certain conditions even the longest-lived transuranic nuclides would decay in place if the analogue concept is valid (Chapters 7 and 10).

Uranium

Considering the fact that uranium mining and processing has become a major industry, remarkably little is known about the ecological transport mechanisms that govern the movement of this element in the food chains. Data on soil–plant or plant–animal relationships are relatively scarce, although it has been known since the 1940s that the practice of using phosphate fertilizers results in the presence of uranium in food in concentrations up to 8 ng/g (Rodden, 1948; Reid *et al.*, 1977). Garten (1978) reviewed the transport of uranium in food chains and reported soil-to-plant CRs that range from 3×10^{-5} to 9×10^{-2}.

Radium

The behavior of radium in the environment has been the subject of reviews by Williams (1982), McDowell-Boyer *et al.* (1980), and Sheppard (1980). Interest in the subject has been stimulated during the past few years by the potential hazards from ^{226}Ra in mill tailing piles (Chapter 8).

Penna Franca *et al.* (1986) studied the mobilization of ^{228}Ra from the Morro do Ferro in Brazil and reported that the annual mobilization rate by ground water solubilization is of the order of 10^{-7}. One of the principal reasons for the low rate of mobilization was shown to be the tenaciousness with which radium is sorbed on clays and organic materials. Prantl *et al.* (1980) found the uptake of radium-226 in 11 types of root and leafy vegetables grown on soil contaminated with uranium tailings to be 1.1 × 10^{-3}. McDowell-Boyer *et al.* reviewed 12 published reports of the uptake of radium-226 by vegetables, grains, and feed hay and found a wide range of concentration ratios. The CR (dry weights) for grains ranged from 6 × 10^{-2} to 2 × 10^{-5}.

Neptunium

Neptunium is a transuranic actinide element produced in reactors and bombs by neutron reactions with uranium. Its short-lived nuclides, which have half-lives measured in days, are prominent in the early fallout from nuclear weapons explosions, but ^{237}Np, with a half-life of 2.1 × 10^6 years, does not become important until after about 10^5 years, when almost all other nuclides have decayed.

The nuclides of neptunium received very little attention until the publication of ICRP Report 30 (ICRP, 1979), which recommended that the assumed gastrointestinal tract-to-blood absorption factor for ingested Np be increased from 10^{-4} to 10^{-2}; the former value had been in use for many years. Since the half-life of ^{237}Np is one of the longest in high-level nuclear wastes, the increase in the assumed transfer factor affected the risk assessments for waste repositories in the period beyond about 10^5 years (NAS–NRC, 1983a; Thompson, 1982; Cohen, 1982).

There is relatively little information about the behavior of neptunium in the pentavalent state in the environment, but there is evidence that this element is more rapidly transferable from soils than in the quadrivalent state. Since it has been reported that neptunium is reduced to its quadrivalent state under the anaerobic conditions that would be expected to exist in waste repositories (Bondietti and Francis, 1979), the true potential significance of ^{237}Np for long-range risk assessment is uncertain.

The relative ease of transfer of neptunium from soil is illustrated by the report that the concentration ratio (concentration in dry weight of plant divided by concentration in dry weight of soil) is about 0.4, compared to 10^{-4} for Pu and 2 × 10^{-3} for Am and Cm (Schreckhise and Cline, 1980b). However, this ease of transfer has not been demonstrated in animal uptake studies (Thompson, 1982; NCRP, 1986).

Transport in Aquatic Surface Systems

Surface waters are coupled to subsurface aquifers, soils, and the atmosphere. Trace contaminants that somehow find their way into deep underground aquifers may in time reach surface waters and become incorporated into the biosphere. Atmospheric pollutants eventually deposit on soils or surface waters, and the main mechanisms for removal of contaminants from soil involve transport by water by a sequence of processes including surface runoff and leaching into soil water that eventually seeps to streams.

Because warm water is discharged by the condenser cooling systems of some power stations, the question is sometimes raised of possible synergistic ecological effects of temperature and ionizing radiation. No such synergism has been demonstrated over a wide range of temperatures or dose rates, except that in some cases organisms grown in warm water have been shown to absorb radionuclides as much as 50% faster owing to increased growth rates (Harvey, 1970). Ophel and Judd (1966) administered ^{131}I and ^{90}Sr to goldfish (*Carassius auratus*) and found no impairment of ability to withstand near-lethal temperatures at doses of 10^4 rad to bone and 10^5 rad to thyroid. Angelovic *et al.* (1969) reported that the estuarine fish (*Fundulus heteroclitus*) has lowered tolerance to heat and salinity when irradiated, but the administered doses were very high, in the range 2000–6000 rem.

MIXING WITHIN AQUATIC SYSTEMS

It was seen in Chapter 5 that the concentration of an atmospheric pollutant downwind of a source of known strength can be calculated conveniently if a few readily obtainable meteorological data are available. The same diffusion equations can be used under many topographical conditions and, if the source strength is known, one need only measurewind direction and velocity and estimate the degree of atmospheric stability to approximate the downwind concentration within reasonable limits of uncertainty.

Although much progress has been made in recent years (NCRP, 1984a; Peterson, 1983; Jirka *et al.*, 1983), a generalized approach to the dispersion of pollutants introduced into a body of water is not possible in the state of our knowledge. The rate of mixing is dependent on depth of water, type of bottom shoreline configuration, tidal factors, wind, temperature, and the depth at which the pollutant is introduced, among other factors. Each stream, river, bay, lake, sea, and ocean has its own mixing characteristics

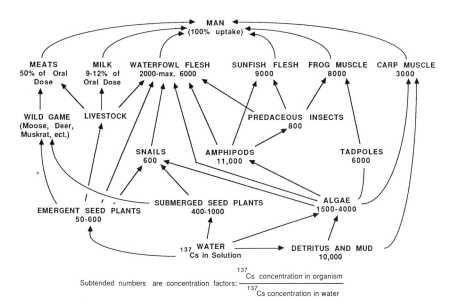

Fig. 6–5. Freshwater food web, illustrating the pathways to humans for cesium-137 in the aquatic environment. (From Pendleton and Hanson, 1958.)

that vary from place to place and from time to time. Hydrologists have developed useful dispersion equations that are valuable for the specific situations for which they are intended, but site-specific parameters are usually required that limit the general applicability of the equations without prior field studies. Some of the problems associated with the use of models in aquatic systems were discussed briefly in Chapter 4.

The mixing process in the aquatic environment is also complicated by the fact that the fate of a pollutant may be dependent on other physical and biological processes (Figs. 6–5 and 6–6). If a pollutant is a suspended solid, it can settle to the bottom, be filtered by organisms, or become attached to plant surfaces. Pollutants in solution can sorb on suspended organic and inorganic solids or can be assimilated by plants and animals. The suspended solids, dead biota, and excreta settle to the bottom and become part of the organic-rich substrate that supports the benthic community of organisms. The sediments more often act as a sink (temporary or permanent) for pollution, but they may also become a source, as when they are resuspended during periods of increased turbulence or are dredged and deposited elsewhere. The sediments may also serve as a secondary source of pollution when desorption occurs. Lentsch *et al.* (1972) have shown that the role of estuarine sediments as a source for Mn

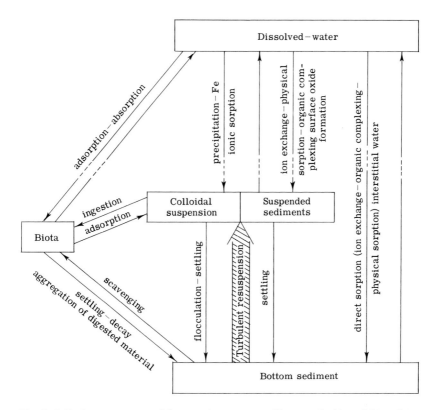

Fig. 6–6. Basic components of the aquatic ecosystem. The complexities of the pathways among the biota are illustrated in Fig. 6–5. (From Hairr, 1974.)

and ^{54}Mn is related to salinity and that the K_d for ^{54}Mn may oscillate with the tidal cycle under estuarine conditions. Figure 6–7 illustrates how the K_d for ^{137}Cs can be influenced by the presence of cations that compete for sorption sites. It is seen that the K_d for ^{137}Cs in the tidal portion of the Hudson River estuary varies inversely with Cl$^-$ over a range of two orders of magnitude (Jinks and Wrenn, 1976; Linsalata *et al.*, 1985b). The chloride ion itself plays no role in competing with ^{137}Cs for sorption sites on particulate matter, but is simply a measure of salinity and the associated presence of cations that are normally present in seawater and compete with Cs for sorption sites. Laboratory studies have shown that K$^+$ is the primary cation in competition with Cs$^+$ for sorption sites (Hairr, 1974).

Prediction of the dispersion of pollutant species that have large K_d values and thus favor the particulate phase is more difficult than for those that remain in solution. Elements that tend to remain in solution include

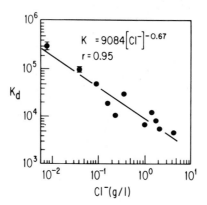

Fig. 6–7. Relationship of ^{137}Cs distribution coefficients (K_d) and chloride concentrations in continuous water samples at Indian Point, New York, in 1971. (From Jinks and Wrenn, 1976.)

Sr, Cr, and Sb. Elements that are easily sorbed on sediments and suspended matter include Cs, Mn, Fe, Co, and the actinides.

If sufficient information can be obtained about physical characteristics of a body of water, it is possible to estimate with some degree of certainty its capacity to receive radioactive nuclides without exceeding the permissible limits of human exposure. However, in the absence of site-specific information, the uncertainties that exist are likely for some time to result in a highly conservative approach (i.e., one that errs on the safe side) to the discharge of radionuclides into the aquatic environment. Proper planning of liquid-waste management practices often requires that dispersion and ecological studies be conducted at each site where a radioactive waste outfall is to be located.

MIXING CHARACTERISTICS OF RECEIVING WATERS

The Oceans

The oceans (Sverdrup *et al.*, 1963; Kinne, 1970) cover an area of 3.6 × 10^8 km^2 and, with an average depth of 3800 m, contain a total volume of 1.37 × 10^9 km^3 (Revelle and Schaefer, 1957). Bordering the oceans are the continental shelves, which skirt most of the coastlines to a depth of about 150 m and in some places extend seaward for more than 150 km.

The near-surface water of the oceans, to a depth which varies geographically from 10 to 200 m, is a region in which rapid mixing occurs as a result of wind action. Because of this mechanical mixing, the vertical gradients of temperature, salinity, and density are nearly uniform.

About 75% of the oceans is cold, deep water at a temperature of 1–4°C and a salinity of 3.47%, but between the surface water and the deep water is an intermediate zone characterized by decreasing temperature and increasing salinity and density with depth. The fact that the salinity increases with depth reduces vertical motions because lighter water overlies dense water. The intermediate zone thus tends to restrict exchange between the surface waters and the deep waters (Revelle *et al.*, 1956; Pritchard *et al.*, 1971). The zone where the density change is greatest, sometimes called the *thermocline* (when due to temperature) or *pycnocline* (when due to density), may be as much as 1000 m in depth, but is usually less.

The characteristic currents of the ocean surface are due primarily to wind action and tend to be related to the surface wind patterns. The movement of surface water has been shown to be as much as 144 km/day in the Florida current and 66 km/day in the Kuroshio current of the western Pacific (NAS–NRC, 1957b). Radioactivity from Bikini atoll was found to drift westward at a rate of about 14 km/day after tests of nuclear weapons in 1954 (Miyake and Saruhashi, 1960). The major surface currents of the oceans are shown in Fig. 6–8.

A number of studies of the manner in which radioactive substances diffuse vertically and horizontally in the mixed layer have been made in connection with the U.S. weapons testing program in the Marshall Islands (Chapter 13), and in the Irish Sea, where the British have conducted extensive studies of the fate of radioactive wastes discharged from Sellafield (formerly known as Windscale) (Chapter 11).

Studies of mixing of radioactive fallout in the Pacific indicated a persistent holdup of radioactivity near the surface for as long as 5 years after cessation of atmospheric tests of nuclear weapons (Volchok and Kleinman, 1971). Strontium-90 and ^{137}Cs have been the nuclides most thoroughly studied, and the ratio of the two nuclides has been constant with respect to time, depth, and sampling location. The complexities of the subject are well illustrated in the report by Bowen *et al.* (1980) of the vertical and horizontal distribution of fallout radionuclides in the Pacific.

After a series of tests in the Marshall Islands in 1954, extensive surveys were made of the spread of radioactivity in the northern Pacific. The general course of the contamination was initially in a westerly direction to the region of the Asiatic mainland, where the contamination turned north into the Kuroshio current. The data from these surveys have been summarized by Miyake and Saruhashi (1960) and are illustrated in Fig. 6–9.

Fission products introduced into surface waters near Bikini in the Marshall Islands (Folsom and Vine, 1957) were found to have moved 225 km in 40 days and to have diffused during this time only to a depth of 30–60 m. The horizontal area was found to be about 40,000 km^2. The dilution

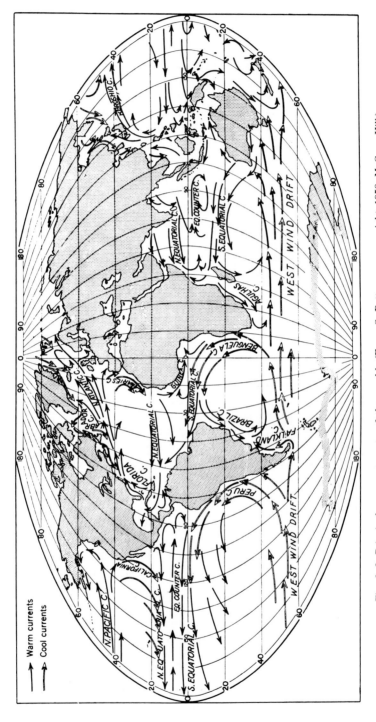

Fig. 6–8. Principal ocean currents of the world. (From S. Petterssen; copyright 1958, McGraw-Hill.)

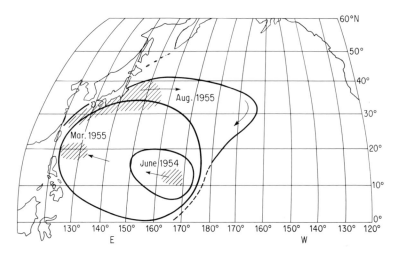

Fig. 6–9. Horizontal dispersion of nuclear weapons debris in the western Pacific Ocean after tests by the United States in the Marshall Islands in 1954. (From Miyake and Saruhashi, 1960.)

was such that had 1000 Ci been introduced, the average concentration at the end of 40 days would have been 1.5×10^{-10} µCi/ml.

The mechanisms and rates of dispersion of water from the deep ocean bottom can be inferred from measurements of the vertical distributions of ^{226}Ra and ^{230}Th, both of which are produced from decay of ^{238}U in the bottom sediments. From various data, Koczy (1960) developed the model of vertical diffusion shown in Fig. 6–10. Dissolved substances released from the ocean floor diffuse slowly through a friction layer 20–50 m in depth, within which the rates of mixing are controlled by molecular diffusion. Mixing is most rapid (3–30 cm²/sec) just above the friction layer and decreases rapidly with height above the ocean floor to a level about 1000 m below the surface, where a secondary minimum ($\sim 10^{-2}$ cm²/sec) is thought to exist. Diffusion rates then increase as one approaches the mixed layer, where the diffusion coefficients range from 50 to 500 cm²/sec.

Koczy estimated the vertical velocity in the Atlantic Ocean to be between 0.5 and 2 m/year at depths between 750 and 1750 m. If these values apply at greater depths, a radioactive solution placed at a depth of 3000 m would not appear in the surface water for more than 1000 years. Using these data, he estimated that 10^9 Ci of ^{90}Sr could be deposited at a depth of 300 m below the thermocline without exceeding the maximum permissible concentration (MPC) for ^{90}Sr in the mixed layer.

The vertical motions of the oceans have also been studied by Pritchard *et al.* (1971), using vertical profiles of ^{14}C concentrations. A vertical ve-

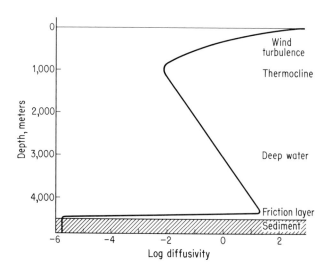

Fig. 6–10. Vertical diffusion from the ocean deeps according to Koczy (1960).

locity of about 6.6 m/year was found to be typical in the northeast Pacific Ocean to a depth of about 1000 m.

Pritchard (1960) had earlier calculated the rate at which a radioactive substance would diffuse to the surface if released from the ocean depths. He used a simplified model in which a 5-m layer of contamination was assumed to rise by 5 m each year and its position occupied by a new contaminated layer. No exchange between the 5-m layers was assumed, but horizontal diffusion was assumed to occur, so that as the radioactivity rises, each layer has a greater horizontal spread and, therefore, a lower concentration than the layer beneath.

The horizontal velocity for the deep ocean was not known but was assumed by Pritchard to lie between 2×10^{-3} and 2×10^{-4} m/sec, so that horizontal spread of the material over most of the area of the North Atlantic would occur between 40 and 400 years, which is somewhat less than the 600 years required to rise from a depth of 4000 m to a depth of 1000 m. Based on Pritchard's model, at the end of 600 years the radioactivity released during a 1-year period would be contained in a layer 5 m thick having a horizontal area of 3×10^{13} m^2. From this it was calculated that 5×10^{11} Ci of ^{90}Sr could be placed on the bottom each year at a depth of 4000 m without exceeding a concentration of 10^{-9} μCi/ml at the base of the 1000-m layer. An alternative estimate, based on the assumption that rapid mixing occurs in the bottom 1000 m, reduces Pritchard's estimate to 9×10^8 Ci, which is in close agreement with the figure suggested by Koczy.

The above estimates are based on relatively few data. New information is becoming available as a result of the international research program being undertaken to study the feasibility of using the deep ocean sediments as a repository for high-level radioactive wastes (Chapter 11) (Robinson and Marietta, 1985).

Rivers, Estuaries and Coastal Waters

Knowledge of the behavior of trace substances in rivers, estuaries, and coastal waters is important because these aquatic systems are major receptors of effluents from industrial plants and municipalities (Fisher *et al.*, 1979). The term *estuary* is usually applied to the tidal reaches of a river; it is a semienclosed coastal body of water which has a free connection to the open sea and within which the marine water is measurably diluted with fresh water from land drainage (Pritchard, 1967).

The estuarine waters, in which tidal action brings about mixing of salt and fresh water, are of special importance because of their high biological productivity. Not only are shellfish frequently harvested in estuarine waters in large quantities, but the waters also serve as the nursery grounds for many species of fish that later move to offshore waters, where they are harvested (Reid and Wood, 1976).

Each estuary has its own physical characteristics which must be studied in detail on an individual basis. The kinds of studies that can be made are exemplified by an investigation of diffusion and convection in the Delaware River basin (Parker *et al.*, 1961). A scale model of the Delaware basin was constructed at the United States Army Waterways Experiment Station at Vicksburg, Mississippi (Fig. 6–11). The model was 1200 m long and about 200 m wide. Figure 6–12 illustrates the type of information obtained by dye studies over a period of 58 tidal cycles (about 1 month). In this study, which involved instantaneous injection of a given dose, the concentration at the end of 58 tidal cycles remained at approximately 1% of the maximum concentration during the initial tidal cycle. The conditions of the experiment were conservative: there was no radioactive decay, sedimentation, or biological uptake. Thus, only the diminution in concentration due to mixing was measured.

The most extensive set of river mixing measurements in the United States is probably that made on the Columbia River, where the large plutonium-producing reactors at Hanford used river water for cooling and discharged traces of induced radioactivity, notably ^{32}P and ^{65}Zn, in the effluents. Studies of these two isotopes have been made in the water, sediments and biota of the river. An example of the manner in which dilution of the Hanford effluents takes place is shown in Fig. 6–13 (Foster, 1959).

Many models have been published to describe the dispersion of pollu-

Fig. 6–11. Scale model of the Delaware River, one of several models of rivers, estuaries, and bays by which flow characteristics are studied by the U.S. Army Corps of Engineers at their laboratory in Vicksburg, Mississippi. (Courtesy of U.S. Army Corps of Engineers.)

tants introduced into aquatic systems. The models take into consideration the location of the source, the form and rate of input of the pollutant, removal and transformation mechanisms, and transport processes in the water and sediments (Peterson, 1983; NCRP, 1984a). The aquatic transport and diffusion equations provide estimates of the radionuclide concentrations within a water body, the rate of deposition and accumulation of radionuclides on the shoreline and bottom, and movement of the radionuclides through biotic pathways to humans (Jirka *et al.*, 1983). The methods by which the models and computer codes are developed are discussed elsewhere (Little, 1983). The factors with which the models must deal are numerous, as can be seen from Fig. 6–14, which identifies some of the variables that must be included in the development of any mathematical model or computer code. It has been noted elsewhere (NCRP, 1984a) that in spite of all the measurements that have been made in the most thoroughly studied rivers of the United States during the past 30 years, there are insufficient data to validate the diffusion models that have been published.

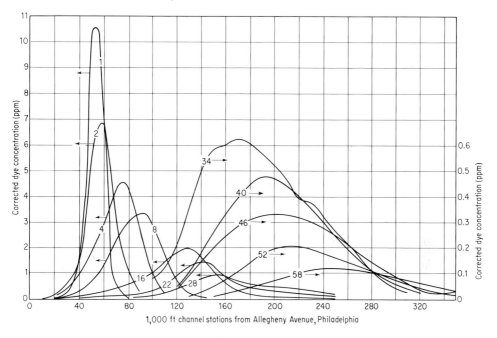

Fig. 6–12. Longitudinal distribution of contaminants after a designated number of tidal cycles in the Delaware River near Philadelphia. (From Parker *et al.*, 1961; reproduced from *Health Physics*, Vol. 6, by permission of the Health Physics Society.)

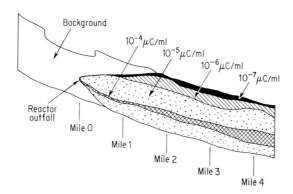

Fig. 6–13. Horizontal mixing of radioactive reactor effluents in the Columbia River. (From Foster, 1959.)

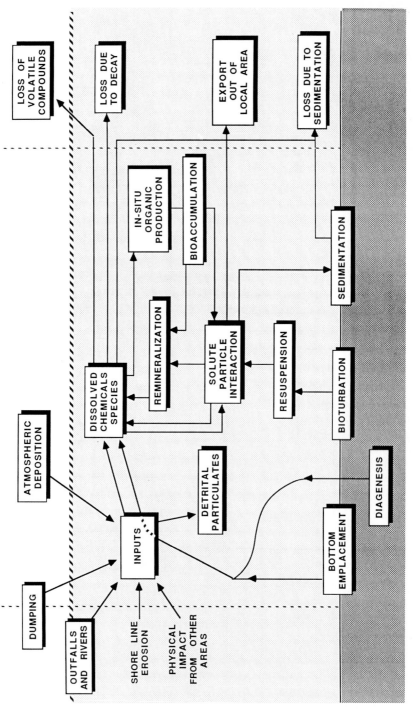

Fig. 6-14. Possible transformations of a pollutant in the water column. (From National Oceanic and Atmospheric Administration, 1979.)

Biological Uptake of Radionuclides

In an aquatic system, the supply of basic compounds such as carbon dioxide and of elements such as oxygen, calcium, hydrogen, and nitrogen is either contained in solution or held in reserve in the bottom sediments (Reid and Wood, 1976). These nutrients are absorbed and metabolized through the utilization of solar energy by two main types of food-producing organisms: rooted or large floating plants and minute floating plants called phytoplankton. Since food production by the higher plant forms is limited to relatively shallow water, the phytoplankton have the main responsibility for converting the mineral resources of the aquatic environment into food for the higher organisms.

The phytoplankton serve as food for small zooplankton, which, in turn, serve as the basic nourishment of several higher trophic levels. The phytoplankton also serve as food for certain filter-feeding fish and bottom-dwelling animals. Sedimentation of the excrement of aquatic animals and the action of organisms which decompose dead plants and animals eventually complete the cycle by returning the nutrient elements to abiotic forms, in which they again become available to the phytoplankton (Lowman *et al.*, 1971). Some of the complexities of the aquatic ecosystem are illustrated in Fig. 6–5.

Some elements, including copper, manganese, zinc, and iron, that may or may not be essential to life processes are present in varying amounts in water and sediments. The aquatic organisms sequester these trace elements to varying degrees, and this phenomenon must be taken into consideration when any decision is made about the rate at which given radionuclide can be released to a body of water.

Onishi *et al.* (1981) thoroughly reviewed the transport of radionuclides in aquatic systems. The ratio of the concentration of an element in the organism to the concentration in the water is known as the concentration factor (CF). This ratio should be measured under equilibrium conditions, since an organism that has just entered a contaminated environment would obviously provide an inappropriate sample and would yield a spuriously low CF.

The CF can also be greatly influenced by the presence of chemical congeners in the water. One would expect the CF for ^{90}Sr to be greatly affected by the concentration of Ca in the water. Because of the relatively high concentration of K in seawater, the CF for ^{137}Cs in freshwater biota is orders of magnitude higher than in marine or estuarine organisms.

Considering all the factors outlined above, it is not surprising that for each element of interest there is a wide range of CFs reported in the literature. The variability could undoubtedly be reduced if reported CFs could be limited to those in which the organism was in equilibrium with its environment and if other sources of variability could be taken into consideration. In most cases the reports in the literature do not provide sufficient information to permit such adjustments to be made (NCRP, 1984a). If the reported data could be normalized to either wet weight or dry weight, the variability would be reduced by a factor of 5 (Blaylock, 1982).

The Nuclear Regulatory Commission has issued recommended CFs for use in dose assessment (U.S. NRC, 1977b), but for most elements it is difficult to justify single recommended values in view of the great variability in the CFs reported in the literature.

The significance of the presence of radionuclides in marine and freshwater foods depends on the part of the organism in which the radionuclide is located. A radionuclide is more important in risk assessment if it concentrates in an organ that is consumed by humans than if it deposits in the portion that is not eaten. Thus, although clams, oysters, and scallops concentrate ^{90}Sr, as do certain crabs, this element is stored in the shell, which is not ordinarily consumed. On the other hand, ^{65}Zn and ^{60}Co concentrate in the edible tissues of seafood. A CF obtained from analysis of whole fish can be an order of magnitude higher than a CF based on analysis of the fish muscle (NCRP, 1984a).

In British studies of the fate of radioactive effluents from Windscale (now called Sellafield) (Dunster, 1958; Preston and Jeffries, 1969), the factor which limited the quantity of waste discharged into nearby coastal waters was the accumulation of radioactivity in seaweed harvested in Cumberland about 20 km from the point of discharge. The local inhabitants, who regularly consumed substantial quantities of seaweed, were the limiting factor in determining the maximum amount of radioactivity that could be discharged into this particular environment.

Another example in which local factors influence the significance of river water contamination was found at Hanford, where Columbia River water irrigates thousands of square miles of land downstream from the Hanford reservation, resulting in ^{65}Zn contamination of farm produce (Davis *et al.*, 1958). The concentration in beef and cow's milk was particularly noteworthy.

ROLE OF THE SUSPENDED SOLIDS AND SEDIMENTS

The sediments play a predominant role in aquatic radioecology by serving either as a sink or as a temporary repository for radioactive substances, which can then pass by way of the bottom-feeding biota or by resuspension or dissolution to the higher trophic levels (Duursma and Gross, 1971).

Figure 6–15 (Linsalata *et al.*, 1986d) illustrates the high degree of con-

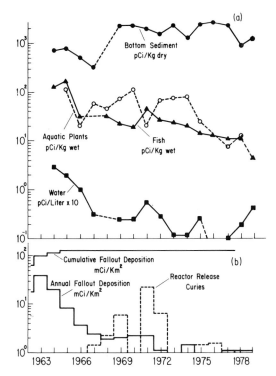

Fig. 6–15. Annual average ^{137}Cs concentrations in water (<0.45 μm), surface sediment (0–5 cm), rooted vegetation, and various indigenous fish species collected from the freshwater reaches of the Hudson River estuary. The apparent increase in ^{137}Cs concentration in bottom sediments (1967–1969) was probably a result of changes in the sampling locations. (From Linsalata *et al.*, 1986d.)

centration of ^{137}Cs that takes place in Hudson River sediments ($K_d = 10^4$–10^5) relative to that in fish, rooted plants, or dissolved in water. These data were collected in the northern (freshwater) reaches of the Hudson River estuary, beyond the range of influence of a nuclear power reactor located at Indian Point. Shown in Fig. 6–15 are (a) the annual mean concentrations of this nuclide in filtered river water, surficial (0–5 cm) bottom sediments, aquatic plants, and indigenous fish, and (b) the rate of fallout deposition and reactor discharge of ^{137}Cs. In general, the long-term trend of ^{137}Cs concentrations in fish (and to a lesser extent in plants) tends to follow the concentrations in water rather closely (simple correlation coefficient $r = 0.50$). However, between 5 and 20% of the ^{137}Cs content of fish has, for various times, locations, and species, been attributed to a sedimentary source (Wrenn *et al.*, 1972; Jinks, 1975).

Radionuclides introduced into a body of water reach the bottom sediment primarily by sorption on suspended solids that later deposit on the bottom. The deposited remains of biota that have absorbed or adsorbed pollutants may also be an important source.

In studies of the Clinch River below Oak Ridge, Tennessee, the amounts of radioactivity contained by the suspended solids were found to be variable (Parker *et al.*, 1966), which is not surprising considering that the load and composition of particulate matter varies from place to place in a river. The tendency of the sediments to remove pollutants depends on their capacity for sorption and ion exchange as well as the salinity of the overlying water. It was estimated that of the total quantity of radioactive materials released into the river during a 20-year period, the sediments contained 21% of the ^{137}Cs, 9% of the ^{60}Co, 0.4% of the ^{106}Ru, about 25% of the rare earth radionuclides, and about 0.2% of the ^{90}Sr (Pickering *et al.*, 1966). The depth distribution of ^{137}Cs in cores several feet deep were found to be well correlated with annual releases to the river. This was found not to be so for ^{60}Co. In other river systems the sediments might be less quiescent owing to stirring or scouring during periods of high runoff.

Lentsch *et al.* have studied the behavior of ^{54}Mn released into the Hudson River from reactors located at Indian Point. The manganese content of several species of rooted aquatic plants has been found to be proportional to the dissolved manganese concentration in water, but this in turn has been shown to be highly variable owing to the periodic intrusion of salt water, which released manganese bound in the river sediments. In this investigation, the ^{54}Mn was found to behave similarly to stable manganese present in the river system (Lentsch *et al.*, 1972).

Chapter 7

Natural Radioactivity

Natural radioactivity originates from extraterrestrial sources as well as from radioactive elements in the earth's crust. About 340 nuclides have been found in nature, of which about 70 are radioactive and are found mainly among the heavy elements. All elements having an atomic number greater than 80 possess radioactive isotopes, and all isotopes of elements heavier than number 83 are radioactive.

The radioactivity of the earth now includes the primordial radionuclides, whose half-lives are sufficiently long that they have survived the interval since their creation, and the secondary radionuclides which are derived by their radioactive decay. A much larger number of radioactive isotopes than now exist were produced when the matter of which the universe is formed first came into being several billion years ago, but most of them have decayed out of existence. The radionuclides which now exist are those that have half-lives at least comparable to the age of the universe. Radioisotopes with half-lives of less than about 10^8 years have become undetectable in the 30 or so half-lives since their creation, whereas radionuclides with half-lives greater than 10^{10} years have decayed very little up to the present time.

In most places on earth the natural radioactivity varies only within relatively narrow limits, but in some localities there are wide deviations from normal levels owing to the presence of abnormally high concentrations of radioactive minerals in the local soils.

TABLE 7–1a

SINGLY OCCURRING NATURAL RADIONUCLIDES PRODUCED BY COSMIC RAYS[a]

Radionuclide	Half-life	Tropospheric concentration (pCi/kg air)	Principal radiations and energies (MeV)	Observed average concentrations in rainwater (pCi/liter)
^{3}H	12 years	3.2×1.0^{-2}	β^- 0.0186	—
^{7}Be	53 days	0.28	γ 0.477	18.0
^{10}Be	1.6×10^6 years	3.2×1.0^{-8}	β^- 0.555	—
^{14}C	5730 years	3.4	β^- 0.156	—
^{22}Na	2.6 years	3.0×1.0^{-5}	β^+ 0.545, γ 1.28	7.6×10^{-3}
^{24}Na	15.0 hr	—	β^- 1.4, γ 1.37, 2.75	0.08–0.16
^{32}P	14 days	6.3×1.0^{-3}	β^- 1.71	"a few"
^{33}P	24 days	3.4×1.0^{-3}	β^- 0.246	"a few"
^{35}S	88 days	3.5×1.0^{-3}	β^- 0.167	0.2–2.9
^{36}Cl	3.1×10^5 years	6.8×1.0^{-9}	β^- 0.714	—
^{38}S	2.87 hr	—	β^- 1.1, γ 1.88	1.8–5.9
^{38}Cl	37 min	—	β^- 4.91, γ 1.60, 2.17	4.1–67.6
^{39}Cl	55 min	—	β^- 1.91, γ 0.25, 1.27, 1.52	4.5–22.5

[a]Perkins and Nielsen (1965). Reproduced from *Health Physics*, vol. 11, by permission of the Health Physics Society.

Naturally Occurring Radioactive Substances

The naturally occurring radionuclides can be divided into those that occur singly (Tables 7-1a and 7-1b) and those that are components of three chains of radioactive elements: (1) the uranium series, which originates with ^{238}U (Table 7–2); (2) the thorium series, which originates with ^{232}Th (Table 7–3); and (3) the actinium series, which originates with ^{235}U (Table 7–4). In addition, there are singly occurring radionuclides of both cosmic and terrestrial origin. The three families of radioactive heavy elements account for much of the background radiation to which humans are exposed. A fourth family, the neptunium series, which originated in the parent element ^{241}Pu, is known to have existed at one time, but this nuclide has a half-life of only 14 years and existed only briefly after its formation. Other members of that series also have relatively short half-lives. The only surviving member of the neptunium family is the nearly stable nuclide ^{209}Bi, which has a half-life estimated to be about 2×10^{18} years.

In nature, ^{235}U and a few other nuclides of uranium and thorium undergo fission spontaneously or as a result of interactions with neutrons that

TABLE 7–1b

<small>Singly Occurring Natural Radionuclides of Terrestrial Origin[a]</small>

Radio-nuclide	Abun-dance (%)	Half-Life (years)	Principal radiations: energy (MeV) and yield (%)	Specific activity (elemental) (pCi/g)
^{40}K	0.012	1.26×10^9	β^- 1.33, 89% γ with EC 1.46, 11%	855
^{50}V	0.25	6×10^{15}	γ with β^- 0.78, 30% γ with EC 1.55, 70%	3.0 $\times 10^{-3}$
^{87}Rb	27.9	4.8×10^{10}	β^- 0.28, 100%	2.4 $\times 10^4$
^{115}In	95.8	6.0×10^{14}	β^- 0.48, 100%	4.98
^{123}Te	0.87	1.2×10^{13}	EC	2.11
^{138}La	0.089	1.12×10^{11}	β^- 0.21, 80% γ with EC (0.81, 1.43), 70%	20.7
^{142}Ce	11.07	$>5 \times 10^{16}$	(α)	5.6 $\times 10^{-3}$
^{144}Nd	23.9	2.4×10^{15}	α 1.83	0.25
^{147}Sm	15.1	1.05×10^{11}	α 2.23	3.5 $\times 10^3$
^{148}Sm	11.27	$>2 \times 10^{14}$	—	1.37
^{146}Sm	13.82	$>1 \times 10^{15}$	—	0.33
^{152}Gd	0.20	1.1×10^{14}	α 2.1	4.3 $\times 10^{-2}$
^{156}Dy	0.052	$>1 \times 10^{18}$	—	1.2 $\times 10^{-6}$
^{174}Hf	0.163	2×10^{15}	α 2.5	1.68 $\times 10^{-3}$
^{176}Lu	2.6	2.2×10^{10}	β^- 0.43 γ 0.089, 0.203, 0.306	2.4 $\times 10^3$
^{180}Ta	0.012	$>1 \times 10^{12}$	—	0.239
^{187}Re	62.9	4.3×10^{10}	β^- 0.003	2.8 $\times 10^4$
^{190}Pt	0.013	6.9×10^{11}	α 3.18	0.36

[a] Data assembled from Lederer *et al.* (1967).

originate from cosmic rays or other natural sources. The half-life of ^{235}U owing to spontaneous fission is between 10^{15} and 10^{16} years, which means that decay by this process proceeds at a rate less than 10^{-7} of that due to α emission. Others of the heavy nuclides undergo spontaneous fission with half-lives that range from 10^{14} to 10^{20} years (Rankama, 1954).

Many transuranic elements, such as plutonium, neptunium and americium, which now exist because they have been produced artificially (Seaborg, 1958), must have existed in nature at one time, but their half-lives are so short that they disappeared long ago. However, some of the transuranic elements are produced in minute amounts by naturally occurring neutrons that result in fission of uranium isotopes. Plutonium-239 has been detected in pitchblende in a ratio to ^{238}U of 10^{-11} to 10^{-13}, and ^{237}Np has been identified in uranium minerals in a ratio to ^{238}U of 1.8 μ 10^{-12}.

TABLE 7–2

NUCLIDES OF THE URANIUM SERIES

Isotope	Relative isotopic abundance (%)	Half-life	Radiation	Energy (MeV)	Percent yield
^{238}U	99.28	4.5×10^9 years	α	4.20	75
				4.15	23
			γ	0.048	23
^{234}Th	—	24 days	β	0.192	65
				0.100	35
			γ	0.092	4.0 (doublet)
^{234m}Pa	—	1.2 min	β	2.29	98
				1.53	<1
				1.25	<1
			γ (IT)	0.39	0.13
			γ	0.817	4
^{234}U	0.0057	2.5×10^5 years	α	4.77	72
				4.72	28
			γ	0.093	5
^{230}Th	—	8.0×10^4 years	α	4.68	76
				4.61	24
				4.51	0.35
			γ	0.068	0.6
				0.253	0.02
^{226}Ra	—	1622 years	α	4.78	94.3
				4.59	5.7
			γ	0.186	4
				0.26	0.007
^{222}Rn	—	3.8 days	α	5.48	100
			γ	0.510	0.007
^{218}Po	—	3.05 min	α	6.0	99+
^{214}Pb	—	26.8 min	β	0.72	100
			γ	0.053	1.6
				0.242	4
				0.295	19
				0.352	36
^{218}At	—	1.5–2.0 sec	α	6.70	94
				6.65	6
^{214}Bi	—	19.7 min	β	3.26	19
				1.51	40
				1.00	23

TABLE 7–2 (*continued*)

Isotope	Relative isotopic abundance (%)	Half-life	Radiation	Energy (MeV)	Percent yield
				1.88	9
			α	5.52	0.008
				5.45	0.011
				5.27	0.001
^{214}Po	—	1.64×10^{-4} sec	α	7.68	100
			γ	0.799	0.014
$^{210'}$Tl	—	1.3 min	β	1.9	56
				1.3	25
				2.3	19
			γ	0.296	80
				0.795	100
				1.31	21
^{210}Pb	—	22 years	β	0.015	81
				0.061	19
			γ	0.0465	4
^{210}Bi	—	5.0 days	β	1.17	99+
			α	5.0	5×10^{-5}
^{210}Po	—	138 days	α	5.3	100
			γ	0.80	0.0011
$^{206'}$Tl	—	4.2 min	β	1.51	100
^{206}Pb	25.2	Stable	—	—	—

URANIUM

The uranium normally found in nature consists of four isotopes having mass numbers 230, 234, 235, and 238. Uranium-238 is present in the amount of 99.28% and is usually in equilibrium with ^{234}U, which is present in the amount of 0.0058%. Uranium-235, the parent isotope of the actinium series, is present in the amount of 0.71%. Uranium-230, which is also a member of the ^{238}U series, has a short half-life (20.8 days).

Uranium is found in all rocks and soils. Typical concentrations are listed in Table 7–5, which shows that the acid igneous rocks contain concentrations of the order of 3 ppm, about 100 times greater than the ultrabasic igneous rocks but considerably less than the phosphate rocks of Florida and southeastern Idaho and neighboring areas, which contain as much as 120 ppm and which have been considered as a commercial source of uranium (Clegg and Foley, 1958; NCRP, 1975c). The high uranium content

TABLE 7–3

NUCLIDES OF THE THORIUM SERIES

Isotope	Relative isotopic abundance (%)	Half-life	Radiation	Energy (MeV)	Percent yield
^{232}Th	100	1.4×10^{10} years	α	4.01	76
				3.95	24
			γ	0.055	24
^{228}Ra	—	6.7 years	β	0.055	100
^{228}Ac	—	6.13 hr	β	2.18	10.0
				1.85	9.6
				1.72	6.7
				1.11	53.0
				0.64	7.6
				0.46	13
			γ	0.058	53.0
				0.129	5.2
				0.184	<1
^{228}Th	—	1.9 years	α	5.42	71
				5.34	28
			γ	0.083	1.6
^{224}Ra	—	3.64 days	α	5.68	94
				5.45	6
				5.19	0.4
			γ	0.241	3.7
^{220}Rn	—	55 sec	α	6.28	100
			γ	0.50	0.07
^{216}Po	—	0.16 sec	α	6.77	100
^{212}Pb	—	10.6 hr	β	0.33	~88
				0.57	~12
			γ	0.176	<1
				0.238	47
				0.300	3.2
^{212}Bi	—	60.5 min	β	2.25	64
			α	6.086	9.8
				6.047	25.1
			γ's with β	1.81	1.0
				1.61	1.8
				1.03	2
				0.83	13
				0.72	7
			γ's with α	0.040	2
				0.288	0.5
				0.46	0.8
					(complex)

TABLE 7–3 (*Continued*)

Isotope	Relative isotopic abundance (%)	Half-life	Radiation	Energy (MeV)	Percent yield
²¹²Po	—	3.04×10^{-7} sec	α	10.55	<1
				8.785	99
²⁰⁸Tl	—	3.1 min	β	1.80	100
			γ	2.61	100
				0.86	12
				0.58	86
				0.51	23
²⁰⁸Pb	52.3	Stable	—	—	—

of phosphate rocks is reflected in correspondingly high uranium concentrations in commercial phosphate fertilizers.

The uranium content of air in New York State has been found to range from 0.035 to 0.47 fCi/m³ and to be correlated with the concentration of suspended particulates (McEachern *et al.*, 1971). Airborne soil and possibly coal fly ash are the most likely sources.

Uranium occurs in traces in many commercial products. Pre-World War II samples of steel analyzed by Welford and Sutton (1957) contained uranium in the range of 0.01 to 0.2 ppm. Surprisingly, photographic emulsions and other photographic materials contain from 0.2 to 1 ppm of uranium (Smith and Dzuiba, 1949).

Because uranium occurs in soils and fertilizers, the element is present in food and human tissues. On average, the annual intake of uranium from all dietary sources is about 0.14 mCi. The intake of uranium from tap water is negligible by comparison, but a few exceptions have been reported, notably in France, Finland, and the Soviet Union (Hess *et al.*, 1985; UNSCEAR 1982). The skeleton is estimated to contain about 25 µg of uranium, equivalent to about 8 pCi. The dose to the skeleton, which receives a higher dose from uranium than any other organ, is about 0.3 mrem/year (UNSCEAR, 1982; ICRP, 1975; Wrenn *et al.*, 1985).

RADIUM-226

Radium-226 and its daughter products are responsible for a major fraction of the dose received by humans from the naturally occurring internal emitters. Referring to Table 7–2, it is seen that ²²⁶Ra is an α emitter that

TABLE 7–4

NUCLIDES OF THE ACTINIUM SERIES

Isotope	Relative isotopic abundance (%)	Half-life	Radiation	Energy (MeV)	Percent yield
^{235}U	0.72	7.1×10^8 years	α	4.32	3
				4.21	5.7
				4.58	8 (doublet)
				4.5	1.2
				4.4	57
				4.37	18
			γ	0.110	2.5
				0.143	11.0
				0.163	5.0
				0.185	54.0
				0.205	5.0
^{231}Th	—	25.64 hr	β	0.302	52
				0.218	20
				0.138	22
			γ	0.026	2
				0.085	10 (complex)
^{231}Pa	—	3.25×10^4 years	α	5.00	24
				4.94	22
				5.02	23
				5.05	10
			γ	0.027	6
				0.29	6 (complex)
^{227}Ac	—	21.6 years	β	0.046	99
			α	4.95	1.2 (doublet)
			γ	0.070	0.08
^{227}Th	—	18.2 days	α	6.04	23
				5.98	24
				5.76	21
				5.72	14 (doublet)
			γ	0.050	8
				0.237	15 (complex)
				0.31	8 (complex)

TABLE 7–4 (*continued*)

Isotope	Relative isotopic abundance (%)	Half-life	Radiation	Energy (MeV)	Percent yield
^{223}Fr	—	22 min	β	1.15	99+
			α	5.35	4×10^{-3}
			γ	0.050	40
				0.080	13
				0.234	4
^{223}Ra	—	11.4 days	α	5.75	9
				5.71	54
				5.61	26
				5.54	9
			γ	0.149	10 (complex)
				0.270	10
				0.33	6 (complex)
^{219}Rn	—	4.0 sec	α	6.82	81
				6.55	11
				6.42	8
			γ	0.272	9
				0.401	5
^{215}Po	—	1.77×10^{-3} sec	α	7.38	99+
			β	—	2.3×10^{-4}
^{211}Pb	—	36.1 min	β	1.36	92
				0.95	1.4
				0.53	6
			γ	0.405	3.4
				0.427	1.8
				0.832	3.4
^{215}At	—	$\sim 10^{-4}$ sec	α	8.00	100
^{211}Bi	—	2.16 min	α	6.62	84
				6.28	16
			β	—	0.27
			γ	0.35	14
^{211}Po	—	0.52 sec	α	7.45	99
				6.89	0.5
			γ	0.57	0.5
				0.90	0.5
^{207}Tl	—	4.79 min	β	1.44	100
			γ	0.870	0.16
^{207}Pb	21.7	Stable	—	—	—

TABLE 7–5

AVERAGE URANIUM CONCENTRATION IN VARIOUS ROCKS[a]

Rock type	Uranium concentration (ppm)
Acid igneous	3.0
Intermediate igneous	1.5
Basic igneous	0.6
Ultrabasic igneous	0.03
Meteorites	0.003
Phosphate rock (Fla.)	120
Phosphate rock (N. Africa)	20–30
Bituminous shale (Tenn.)	50–80
Normal granite	4
Limestones	1.3
Other sedimentary rocks	1.2

[a]From Lowder and Solon (1956).

decays, with a half-life of 1622 years, to ^{222}Rn, with a half-life of 3.8 days. The decay of radon is followed by the successive disintegration of a number of short-lived α- and β-emitting progeny. After six decay steps, in which isotopes are produced that range in half-lives from 1.6×10^{-4} sec to 26.8 min, ^{210}Pb is produced, which has a half-life of 22 years. This nuclide decays through ^{210}Bi to produce ^{210}Po (half-life 238 days), which decays by alpha emission to stable ^{206}Pb.

TABLE 7–6

AVERAGE RADIUM, URANIUM, THORIUM, AND POTASSIUM CONTENTS IN VARIOUS ROCKS[a]

Type of rock	^{226}Ra (pCi/g)	^{238}U (pCi/g)	^{232}Th (pCi/g)	^{40}K (pCi/g)
Igneous	1.3	1.3	1.3	22
Sedimentary				
Sandstones	0.71	0.4	0.65	8.8
Shales	1.08	0.4	1.1	22
Limestones	0.42	0.4	0.14	2.2

[a] Adopted from UNSCEAR (1958).

TABLE 7–7

AVERAGE RADIUM CONTENT OF VARIOUS ROCKS AND SOILS[a]

No. of specimens	Classification	Specimen, average value (pCi/g)
1	Quartz-mica schist	0.20
1	Quartzite	0.20
4	Limestones	0.29
1	Sandstone	0.32
2	Glacial sand and rubble	0.38
4	Gravels	0.41
2	Soils	0.73
1	Kaibab limestone	0.97
5	Granites	1.02
2	Peruvian lavas	2.06
23	All specimens	0.70

[a]From Evans and Raitt (1935).

Radium, being an α emitter, does not add directly to the γ activity of the environment, but does so indirectly through its γ-emitting descendants.

Radium-226 Content of Rocks and Soils

Radium-226 is present in all rocks and soils in variable amounts. Igneous rocks tend to contain somewhat higher concentrations than sandstones and limestones. Rankama and Sahama (1950) give a mean concentration of 0.42 pCi/g in limestone and 1.3 pCi/g in igneous rock, as listed in Table 7–6. Evans and Raitt (1935), who measured the ^{226}Ra content of rock and soil specimens from the sites of cosmic-ray observations made by Millikan and co-workers, give the values shown in Table 7–7, which are somewhat lower than the average values given by Rankama and Sahama.

Radium-226 in Water

The radium content of public water supplies has been reviewed comprehensively by Hess *et al.* (1985), who described the geological and geochemical factors that influence the concentration of the two principal radium isotopes, ^{226}Ra and ^{228}Ra, which are progeny of uranium and thorium, respectively. There is more ^{232}Th than ^{238}U in nature on an activity basis, but there are geochemical factors that cause local concentrations of uranium, which often results in greater amounts of ^{226}Ra relative to ^{228}Ra.

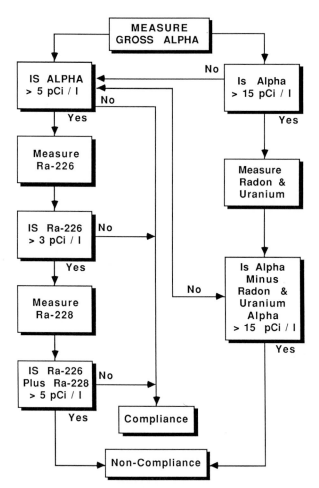

Fig. 7–1. Flowchart for measuring alpha activity of water according to EPA procedures. (From Lappenbusch and Cothern, 1985; reproduced from *Health Physics*, vol. 48, by permission of the Health Physics Society.)

Thus, it is generally assumed that the ratio ^{226}Ra : ^{228}Ra is greater than unity, although until recently most of the reported measurements have been of ^{226}Ra, with fewer measurements of ^{228}Ra.

The radium content of surface waters is low (0.1–0.5 pCi/liter) compared to most ground waters. Dissolved radium sorbs quickly to solids and does not migrate far from its place of release to ground water. It has been suggested that radium transport in ground water is even less than that of ^{222}Rn, which has a half-life of only 3.8 days (King *et al.*, 1982; Kirshnaswami *et al.*, 1982).

TABLE 7–8

SUMMARY OF [228]Ra AND [226]Ra DISTRIBUTION IN GROUND WATER BY AQUIFER TYPE FOR THE ATLANTIC COASTAL PLAIN AND PIEDMONT PROVINCES[a]

Aquifer type	Number of samples	Ra-228 (pCi/liter)		Ra-226 (pCi/liter)	
		Geometric mean	Range	Geometric mean	Range
Igneous rocks (acidic)	42	1.39	0.0–22.6	1.80	0.0–15.9
Metamorphic rocks	75	0.33	0.0–3.9	0.37	0.0–7.4
Sand	143	1.05	0.0–17.6	1.36	0.0–25.9
Arkose	92	2.16	0.0–13.5	2.19	0.0–23.0
Quartzose	50	0.27	0.0–17.6	0.55	0.0–25.9
Limestone	16	0.06	0.0–0.2	0.12	0.0–0.3

[a]Hess *et al.* (1985). Reproduced from *Health Physics*, vol. 48, by permission of the Health Physics Society.

Interest in the radium content of water supplies has increased during the past few years, following the publication in 1979 of a proposed EPA drinking water standard for total radium of 5 pCi/liter (Lappenbusch and Cothern, 1985). Compliance is determined by the sequence of steps illustrated in Fig. 7–1. Surveys of water supplies from many states (Cothern and Lappenbusch, 1985) show that the radium limit is exceeded in many communities that obtain water from wells, including communities of about 600,000 persons in Illinois, Iowa, Missouri, and Wisconsin (Lucas, 1982). About 75% of the supplies that exceeded 5 pCi/liter were located in two areas of the United States: (1) the Piedmont and Coastal Plain areas of the Middle Atlantic states, and (2) the North Central states of Minnesota, Iowa, Illinois, Missouri, and Wisconsin. The findings in the Atlantic Coastal Plain and Piedmont areas are summarized in Table 7–8. There is wide variability, with [226]Ra concentrations being generally higher than the [228]Ra concentrations. The concentration of [226]Ra is in some cases as high as 25 pCi/liter, with [228]Ra concentrations up to about 17 pCi/liter. It has been estimated that an individual who consumes 2 liters/day of water that contains 5 pCi [226]Ra/liter would, at the end of a lifetime, be receiving a bone dose of 150 mrem/year. The dose would be increased by the presence of [228]Ra. In rare cases where the concentration of [226]Ra is as high as 25 pCi/liter, the bone dose can approach nearly 1 rem/year.

The EPA regulations, which limit the [226]Ra content of potable water to 5 pCi/liter, may require that some well water supplies be treated. A study of the effectiveness of water treatment methods for the removal of radium was undertaken by Brinck *et al.* (1978) in areas of Iowa and Illinois where the EPA limit was exceeded. Four different water treatment methods were studied. Reverse osmosis and sodium ion exchange process were generally about 92% effective. The removal efficiency of the lime–soda ash softening process varied from 75 to 95%. Systems designed to remove iron only were found to have removal efficiencies from 11 to 53%. A detailed analysis of the costs and effectiveness of various methods of removing radioactivity from drinking water is given by Reid *et al.* (1985).

Little variation is found in the ^{226}Ra content of Atlantic Ocean surface water outside Antarctica (Broecker *et al.*, 1976). Eighty samples analyzed by Broecker and co-workers averaged 0.03 pCi/liter. Deep ocean water contains somewhat greater amounts of ^{226}Ra, which were explained by the inflow of O_2-deficient bottom water from the polar regions.

Radium-226 in Food

Radium is chemically similar to calcium and is absorbed from the soil by plants and passed up the food chain to humans. Because the radium in food originates from soil and the radium content of soil is variable, there is considerable variability in the radium content of foods. In addition, it is reasonable to expect that chemical factors such as the amount of exchangeable calcium in the soil will determine the rate at which radium will be absorbed by plants.

One of the earliest attempts to estimate the radium content of food was undertaken by Mayneord and associates (1958, 1960), who made α-radiation measurements of ashed samples of foods and differentiated the α activities of the thorium and uranium series by counting the double α pulses from the decay of ^{220}Rn and ^{212}Po, whose disintegrations are separated by only the 0.158-sec half-life of ^{212}Po. These early measurements served to approximate the total ^{226}Ra and ^{228}Ra content of foods (Table 7–9) and were highlighted by the fact that Brazil nuts were found to be much more radioactive than other foods. This was later investigated by Penna Franca *et al.* (1968), who showed that the anomaly is due to the tendency of the Brazil nut tree (*Bertholletia excelsa*) to concentrate barium, a chemical congener of radium. Penna Franca found the radium content of Brazil nuts to range between 273 and 7100 pCi/kg, with only 3 of 15 samples assaying less than 1000 pCi/kg. The radioactivity was about equally divided between ^{226}Ra and ^{228}Ra and was not related to the radium or barium content of the soil in which the tree was grown. The radium concentration of Brazil nuts is on the order of 1000 times greater than the radium concentration in the foods that make up the average diet in the United States.

Fisenne and Keller (1970) estimated the ^{226}Ra intake of inhabitants of New York City and San Francisco to be 1.7 and 0.8 pCi/day, respectively.

TABLE 7–9

ALPHA RADIOACTIVITY OF FOODS[a]

Foodstuff	Maximum α activity observed per 100 g (pCi)
Brazil nuts	1400
Cereals	60
Teas	40
Liver and kidney	15
Flours	14
Peanuts and peanut butter	12
Chocolates	8
Biscuits	2
Milks (evaporated)	1–2
Fish	1–2
Cheeses and eggs	0.9
Vegetables	0.7
Meats	0.5
Fruits	0.1

[a]From Mayneord *et al* (1958).

However, this twofold difference is not reflected by differences in the ^{226}Ra content of human bone from the two cities, which were found (Fisenne *et al.*, 1981) to be 0.013 pCi ^{226}Ra/g of bone ash for both cities. Studies of this kind involve highly sophisticated and sometimes uncertain food and bone sampling techniques which may involve errors that can obscure differences of a factor of 2. The ^{226}Ra content of the New York City diet is shown in Table 7–10. The relatively large contribution from eggs seems curious and has not been explained.

Radium-226 Content of Human Tissues

A number of investigators in various parts of the world have provided measurements of the ^{226}Ra content of bone, which contains about 80% of the total body radium. The data are summarized in Fig. 7–2, in which the concentrations range from about 0.008 to 0.10 pCi/g Ca, which results in a total body burden of about 8 to 100 pCi in an adult skeleton containing 1000 g of Ca (NCRP, 1984b). The median body burden is 23 pCi.

Estimates of the dose delivered by ^{226}Ra and its daughter products require knowledge of the fraction of radon retained by the tissue in which the radium is deposited, since most of the dose is due to alpha emissions from the decay of radon and its daughters. The dose estimate is also

TABLE 7–10

^{226}Ra in New York City Diet[a]

Diet category	kg year	g Ca year	1966 pCi kg	1966 pCi year	1968 pCi kg	1968 pCi year	Average pCi kg	Average pCi year
Dairy products	200	216.0	0.25	50	0.30[b]	60	0.25	50[c]
					0.19[b]	38		
Fresh vegetables	48	18.7	0.50	24	1.6	77	1.1	53
					1.6	77		
Canned vegetables	22	4.4	0.65	14	0.68	15	0.67	15
Root vegetables	10	3.8	1.4	14	1.2	12	1.3	13
Potatoes	38	3.8	2.8	106	1.7	65	2.3	87
Dry beans	3	2.1	1.1	3.3	0.98	2.9	1.0	3
Fresh fruit	59	9.4	0.43	25	0.20	12	0.32	19
Canned fruit	11	0.6	0.17	1.9	0.15	1.7	0.16	1.8
					0.16	1.8		
Fruit juice	28	2.5	0.42	12	0.90	25	0.66	18
Bakery products	44	53.7	2.8	123	1.7	75	2.3	101
Flour	34	6.5	1.9	65	2.3	78	2.1	71
Whole grain products	11	10.3	2.2	24	2.7	30	2.5	28
Macaroni	3	0.6	2.1	6.3	1.4	4.2	1.8	5.4
Rice	3	1.1	0.76	2.3	3.3	9.9	2.0	60
Meat	79	12.6	0.01	0.8	0.02	1.6	0.02	1.6
Poultry	20	6.0	0.76	15	0.10	2.0	0.44	8.8
					0.11	2.2		
Eggs	15	8.7	6.1	92	14	210	10	150
					14	210		
Fresh fish	8	7.6	0.67	5.4	1.1	8.8	0.89	7.1
Shellfish	1	1.6	0.80	0.8	0.90	0.9	0.85	0.9
Yearly intake		370		584.8		680.2		639.6
Daily intake pCi/g Ca				1.6		1.8		1.7

[a] Fisenne and Keller (1970).
[b] Two different samples.
[c] Average of three samples.

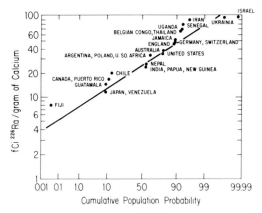

Fig. 7–2. Geographic distribution of measured ^{226}Ra concentration in human bone. (From National Council on Radiation Protection and Measurements, 1984b.)

complicated by the fact that the alpha energy is deposited at loci determined by the pattern of deposition within the tissue. UNSCEAR has used a radon retention factor of 0.33 and has assumed that the radium is uniformly deposited throughout the bone. On this basis, the average annual effective dose equivalent to bone from deposited radium is about 7 mrem, using a quality factor of 10 for the dose from alpha emissions (UNSCEAR, 1982).

THORIUM-232

The thorium content of various rocks, as reported by Faul (1954), indicates a range of 8.1 to 33 ppm for igneous rocks, with a mean value of 12 ppm. Rankama and Sahama (1950) reported that the concentration in sandstone is 6 ppm, intermediate between limestone and the igneous rocks. The thorium content of igneous rocks is thus about four times the uranium content, but since the specific activity of ^{232}Th is 0.11 pCi/g, compared to 0.33 pCi/g ^{238}U, the radioactivity due to the two nuclides is more nearly 1 : 1. Later in this chapter it will be seen that there are areas of the world in which the ^{232}Th content of rocks and soils is very much greater than normal.

The characteristics of the thorium series are different from those of the uranium series in a number of respects (see Tables 7–2 and 7–3).

1. ^{228}Ra has a shorter half-life than ^{226}Ra (5.8 compared to 1620 years).

2. ^{228}Ra is a β emitter that decays to α-emitting ^{228}Th, which has a half-life of 1.9 years. ^{228}Th, in turn, decays through a series of α emitters

including the noble gas ^{220}Rn (thoron), which has a half-life of only 54 sec, compared to 3.8 days for ^{222}Rn. Because of its short half-life, thoron has less opportunity than radon to diffuse from the matrix in which it is found.

3. The solubility of ^{228}Ra in soil is comparable with that of ^{226}Ra, but the dose rate to an organism from assimilated $^{-28}$Ra, a β emitter, is time-dependent because the dose depends on ingrowth of α-emitting ^{228}Th and its short-lived descendants.

4. In the ^{228}Ra chain, there is no long-lived "stopping" nuclide comparable to ^{210}Pb ($T_{1/2}$ = 22 years). The longest-lived nuclide beyond ^{228}Th is ^{212}Pb, with a half-life of 10.6 hr. The dosimetry and radiochemistry of the thorium series tend to be complicated by these characteristics (Fresco *et al.*, 1952).

Because of its relative insolubility and low specific activity, ^{232}Th is present in biological materials only in insignificant amounts. Linsalata *et al.* (1986a) have reported ^{232}Th concentrations in the edible portions of washed vegetables grown in silty clay and peaty soils of New York State to range between 0.001 pCi/kg (fresh weight) for carrots and 0.093 pCi/kg for a variety of squash. The mean value for 25 vegetable samples including potatoes, corn kernels, carrots, beans, and squash was 0.018 ± 0.022 pCi/kg (fresh weight). Wrenn *et al.* (1981) measured ^{232}Th, ^{230}Th, and ^{228}Th in the tissues of residents of New York and Colorado. Thorium was found to be present in the highest concentrations in pulmonary lymph nodes and lung, indicating that the principal source of human exposure is inhalation of suspended soil particles. Because thorium is removed from bone very slowly, the concentrations of both ^{230}Th (which is found in the ^{238}U decay series) and ^{232}Th were found to increase with age.

RADIUM-228 (MESOTHORIUM)

Although ^{228}Ra (referred to as mesothorium in some earlier writings) frequently occurs in soil and water in approximately a 1 : 1 ratio to ^{226}Ra, there is surprisingly little information about its occurrence in foods or in human tissues. Systematic ^{228}Ra measurements in food and water have not been made on a scale comparable to those of ^{226}Ra, but such data as do exist suggest that under normal circumstances the ^{228}Ra content of food, water, and human tissues is one-half to one-fourth of the ^{226}Ra content (UNSCEAR, 1966).

RADON-222 AND RADON-220 (THORON)

When ^{226}Ra decays by α emission, it transmutes to its daughter ^{222}Rn, an inert gas with a half-life of 3.8 days. Similarly, ^{224}Ra, which is a descendant of ^{232}Th, decays by α emission to 54-sec ^{220}Rn, commonly known as thoron.

The 3.8-day radon isotope has a greater opportunity than the nuclei of shorter-lived thoron to escape to the atmosphere. The mechanisms by which radon diffuses from soil into the atmosphere have been discussed by Tanner (1964, 1980).

The average radon concentration in Washington, D.C., has been shown by Lockhart (1964) to be more than 100 times greater than the average concentration in ice-covered Little America and more than 12 times the values observed at Kodiak, Alaska. The same investigator reported wide variability from day to day. For example, the mean daily concentrations varied more than a hundredfold in Washington, D.C., during 1957 (Lockhart, 1958). Gesell (1983) has reviewed the reported data from various parts of the United States and estimates the annual average concentration to range from 0.016 pCi/liter in Kodiak, Alaska, to 0.75 pCi/liter in Grand Junction, Colorado. The average annual concentrations showed less variation at four more normal localities (0.22–0.30 pCi/liter).

Other reports from several countries indicate that the average concentrations of radon in outdoor air may be taken normally to be in the range 0.1–0.5 pCi/liter. A number of investigators have observed periodicity in hour-to-hour observations of the radon and thoron content of outdoor air (UNSCEAR, 1982). Maximum concentrations are observed in the early hours and the lowest values are found in the late afternoon, when the concentrations are about one-third the morning maxima (Gold *et al.*, 1964).

It is likely that the variations at any given locality are dependent on meteorological factors that influence both the rate of emanation of the gases from the earth and the rate of dilution in the atmosphere (Wilkening, 1982). Thus, the rate of emanation from soil may increase during periods of diminishing atmospheric pressure and decrease during periods of high soil moisture, although the evidence is not consistent on this point. It is also likely that the history of an air mass for several days prior to observation influences its radon and thoron content (Barreira, 1961). Passage of the air over oceans, and possibly precipitation, would tend to reduce the concentration of these gases, whereas periods of temperature inversion should cause the concentrations to increase by limiting the volume of the atmosphere within which dilution can take place. The gases can be expected to be present in greater amounts over large masses of igneous rock than over large bodies of water or over sedimentary formations.

Because the daughter products of radon and thoron are electrically charged when formed, they tend to attach themselves to inert dusts, which are normally present in the atmosphere. If the radioactive gases coexist with the dust in the same air mass for a sufficiently long time, the parents and their various daughters will achieve radioactive equilibrium. The growth of the ^{222}Rn daughters approaches an equilibrium in about 2 hr and beyond that time further growth would slow because of the presence of 22-year ^{210}Pb. Blifford and associates (1952) investigated the relationships between the concentrations of radon and its various decay products in the normal atmosphere and found, as would be expected, that the atmosphere is markedly depleted in ^{210}Pb relative to the precursors of this isotope. This is because the inert dust of the atmosphere, the radon, and the radon daughters coexist long enough under normal circumstances for equilibria to be reached between radon and the shorter-lived daughters. Since the radionuclide with the longest half-life prior to ^{210}Pb is 26.8-min ^{214}Pb, equilibrium is reached in about 2 hr. The ^{210}Pb, which has a 22-year half-life, would take about 100 years to reach equilibrium. Various mechanisms exist for removing dust from the atmosphere, and the ratio of ^{210}Pb to its shorter-lived ancestors was shown by Blifford to be indicative of the length of time the dust resides in the atmosphere. He concluded by this method of analysis that the mean life of the atmospheric dust to which the radon daughters are attached is 15 days.

Wilkening (1964) found the concentration of ^{222}Rn daughters in the atmosphere to be depleted during passage of a thunderstorm. He attributed this to the action of electric fields, which changed from a normal value to about 1.8 to -340 V/cm. Deposition of the daughter products during rainstorms may temporarily increase the gamma background.

The thorium series below thoron (^{220}Rn) has no long-lived member. The equilibrium between thoron and its daughters will be achieved at a rate governed by the time required for the buildup of ^{212}Pb (half-life 10.6 hr).

The natural radioactivity of atmospheric dust, owing primarily to the attached daughters of radon, can be demonstrated readily. When air is drawn through filter media, the radon daughters attached to the filtered atmospheric dust cause both the α and β activity of the filter media to rise. Curve A of Fig. 7–3 illustrates the manner in which the increase in α radioactivity occurs in the case of normal air containing 0.05 pCi/liter of radon in equilibrium with its daughter products. The rise in α activity increases for about 2 hr, at the end of which time the accumulated daughters decay at a rate compensated by the newly deposited daughters. The radioactivity of the filter media will not increase beyond this eqiulibrium unless the rate of air flow or the concentration of radon is increased. When air flow ceases, the α radioactivity of the filter will diminish, as shown in curve B of Fig. 7–3, with an effective half-life of about 40 min.

Thus, the adsorbed radon daughters have the effect of endowing the

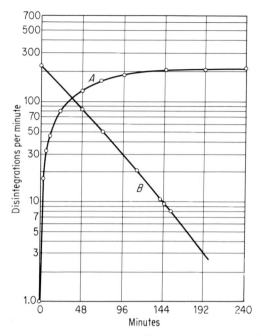

Fig. 7–3. (Curve A) Buildup of the α activity on a filter paper through which is drawn 1 cfm of air containing 5×10^{-14} Ci/liter of radon in equilibrium with its daughter products up to but not including ^{210}Pb. (Curve B) Decay of the accumulated α activity after cessation of flow.

ordinary dusts of the atmosphere with apparent radioactivity. Wilkening (1952) found that the radon daughters tend to distribute themselves on atmospheric dust in a manner which depends on the particle size of the dust and that the bulk of the activity is contained on particles having diameters less than 0.035 μm. Anderson and associates (1954; Anderson and Turner, 1956) found a close correlation between the concentration of radon daughters in the atmosphere and the concentration of suspended solids. This suggests that the dust concentration of the atmosphere, like the radon concentration, is an indication of the history of an air mass: air that has been recently over the oceans would be depleted in both radon and dust.

Thus, as discussed in Chapter 2, the dose from the radon series is delivered primarily by the daughter products adsorbed on inert dust that deposits in the lung. When air that contains radon or thoron in partial or total equilibrium with their daughter products is inhaled, the inert gases are largely exhaled immediately. However, a fraction of the dust particles is deposited in the lung, with the place of deposition and the manner of clearance from the lung dependent on the factors discussed in Chapter 2. With each breath, additional inert dust is deposited until radioactive equilibrium is reached, at which point the amount of activity deposited in unit time equals the amount eliminated from the lung by the combination of physiological clearance and radioactive decay. In the case of radon in equilibrium with its daughters, the total energy dissipation in the lungs from the daughter products is about 500 times greater than that derived from radon itself. Based on the models developed by Altshuler *et al.* (1964) and Jacobi (1964), the dose to the basal cells of the epithelium of segmental and lobar bronchi could be as high as 2000 mrem/year from the radon daughters normally present in the outdoor atmosphere (UNSCEAR, 1982).

Blanchard and Holaday (1960) have shown that, as in the case of radon, the dose delivered to the lung by thoron is about one-thousandth of the dose from the daughter products with which it is in equilibrium. However, the dose to the lung from thoron and its daughters does not add significantly to the dose received from the radon series.

Radon is partially absorbed when inhaled, and the decay products deliver a dose to the whole body that has been estimated at 3.0 mrem/year from a mean atmospheric radon concentration of 0.5 pCi/liter (UNSCEAR, 1966).

Radon dissolved in potable water is another source of human exposure, mainly because the radon is released from solution at the tap and enters the home atmosphere (Prichard and Gesell, 1983; Cross *et al.*, 1985). The highest concentrations of radon in drinking water have been reported from Maine and New Hampshire, where concentrations as high as 10^5 pCi/liter have been found (Smith *et al.*, 1961; Hess *et al.*, 1980). Well waters in the central United States ranged from 117 to 287 pCi/liter with a mean of 197 pCi/liter. Cross *et al.* (1985) suggested that a concentration of no less than 10,000 pCi Rn/liter could be supported as a standard for radon in drinking water, based on the comparability of the risk to that resulting from other standards that have been established.

Indoor Radon

One of the most surprising developments during the past decade has been the finding that in many homes the concentration of ^{222}Rn (and radon daughter products) is so high as to involve risks far greater than those from many other pollution hazards that have attracted attention. The problem exists mainly in residential dwellings because the radon originates from the soil and ground water, which have their greatest effects on one- or two-story buildings. The building materials themselves are a minor source of radon compared to soil and ground water, except where the materials contain relatively high concentrations of radium. This is true, for example, where gypsum board has been manufactured as a by-product of phosphate mining.

Interest in indoor radon developed worldwide in the mid-1970s, when convenient instrumentation became available. This happened to coincide with the emphasis placed on the need to conserve energy by weatherproofing homes, thereby reducing the rate of infiltration of outside air. It was widely believed that the high radon concentrations found were the result of the energy conservation programs, since the radon concentrations would be expected to be inversely proportional to the rate of ventilation, which, in the U.S. homes, ranges from 0.5 to 1.5 air changes per hour. However, recent studies have shown that the ventilation rate does not have a major influence on the radon concentration in homes because the rate varies only within a factor of about 3, whereas other variables can have a more pronounced effect. Steinhausler (1975) has shown that meteorological factors in particular can influence the indoor concentrations of radon and its daughters.

Radon can enter the indoor atmosphere in a number of ways, including diffusion from materials of construction or diffusion from soil through breaches in the foundation. However, there is evidence that diffusion of radon from soil is a minor source compared to the movement of soil gases directly through the foundation as a result of slight pressure differentials that can result from barometric changes, temperature differentials, or wind velocity. Water supplies ordinarily make a small contribution to the indoor radon concentration, but can be the predominant source in areas where the radon content of ground water is unusually high. The approximate contributions of various sources to the indoor radon concentrations are given in Table 7–11 (Nero, 1985).

In a study of the indoor radon concentrations in an area of Maine in which the ground water contained up to about 50,000 pCi ^{222}Rn/liter, it was found that the concentrations of ^{222}Rn in water and air in single-family dwellings were weakly correlated ($r = 0.50; N = 70$), with a regression slope of 1.3 pCi/liter of air per 10,000 pCi/liter of water (Hess *et al.*, 1983). Based on other measurements (Hess *et al.*, 1981), it was estimated that the contribution of ground water to the ^{222}Rn concentration in indoor air was 0.8 ± 0.2 pCi/liter per 10,000 pCi/liter of water, standardized for one air change per hour. Thus, the contribution of ground water to indoor air may be approximated by the ratio 1.0 pCi radon/liter of air per 10,000 pCi radon/liter of water.

There has not as yet been a systematic national survey of the extent of indoor exposure, but measurements in a total of 552 homes have been published in 19 different reports to date. These data have been aggregated

TABLE 7–11

APPROXIMATE CONTRIBUTION OF VARIOUS SOURCES TO INDOOR
RADON CONCENTRATIONS[a]

Source	Single-family homes (pCi/liter)	Apartments (pCi/liter)
Soil (estimates based on flux measurements	1.5	<1
Public water supplies	0.01[b]	0.01[b]
Building materials	0.05	0.1
Outdoor air	0.2	0.2

[a]From Nero (1985).

[b]Applies to 80% of the population that are served by public supplies. The contributions may average about 0.4 pCi/liter in homes using private wells, with higher concentrations in some areas.

by Nero (1985) with the results shown in Fig. 7–4. The measurements conform to a lognormal distribution with a geometric mean of 0.96 ± 2.84 pCi/liter. The distribution has a long tail, with 1–3% of the houses exceeding 8 pCi/liter, which approaches the 2 WLM/year remedial action limit recommended by the NCRP (1984b).

A linear relationship between radon exposure and incidence of lung cancer has been observed among uranium miners exposed to more than 100 WLM. Estimates of the lung cancer risk from radon inhalation vary from 1.5–45×10^{-6} per lifetime per WLM. NCRP (1984b) adopted a round-

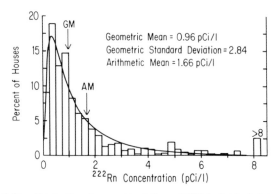

Fig. 7–4. Probability distribution from aggregation of 552 individual data in 19 sets. The smooth curve is the lognormal functional form corresponding to the indicated parameters, calculated directly from the data. (From Nero, 1985.)

ed estimate of 10^{-5} for the purpose of risk computation. The question arises of whether the risk estimates based on experience with uranium miners are applicable to the general population exposed to indoor radon. To some extent the conditions are different. The particle size of the dust to which the radon daughters are attached, differences in the unattached fractions of the radon daughters, the extent to which daughter product equilibrium has been achieved, the radiosensitivity of children compared to adults, and even differences in breathing rates affect the dose estimates and risk coefficients. These factors have been examined by Harley (1984), who concluded that the risk estimates based on experience with miners are applicable to the general population exposed to indoor radon. When the risk coefficients are applied to the currently estimated exposure distributions, it can be estimated that there may be annually 5,000 to 10,000 cases of lung cancer due to radon exposure. Since there are about 80,000 cases of lung cancer per year in the United States, exposure to environmental radon and its daughter products may be responsible for 6–12% of the total number of reported cases. These estimates should be regarded as highly provisional because of the many uncertainties referred to earlier.

Methods of Reducing Indoor Radon Concentrations

Several methods are available to reduce the concentration of radon in indoor air (NCRP, 1984b; Moeller and Fujimoto, 1984). As noted earlier, it was originally believed that reduced ventilation rates in homes as a result of weatherproofing were a major factor in increasing the concentration of radon. This, however, has been shown not to be the case: the overriding factor is the rate at which radon enters the home with soil gases, except in places where homes are supplied by water that has a high radon content.

Nero (1985) has estimated that as many as 1 million U.S. homes have concentrations that exceed the 2 WLM/year limit recommended by NCRP. This is a matter that raises unprecedented questions of public policy, although the NCRP recommendations do not have the force of law. Should the establishment of regulations be left to the states or should they be preempted by the federal government? Should people be required to reduce the radon levels to accepted standards in buildings they own? Should there be different policies in owner-occupied buildings than in rented buildings? Will prospective buyers or real estate require that radon levels be certified? Will the radon levels within a building be an important factor in determining its value? These and many other questions are just beginning to be discussed.

Radon in Natural Gas

A radioactive gas that had properties similar to those of the "radium emanation" was first separated from petroleum in 1904 (Burton, 1904).

Thus, it is not surprising that radon is also present in natural gas. It is found at the wellhead in concentrations that average about 40 pCi/liter, but samples from some fields contain more than 1000 pCi/liter (Johnson *et al.*, 1973; Gesell, 1975). Natural gas at the wells contains from 55 to 98% methane and a much smaller percentage of other heavy hydrocarbons (ethane, propane, butane), as well as carbon dioxide, nitrogen, helium, and water vapor. The gas is blended and processed to produce liquefied petroleum gas (LPG), a product that consists mainly of propane with lesser amounts of ethane. The boiling point of radon is close to that of propane and ethane, which has the effect of increasing the radon concentration in LPG while reducing the concentration in the methane-rich gas pumped into the pipelines. The long-lived daughters of ^{222}Rn (^{210}Pb and ^{210}Po) tend to accumulate on the interior surfaces of the LPG plant machinery and constitute a potential source of expousre of maintenance personnel (Gesell *et al.*, 1975; Summerlin and Prichard, 1985).

Since radon has a half-life of only 3.8 days, exposure of the consumers is influenced by storage time of LPG as well as the pipeline transit time. In 1977 it was estimated that radon from natural gas results in an average dose of about 5 mrem to about 125 million persons living in homes in which natural gas is used (NCRP, 1977c).

LEAD-210 AND POLONIUM-210

Lead-210 (sometimes called by its historic designation, radium D) is a 22-year β emitter separated from its antecedent ^{222}Rn by six short-lived α and β emitters (see Table 7–2). The longest-lived radionuclide between ^{222}Rn and ^{210}Pb is ^{214}Pb, which has a half-life of only 26.8 min. The ^{210}Pb decays to 138-day ^{210}Po via the intermediate ^{210}Bi, which has a 5-day half-life. Thus, following the decay of 3.8-day radon in the atmosphere, ^{210}Pb is produced rapidly, but its long half-life allows very little to decay in the atmosphere before it precipitates to the earth's surface, mainly in rain or snow.

The ^{210}Pb content of the atmosphere has been found to vary from 4.8×10^{-3} to 26×10^{-3} pCi/m^3, with the lowest values at island stations such as San Juan and Honolulu and the higher values in the interior of the United States (Magno *et al.*, 1970). Since the mean residence time of dust suspended in the troposphere is 15–20 days, there is little time for ^{210}Po to be formed in suspended dust.

Radioactive disequilibria are found in the upper profiles of rocks and soils from which ^{222}Rn diffuses. Atmospheric transport and deposition of ^{210}Pb cause ^{210}Pb to be distributed in a more uniform pattern than the ^{226}Ra from which it is derived. It would also be expected that broad-leaved plants on which ^{210}Pb is deposited would be enriched in this radionuclide. This, in fact, is observed.

The ^{210}Pb/^{210}Po ratio depends on the length of time ^{210}Pb exists within a matrix and whether the polonium is selectively removed from its site of prc᎐ ᎐ction by chemical or biological mechanisms. Since ^{210}Po has a half-life of only 138 days, appreciable ingrowth in vegetation can take place during a single growing season, and additional buildup can occur during food storage, with equilibrium being reached in about 1 year. When ^{210}Pb is absorbed into the body, ingrowth of ^{210}Po can occur because ^{210}Pb has deposited in the skeleton, from wi᎐᎐ h it leaves slowly, with a half-life of about 10^4 days.

Jaworowski (1967) reported that rainwater contains from 1 to 10 pCi/liter of ^{210}Pb with a mean of about 2 pCi/liter. In an area having 1 m of rainfall per annum, this would indicate a fallout of about 2 mCi/km^2 per year. UNSCEAR (1982) estimates that the average daily intake of ^{210}Pb from dietary sources is about 3 pCi under normal circumstances. Holtzman (1980) found lower values for U.S. residents, a mean of 1.4 pCi, with a relatively narrow range of 1.3–1.6 pCi/day.

The concentration of ^{210}Pb in a "standard" diet was found by Magno *et al.* (1970) to be not significantly different between locations in the United States and averaged 0.80 pCi/kg. His values were in good agreement with those reported by Morse and Welford (1971), which ranged from 0.70 to 1.0 pCi ^{210}Pb/kg in eight U.S. cities. Using the ICRP model for lead transfer from the gastrointestinal tract to blood, it can be estimated that about 0.2 pCi/day of ^{210}Pb reaches the bloodstream of average inhabitants of the United States. According to Morse and Welford, air and food contributed about equally to the blood level. The ^{210}Po and ^{210}Pb contents of the standard diet are equal, indicating that in most foods there has been sufficient time for this nuclide to reach equilibrium with ^{210}Pb.

There are two notable groups in which the dose from ^{210}Po is apt to be exceptionally high: cigarette smokers and residents of the northlands who subsist on reindeer that consume lichens.

Marsden and Collins (1963) originally noted the presence of alpha activity in tobacco, following which Radford and Hunt (1974) measured the concentrations of ^{210}Po in cigarettes, and Little *et al.* (1965) demonstrated that ^{210}Po was present in the lungs of cigarette smokers. Subsequent investigators have studied this phenomenon (Cohen *et al.*, 1980) and concluded that the dose to the basal cells or the bronchial epithelium in cigarette smokers is increased by 2–12 mrem/year. Autoradiographs of the bronchial epithelium from one smoker showed the ^{210}Po to be highly localized. If the ^{210}Po remained fixed in position, the dose to the tissue surrounding the "hot spot" would be much higher than the average dose to the basal calls of the bronchial epithelium. Based on α-track measurements of autoradiographs of the lung tissue of 13 smokers, Rajewsky and Stahlhofen (1966) had previously estimated that the basal cells of the subsegmental bronchi of cigarette smokers may receive as much as 86 mrem/year and the basal cells of the terminal bronchi may receive as much as 150 mrem/year. These values are higher than those reported by Cohen *et al.* (1980).

Compared to nonsmokers, about twice as much ^{210}Pb and ^{210}Po are found in the ribs (Holtzman and Ilcewicz, 1966). Ribs from smokers contained 0.28 pCi ^{210}Pb/g ash and 0.25 pCi ^{210}Po/g ash. The lungs contained 5.9 pCi ^{210}Po/kg wet. From these data, Holtzman estimated that the dose to the total skeleton is elevated by about 30% in cigarette smokers and the dose to the cells of the bone surface is increased by about 8%.

Polonium-210 is believed to enter tobacco by ingrowth of ^{210}Pb deposited on tobacco leaves from the atmosphere (Martell, 1974). Direct uptake of ^{210}Po from soil is probably not significant. The transfer coefficient to vegetables grown in ^{210}Po-contaminated alkaline soil has been found to range from about 6×10^{-4} to 10^{-6} (Watters and Hansen, 1970). Hill (1966) found a close correlation between the ^{210}Po and ^{137}Cs contents of human tissues from Canadian subjects, thus strengthening the suggestion that dietary habits that tend to favor broad-leaved vegetables or other foods subject to surface deposition may influence the ^{210}Po content of the tissues. Cesium-137 is known to be absorbed by humans mainly from surface deposition on plants (Chapters 6 and 13). Others have noted that pipe tobacco contains less ^{210}Po than cigarettes, apparently because the tobacco used for smoking pipes is not aged as long as cigarette tobacco (Harley and Cohen, 1980).

The dose from ^{210}Po is increased in Laplanders and some Eskimos who eat reindeer and caribou that feed on lichens that absorb trace elements in the atmosphere (Beasley and Palmer, 1966). Kauranen and Miettinen (1969) and Persson (1972) found the ^{210}Po content of Lapps living in northern Finland to be about 12 times higher than that of residents of southern Finland, where more normal dietary regimes exist. These investigators found the liver dose in Laplanders to be 170 mrem/year compared to 15 mrem/year for residents of southern Finland, who do not regularly eat reindeer meat. Unlike other naturally occurring α emitters, ^{210}Po depoists in soft tissues and not bone.

Holtzman (1964) measured the ^{210}Pb and ^{210}Po contents of Illinois potable waters known to be high in radium and found them to be low relative to ^{226}Ra. This seems to indicate a loss of ^{210}Pb owing to chemical precipitation, biological activity, or other factors.

POTASSIUM-40

Of the three naturally occurring potassium isotopes, only ^{40}K is unstable, having a half-life of 1.3×10^9 years. It decays by β emission to ^{40}Ca, followed by K capture to an excited state of ^{40}A and γ-ray emission to the ^{40}A ground state. Potassium-40 occurs to an extent of 0.01% in natural potassium, thereby imparting a specific activity of approximately 800 pCi/g potassium. Representative values of the potassium content of rocks, as summarized by Kohman and Saito (1954), indicate a wide range of values, from 0.1% for limestones through 1% for sandstones and 3.5% for granite.

The potassium content of soils of arable lands is controlled by the use of fertilizers. It is estimated that about 3000 Ci of ^{40}K is added annually to the soils of the United States in the form of fertilizer (Guimond, 1978).

A person who weighs 70 kg contains about 140 g of potassium, most of which is located in muscle. From the specific activity of potassium, it follows that the ^{40}K content of the human body is of the order of 0.1 μCi. This isotope delivers a dose of about 20 mrem/year to the gonads and other soft tissues and about 15 mrem/year to bone. Because of its relative abundance and its energetic β emission (1.3 MeV), ^{40}K is easily the predominant radioactive component in normal foods and human tissues. It is important to recognize that the potassium content of the body is under strict homeostatic control and is not influenced by variations in environmental levels. For this reason, the dose from ^{40}K within the body is constant.

Seawater contains ^{40}K in a concentration of about 300 pCi/liter.

RUBIDIUM-87

Of the two rubidium isotopes found in nature, ^{85}Rb and ^{87}Rb, only the latter is radioactive, with a half-life of 4.8×10^{10} years. Rubidium-87 is a pure β emitter, and it is present in elemental rubidium in the amount of 27.8%, which endows this element with a specific activity of 0.02 pCi/g. Pertsov (1964) quotes Vinogradov in listing the rubidium content of all but highly humic soils as about 0.01%. The ^{87}Rb content of ocean water has been reported to be 2.8 pCi/liter, with marine fish and invertebrates ranging from 0.008 to 0.08 pCi/g wet weight (Mauchline and Templeton, 1964).

It is estimated (UNSCEAR, 1982) that the whole-body dose from ^{87}Rb is 0.6 mrem/year on average.

Natural Radioactivity in Phosphate Fertilizers

Phosphate fertilizers are used in huge amounts everywhere in the world and are essential for food production. The natural resource from which agricultural phosphorus is obtained is phosphate rock, found in sedimentary formations, usually interbedded with marine shales or limestones. The

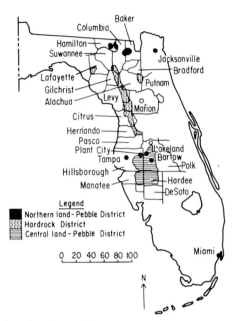

Fig. 7–5. Phosphate deposits in Florida. (From Guimond and Windham, 1980.)

United States is a major phosphate user, with extensive mines located in Florida, North Carolina, Tennessee, Idaho, Montana, Utah, and Wyoming. In recent years, more than 90% of the phosphate rock mined in the United States has come from Florida. The location of the Florida phosphate deposits is shown in Fig. 7–5. They average 4 m in thickness and lie under sand and clay overburdens that vary from about 1 to 10 m in thickness (Roessler *et al.*, 1980). As the surface mining operations proceed, the overburdens, sand tailings, and other waste products are returned to the land.

It has been known since early in this century that the phosphate rocks contain relatively high concentrations of uranium. The concentration of uranium in U.S. phosphate ores ranges from 8 to 400 ppm. The phosphate rocks from the important Florida deposits average 41 ppm, with ^{226}Ra in secular equilibrium. Phosphate rock is mined in huge quantities: it is reported that in 1974 about 26 million tons of ore were sold for fertilizer production in the United States, and that the ore contained about 1000 Ci of ^{226}Ra in equilibrium with ^{238}U (Guimond, 1978).

Several types of phosphate fertilizers are produced, and these are usually blended for application in the field. The radium and uranium tend to separate in the process of producing phosphoric acid, an important step in fertilizer manufacture (Roessler *et al.*, 1979). The radium passes to the gypsum produced as a by-product. A smaller fraction of the radioactivity is rejected in the process and ends up in the mounds of stored waste products.

Elevated concentrations of radon are found within structures located on land relcaimed from phosphate mines (Guimond and Windham, 1980; Roessler *et al.*, 1983). A survey of indoor radon and radon decay product exposures in buildings on or near the Florida phosphate deposits was conducted by Guimond and Windham, who found that 71% of homes built on reclaimed land contained more than 0.01 WL and 23% contained more than 0.05 WL. High levels were also found in structures built over mineral deposits that had not been mined, but relatively few measurements were made. Although these radon concentrations are above the average concentrations found in U.S. homes, it is noteworthy that many homes in areas of the United States contained higher concentrations than are found in the phosphate regions.

The high concentrations of ^{226}Ra also result in increased gamma-radiation exposure in homes built on reclaimed land. Johnson and Bailey (1983) found excess gamma dose to average 17 mrem/year, with value ranging as high as 166 mrem/year.

The radioactivity associated with land reclaimed prior to the 1940s tends to be higher than that in land reclaimed more recently, because methods have been developed for separating the phosphate-rich clay fraction of the deposit. A major fraction of the radium and uranium associated with the phosphate minerals was formerly returned to reclaimed land (Roessler *et al.*, 1980). In 1978 the Florida Department of Health and Rehabilitation Services ruled that some sort of rehabilitation would have to be applied to homes in which the radon concentrations exceeded 0.029 WL. This is about three times the remedial action level suggested by the Surgeon General for application to homes constructed over uranium mill tailings in Grand Junction, Colorado (Chapter 8). It has been estimated that of the approximately 4000 buildings on reclaimed land in 1978, 6–10% would require some kind of corrective action. Roessler *et al.* (1980) found that most of these structures were located on land reclaimed prior to the 1940s.

The contribution of uranium and radium to agricultural lands due to the application of phosphate fertilizers does not significantly affect the dose received from the general population (Kirchmann *et al.*, 1980; Ryan, 1981). However, phosphorus, in the form of mineral phosphate rock, is sometimes added to cattle feed, and this practice can result in increased levels of uranium and radium in cow's milk (Reid *et al.*, 1977). However, continued application of phosphate fertilizers to soil over a period of many years could eventually double the radium and uranium content of the soil, which would result in a corresponding doubling of the dose to bone. Spalding and Sackett (1972) found that the uranium content of North American rivers is higher than in the past, which they attributed to increased runoff of phosphate fertilizers.

Natural Radioactivity in Building Materials

Two important by-products from the processing of phosphate rock are gypsum and calcium silicate slag, both of which are used in the building industry. Use of these products in buildings may increase the gamma-radiation exposure.

The Federal Republic of Germany conducted a survey of 30,000 dwellings and found that, on average, the external radiation exposure was 33% higher within the dwellings than outdoors. Thus, although the building materials absorb the radiation that originates outside the building, exposure within the building is more than compensated by the presence of radionuclides in the materials of construction (Kolb and Schmier, 1978). Data on radioactivity in selected building materials as assembled from various sources by Harley (1978) are given in Table 7–12. Since most people spend 90% of their time indoors, the dose they receive from external natural radiation is increased somewhat. According to Kolb and Schmier, the absorbed dose to the population of the Federal Republic of Germany ranges from 40 to 80 mrad/year indoors (for continuous exposure) compared to 30–60 mrad/year for continuous outdoor exposure. Building materials can also be a source of radon, as was discussed earlier in this chapter.

Natural Radioactivity in Fossil Fuels

Coal contains radionuclides of the uranium and thorium series, as well as ^{40}K. The quantity of radionuclides discharged to the atmosphere per ton of coal consumed depends on the concentration of radionuclides in the coal, the method of combustion, and the efficiency of fly ash recovery. The uranium and thorium concentrations in mined coal have been assembled from various sources by Beck et al. (1980) and are given in Table 7–13, which includes data on most of the coal beds used in the United States.

The mean value for all coals sampled is 1.7 μg/g for uranium, and 4.5 μg/g for natural thorium, which is similar to the average concentrations found in soils and rocks, also shown in Table 7–13. Most investigators have found that the various radionuclides of the uranium and thorium series are in secular equilibrium. Beck et al. (1980) noted that there have

TABLE 7–12

RADIOACTIVITY IN SELECTED BUILDING MATERIALS[a]

Source	Material	^{40}K	^{238}U	^{226}Ra	^{232}Th
Chang *et al* (1974), Taiwan	Wood	90	—	—	—
	Red brick	16	1.2	—	1.8
	Concrete	7	0.9	—	—
Hamilton (1971), United Kingdom	Clay brick	18	3	1.4	1.2
	Silicate brick (gravel)	10	0.2	0.2	0.1
	Granite	30	6	2.4	2.2
	Aerated concrete	19	0.4	2.4	0.4
	Natural gypsum	4	0.4	0.6	0.2
	Concrete block (fly ash)	(6–16)	(1–12)	(0.2–4)	(1.0–1.2)
Wollenberg and Smith (1966a and b), United States	Cement	3.4	1.1	—	0.4
	Silica sand	9	0.3	—	0.5
	Commercial sand	7	0.3	—	0.3
	Red brick	18	—	1.5	1.0
	Silica brick	6	—	0.5	0.4
	Light concrete	14	—	2.0	0.9
	Granite	40	—	3.0	4.5
	Sand	7	—	(<0.4–1)	<0.4
	Cement	4	—	0.7	<0.4
Kolb (1974), West Germany	Granite	34	—	2.8	2.1
	Brick	18	—	1.7	1.8
	Sand, gravel	<7	—	<0.4	<0.5
	Cement	6	—	0.7	<0.6
	Natural gypsum	<2	—	<0.5	<0.3
	Concrete	7	—	0.6	0.6

[a]From Harley (1978).

been exceptions, but did not consider the evidence for disequilibria to be conclusive.

About 90% of the mass of coal is consumed during combustion and the radioactive nuclides tend to concentrate in the nonvolatile fraction or "ash." The ash then undergoes partitioning, according to whether it separates within the furnace and stack or passes with the hot gases to the "fly ash." Modern furnaces in the United States now burn pulverized coal and release 60–85% of the coal ash content into fly ash (Ray and Parker, 1977). In former times, it was common practice to allow the fly ash to escape to the environment, but this is no longer permitted in the United States and other countries, where electrostatic precipitators, bag houses, and scrubbers are employed to reduce the amount of fly ash released with the stack

TABLE 7–13

URANIUM AND THORIUM CONCENTRATIONS IN COAL AS MINED (FRACTION OF DRY WEIGHT)[a]

Region, type	No. of samples	U (μg/g)			Th (μg/g)		
		Range	Geo. mean	Mean	Range	Geo. mean	Mean
Pennsylvania, anthracite	53	0.3–25.2	1.2	1.5	2.8–14.4	4.7	5.4
Appalachian[b]	331	<0.2–10.5	1.0	1.4	2.2–47.8	2.8	4.9
Midwest[b]	143	0.2–43	1.4	3.3	<3–79	1.6	5.2
Northern Great Plains[c,b]	93	<0.2–2.9	0.7	0.9	<2–8.0	2.4	2.7
Gulf Coast, lignite	34	0.5–16.7	2.4	3.2	<3–28.4	3.0	8.3
Rocky Mountains[b,c]	134	<0.2–23.8	0.8	1.6	<3–34.8	2.0	3.6
Alaska[c]	18	0.4–5.2	1.0	1.2	<3–18	3.1	4.4
Illinois Basin	56 (113K)	0.31–4.6	1.3	1.5	0.71–5.1	1.9	2.1
Appalachian	14 (23K)	0.4–2.9	1.3	1.5	1.8–9.0	4.0	4.5
Western	22 (29 K)	0.3–2.5	1.0	1.2	0.62–57	1.8	2.3
Western	19	0.11–3.5	0.85	0.9			
All samples	910 (983 K)		1.04	1.74		2.40	4.47
		Typical Range	Avg.		Typical Range	Avg.	
Soil		0.9–4.0	1.8		2–12	6	
Rocks		0.5–5	2.7		1.6–20	9.6	

[a]From Beck *et al.* (1980).
[b]Bituminous.
[c]Subbituminous.

gases. More than half the power plants in the United States now recover more than 97.5% of their fly ash in scrubbers, electrostatic precipitators, or bag houses (NCRP, 1977c). Because enrichment of some radionuclides occurs in the stack, the concentrations of ^{210}Pb and ^{210}Po in the fly ash are 5–10 times greater than in the original coal (Beck *et al.*, 1980). The emissions of radionuclides from a typical 1000-MW electric power plant in 1972 are given in Table 7–14.

The normal levels of uranium and thorium in the environs are sufficiently high that changes due to the emissions from coal-fired power stations are barely detectable: some investigators (Bedrosian *et al.*, 1970) found no changes in the vicinity of an old, relatively poorly controlled plant, and Beck *et al.* (1980) could find only slight changes in a carefully designed survey of several plants. It has been found, however, that the ^{226}Ra content of snow downwind of coal-fired power plants is higher than

TABLE 7–14

ESTIMATED DOSE EQUIVALENTS TO LUNG OF MAXIMUM
EXPOSED INDIVIDUALS FROM MODEL COAL-FIRED
POWER PLANT RELEASES[a]

| Nuclide | Dose equivalent (mrem/year) | |
	Modern plant	1972 reference plant
^{238}U–^{234}U	0.009	0.46
^{230}Th	0.015	1.55
^{226}Ra	0.003	0.23
^{210}Pb	0.005	0.22
^{210}Po	0.075	3.10
^{232}Th	0.009	0.93
^{228}Th	0.018	1.85
^{228}Ra	0.004	0.38
Total	0.14	8.7

[a]From Beck *et al.* (1980).

the concentration in rainwater (Jaworowski *et al.*, 1975). The concentrations of naturally occurring radionuclides in *airborne* particles collected in the vicinity are elevated above background, and increase the dose to the lung, as shown in Table 7–14. It is seen that there has been a major reduction in the dose equivalent received from a modern plant compared to the 1972 plant, which operated without stringent fly ash control.

The first report on radioactive emissions from coal-burning power plants (Eisenbud and Petrow, 1964) concluded that when the data were normalized for their radiotoxicity relative to the emissions from the first commercially operated pressurized water reactors, which had just begun operation, the dose from the fossil fuel emissions was greater than that from the nuclear reactors. However, the stringent requirements of the Federal Clean Air Act have since resulted in substantial reductions in the lung dose from the fly ash emissions, as shown in Table 7–14.

Induced Radionuclides

A number of radionuclides that exist on the surface of the earth and in the atmosphere have been produced by the interaction of cosmic rays with atmospheric nuclei. The most important of these are tritium (^{3}H), ^{14}C, and ^{7}Be, and of lesser importance are ^{10}Be, ^{22}Na, ^{32}P, ^{33}P, ^{35}S, and ^{39}Cl. The properties of these isotopes and the extent to which they have been reported in various media are listed in Table 7–1.

Carbon-14 is formed by ^{14}N capture of neutrons produced in the upper atmosphere by cosmic-ray interactions. The incident cosmic-ray neutron flux is approximately 1 neutron/sec per square centimeter of the earth's surface, and essentially all these neutrons disappear by ^{14}N capture (Anderson, 1953). The incident neutron flux integrated over the surface of the earth yields the natural rate of production of ^{14}C atoms, which has been estimated to be 0.038 MCi/year (UNSCEAR, 1977) and is believed to have been unchanged for at least 15,000 years prior to 1954, when nuclear weapons testing began to alter the normal ^{14}C inventory to a noticeable extent (NCRP, 1985b).

Carbon-14 of natural origin is present in the carbon of all biota at the historically constant amount of 6 pCi/g C. After death of an organism, the ^{14}C equilibrium is no longer maintained, and the ratio of ^{14}C to ^{12}C diminishes at a rate of 50% every 5600 years, which makes it possible to use the ^{14}C content of organic materials for the purpose of measuring age (Libby, 1952).

Because the ^{14}C originally present in coal and oil has decayed almost completely, the introduction into the atmosphere of carbon from combustion of these fuels tends to reduce the specific ^{14}C activity of atmospheric CO_2. Suess (1958) noted that the $^{14}C:^{12}C$ ratio of tree rings had diminished during the past century, which he attributed to the dilution of atmospheric ^{14}C by combustion of ^{14}C-free fossil fuels. For this reason, the concentration of ^{14}C in atmospheric carbon tends to be lower in urban and industrial areas.

Clayton et al. (1955) and Lodge et al. (1960) have used the ^{14}C content of particulate atmospheric carbon to estimate the fractions of the dust that originate from garbage incineration and combustion of fossil fuels. Incineration of food residues, textiles, paper, and other organic constituents of garbage produces smoke in which the ^{14}C is in contemporary equilibrium, because comparatively little refuse is so old that a significant fraction of the ^{14}C has decayed. Because there has been total decay of the ^{14}C present in the organic matter from which fossil fuels have formed, the relative contributions of the two sources of particulate carbon could be estimated. This technique may become less useful in time because the increasing fraction of plastics contained in municipal refuse is introducing a new source of ^{14}C-free carbon.

The total carbon content of the body is approximately 18%, or 12.6 kg for a 70-kg man, and the ^{14}C body burden from natural sources is thus of the order of 0.1 μCi, but the dose is small owing to the soft quality of the ^{14}C β particles (0.01 MeV). It is estimated that the dose from ^{14}C is 2.4 mrem/year to the skeletal tissues of the body and 0.5 mrem/year to the gonads (UN-SCEAR, 1982).

Tritium (^{3}H), a radioactive isotope of hydrogen, is formed from several interactions of cosmic rays with gases of the upper atmosphere (Suess, 1958). Existing in the atmosphere principally in the form of water vapor, tritium precipitates in rain and snow. Like ^{14}C, it is produced in thermonuclear explosions, and this has increased the atmospheric content of tritium in a manner that will be discussed in a subsequent chapter. The natural production rate of ^{3}H is estimated to be about 0.19 atom/cm^2 sec, corresponding to a steady-state global inventory of about 26 MCi (NCRP, 1979).

The natural concentration of tritium in lakes, rivers, and potable waters was reported to have been 5–25 pCi/liter prior to the advent of weapons testing (UNSCEAR, 1982). The annual absorbed dose from tritium of natural origin is estimated to be about 0.001 mrem/year, uniformly distributed in all tissues.

The other nuclides formed from cosmic-ray interactions with the atmosphere may be potentially useful as tracers for studying atmospheric transport mechanisms, but relatively few observations have been reported.

Natural Sources of External Ionizing Radiation

The dose received from external sources of ionizing radiation originates from cosmic rays and from γ-emitting radionuclides in the earth's crust. The United Nations (UNSCEAR, 1982) estimates the external annual effective dose equivalent from all naturally occurring radiation in "normal" parts of the world to be 30 mrem/year from cosmic sources and 35 mrem/year from terrestrial radiation.

Solon *et al.* (1958) and later Beck (1966) and Beck and dePlanque (1968) made extensive measurements of the natural γ-radiation background in a number of cities throughout the United States. The data of Solon *et al.* were about 30% higher than those reported by Beck, probably because of the greater effect of fallout during the period when their measurements were made. The techniques used by Beck and associates permitted differentiation between fallout and natural radiation. This was not possible at the time Solon *et al.* made their measurements because gamma spectrometry was not as yet practical for use in the field.

The mean exposure in the 124 locations measured by Solon and co-workers was 81 ± 20 mR/year, compared to Beck's mean of 61 ± 23 mR/yr at 210 locations. Solon *et al.* showed that their data were well correlated with barometric pressure, indicating the effect of cosmic sources of radia-

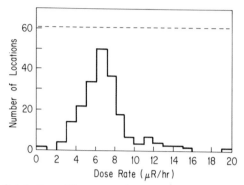

Fig. 7–6. Frequency distribution of the gamma dose rate from natural emitters at 210 different locations in the United States. (From Beck, 1966; reproduced from *Health Physics*, vol. 12, by permission of the Health Physics Society.)

tion. The data gathered by Beck at the principal cities where measurements were made are summarized in Fig. 7–6.

Terrestrial Sources of External Radiation

The terrestrial sources of gamma radiation are ^{40}K and nuclides of the ^{238}U and ^{232}Th series. If the concentrations of these three nuclides in soil are known, the dose can be estimated, using methods developed originally by Hultqvist (1956) and further developed by Beck (1972). The absorbed dose in air, 1 m above soil having unit concentrations of the three nuclides, is given in Table 7–15. If the data of Table 7–15 are adjusted for the typical concentrations of the three nuclides in soil, it is found that ^{232}Th and ^{40}K each contribute 10–12 mrad/year (NCRP, 1975c). The absorbed dose above various kinds of rocks is given in Table 7–16. According to Kohman

TABLE 7–15

Absorbed Dose Rate 1 m above Soil Containing
Unit Concentrations of ^{40}K, ^{238}U, and ^{232}Th,
Including Daughter Products[a]

Nuclide	Soil Concentration	Adsorbed dose rate in air (mrad/year)
^{40}K	1 pCi ^{40}K/g	1.4
^{238}U + daughters	1 pCi ^{238}U/g	13.9
^{232}Th + daughters	1 pCi ^{232}Th/g	21.6

[a]Adapted from Beck (1972).

TABLE 7–16

γ-Radiation Dose Rates from Radium, Uranium, Thorium, and Potassium in Rocks[a]

Type of rock	Dose rate (mrem/year)			
	^{226}Ra	^{238}U	^{232}Th	^{40}K
Igneous	24	26	37	35
Sedimentary				
Sandstones	13	7.7	18	15
Shales	20	7.7	31	36
Limestones	7.7	8.4	4	4

[a] UNSCEAR (1958).

(1959), the similarity of dose rates from the various isotopes listed in Table 7–16 is a coincidence arising from the fact that the various isotopes happen to be present in rocks in amounts that are approximately inversely proportional to their specific activities.

The external γ radiation from radionuclides in the earth's crust is thus influenced by the kind of rock over which the measurements are made. The actual doses to people cover a somewhat narrower range owing to the fact that most people live on soil rather than rock. The soils tend to be less variable in their radioactive content because the igneous rocks, which are high in radioactive content, whether more slowly and, therefore, contribute less radionuclides to soils than the softer sedimentary rocks.

Beck *et al.* (1966) also reported on the variation with time of the natural γ dose rates at two communities in Westchester County, New York, during 1963–1965. The individual observations at one location that averaged 6.9 ± 2.1 μR/hr ranged from 6.0 to 8.2 μR/hr. At the second location, which averaged 7.1 ± 2.1 μR/hr, the range was similar, from 6.0 to 8.3 μR/hr. These investigators attributed the principal source of variation to the effect of soil moisture, which can account for 30% by weight during wet periods. Whereas the soil moisture might be expected to attenuate the radiation from thorium and potassium, the radiation from the uranium series might be expected to increase because soil water would inhibit the diffusion of radon. However, examination of the data published in the report of Beck *et al.* fails to show an inverse correlation of this type.

In addition to the direct ground level measurements and estimates of external exposure that can be made from knowledge of the radionuclide content of soils, exposure estimates can be made from aircraft by using sensitive gamma detectors designed to provide estimates of ground level exposure. During the past 20 years many such surveys have been made by government agencies, either to explore for uranium or to provide information about the levels of radiation in the vicinity of nuclear facilities. These data have been analyzed by Oakley (1972), who made estimates of the population dose distribution in the United States. The data are grouped by three geographic regions: (1) the Atlantic and Gulf coastal plains, for which the mean absorbed dose rate is 23 mrad/year; (2) a portion of the eastern slope of the Rocky Mountains, where the absorbed dose averages 90 mrad/year; and (3) the remainder of the United States, where the average absorbed dose is 46 mrad/year.

COSMIC RADIATION

The primary radiations that originate in outer space and impinge isotropically on the top of the earth's atmosphere consist of 87% protons, 11% α particles, about 1% nuclei of atomic number Z between 4 and 26, and about 1% electrons of very high energy (NCRP, 1975c). An outstanding characteristic of the cosmic radiations is that they are highly penetrating, with a mean energy of about 10^{10} eV and maximum energies of as much as 10^{19} eV. The primary radiations predominate in the stratosphere above an altitude of about 25 km.

It is now known that these radiations originate outside the solar system and that only a small fraction is normally of solar origin. However, the solar component becomes very significant outside the atmosphere following flares associated with sunspot activity that follow an 11-year cycle.

The interactions of the primary particles with atmospheric nuclei produce electrons, γ rays, neutrons, and mesons. At sea level the mesons account for about 80% of the cosmic radiation and electrons account for about 20%. It has been estimated that 0.05% of primary protons penetrate to sea level (Myrloi and Wilson, 1951). With the development of high-altitude aircraft and manned space probes, the dose from primary cosmic radiations has attracted more interest in recent years, and sophisticated radiation measurements are now an important component of the scientific programs using satellites probing into space (Tobias and Todd, 1974).

The dose from cosmic radiation is markedly affected by altitude. The annual cosmic-ray dose is about 28 mrem at sea level. For the first few kilometers above the earth's surface, the cosmic-ray dose rate doubles for each 2000 m increase in altitude (Fig. 7–7). However, for the first 1000 m, the *total* dose rate actually decreases with altitude above the surface, because attenuation of the γ rays from terrestrial sources occurs more rapidly than the increase in cosmic radiation (Schaefer, 1971). Residents of Denver (altitude 1600 m) receive nearly twice the dose at sea level, and in

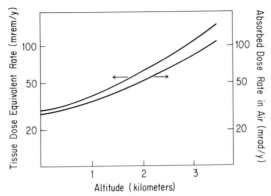

Fig. 7–7. Variation of cosmic-ray dose with altitude above sea level. The charged particle absorbed dose rate in air or tissue is shown in the lower curve and the total dose equivalent rate (charged particles plus neutrons) is shown in the upper curve at 5 cm depth in a 30-cm-thick slab of tissue. (From National Council on Radiation Protection and Measurements, 1975c.)

Leadville, Colorado (altitude 3200 m), the residents receive about 125 mrem/year from cosmic rays, more than four times the annual dose at sea level. The increase in dose rate at higher altitudes is shown in Fig. 7–8, where it is seen that at polar latitudes rates in excess of 1.5 mrem/hr are received at altitudes of 50,000–80,000 ft, the upper limit of high-perfor-

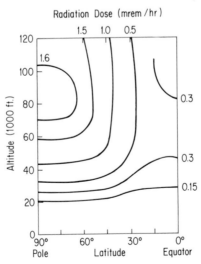

Fig. 7–8. Cosmic radiation field in the earth's atmosphere, from sea level to 120,000 ft. (From H. J. Schaefer; copyright 1971, AAAS.)

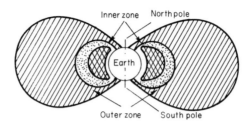

Fig. 7–9. The geomagnetically trapped corpuscular radiations.

mance aircraft such as the SST. On rare occasions, once or twice during the 11-year cycle, a giant solar event may deliver dose equivalents in the range 1–10 rem/hr, with a peak as high as 5 rem during the first hour (Upton *et al.*, 1966). During a well-documented solar flare in February 1956, dose rates well in excess of 100 mrem/hr existed briefly at altitudes as low as 35,000 ft (Schaefer, 1971).

Because of the effect of altitude, the passengers and crew of high-flying aircraft are subject to additional exposure from cosmic rays. A transcontinental flight will result in a dose of about 2.5 mrem, or 5 mrem per round trip (NCRP, 1975c). The cabin attendants and aircraft crew have been estimated to receive an incremental dose of about 160 mrem/year above that received at sea level (NAS–NRC, 1980).

Unusual solar activity is capable of injecting huge quantities of high-energy protons into interplanetary space and can increase the dose to hazardous levels that might require evasive action by aircraft flying above about 13 km. An international commission concerned with high-flying aircraft, as well as the Federal Aviation Administration (FAA) in the United States, has recommended continuous monitoring for solar flares so that high-flying aircraft can take evasive action if necessary (Wilson, 1981).

Above the earth's atmosphere, the dose consists of two main components. One is the dose from highly energetic cosmic radiation trapped in the earth's magnetic field, as illustrated in Fig. 7–9. A second portion is received beyond the earth's magnetic field and is due to the background cosmic radiation, on which may be superimposed very sharp peaks of radiation due to solar flares.

On entering the earth's magnetic field, some of the primary particles are deflected toward the polar regions, resulting in a somewhat lower radiation flux at the equator. This phenomenon becomes more accentuated with altitudes above a few kilometers, as shown in Fig. 7–10. The difference in the dose rate due to geomagnetic latitude varies from 14% at sea level to 33% to 4360 m (Pertsov, 1964).

Cosmic-ray doses to passengers on supersonic and subsonic aircraft for

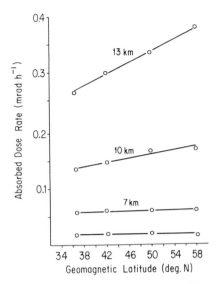

Fig. 7–10. Cosmic radiation absorbed dose index rate at four different altitudes as a function of latitude. (From UNSCEAR, 1977.)

representative flights are given in Table 7–17. Although the dose received per hour is greater when flying in an SST, the dose per trip is less than for subsonic flight because of the reduced flying time.

The geomagnetically trapped radiations consist mainly of protons and electrons produced by backscatter of the primary cosmic-ray beam on the earth's atmosphere and protons of solar origin. Because of differences in the mass-to-charge ratios of protons and electrons, the trajectories for the two particles diverge, giving rise to an inner radiation belt consisting mainly of protons and an outer belt consisting mainly of electrons, as illustrated in Fig. 7–9. The proton region begins about 1000 km above the geomagnetic equator and ends at an altitude of about 3000 km.

When astronauts travel into outer space they are exposed to the intense radiation of the two belts of trapped electrons, the primary cosmic radiation particles and the radiation from solar flares (UNSCEAR, 1982). The largest portion of the dose to the Apollo astronauts was received when they were passing through the earth's radiation belts, and the dose can be influenced by the trajectory through the belts. The Apollo missions lasted from about 5 to 12 days and resulted in average crew doses of 160–1140 mrad. The Skylab missions lasted from 20 to 90 days, and resulted in doses of 1.6–7.7 rad. During the first 11 Shuttle flights, the doses ranged from about 12 to 400 mrad. The doses would be much higher in the event of a

TABLE 7–17

CALCULATED COSMIC-RAY DOSES TO A PERSON FLYING IN SUBSONIC AND SUPERSONIC
AIRCRAFT UNDER NORMAL SOLAR CONDITIONS[a]

| | Subsonic flight at 11 km | | Supersonic flight at 19 km | |
Route	Flight duration (hr)	Dose per round trip (mrad)	Flight duration (hr)	Dose per round trip (mrad)
Los Angeles–Paris	11.1	4.8	3.8	3.7
Chicago–Paris	8.3	3.6	2.8	2.6
New York–Paris	7.4	3.1	2.6	2.4
New York–London	7.0	2.9	2.4	2.2
Los Angeles–New York	5.2	1.9	1.9	1.3
Sydney–Acapulco	17.4	4.4	6.2	2.1

[a]From UNSCEAR (1977), adapted from Wallace (1973).

solar flare. It has been estimated that the dose from a flare that occurred in July 1959 could have been between 40 and 360 rad. Space radiation is undoubtedly an important constraint on long-term space travel.

Dose calculations for the solar flare particles are very difficult owing to the wide range of energies and the complex shielding geometries presented by space capsules. Conversion of cosmic-ray dose estimates from rads to rems is made difficult by the wide range of energies and linear energy transfer (LET). The LET of the incident protons varies from about 0.21 to 58 keV/μm. For high-Z nuclides such as iron, the LET can be as high as 3500 keV/μm. The absorbed dose for the centrally traversed cells varies from 0.07 rad for relatively low-LET protons to 1200 rad for high-LET iron. Estimates of the RBE for the carcinogenic effect of the heavy ions range from 5 for helium to 27 for iron and argon (Fry et al., 1985).

Aerospace activities are only one way in which the dose from nature is increased by technological developments. The dose can also be increased by use of building materials that have high levels of natural radioactivity and, as discussed earlier, by living in houses in which radon and its daughter products accumulate. It has also been mentioned that the dose received from the radon daughters, especially ^{210}Po, can be increased by the practice of smoking cigarettes. Other ways in which the dose from natural sources can be increased include burning natural gas (which may contain radon), mining phosphate minerals (which are often associated with uranium), and injudiciously disposing of uranium mill tailings. The exposures from mill tailings will be discussed in Chapter 8. Other sources of technological enhanced natural radioactivity (Gesell and Prichard, 1975) will be reviewed in Chapter 10.

Areas Having Unusually High Natural Radioactivity

Except for brief mention of the occurrences of high levels of ^{226}Ra in well water, the discussion of natural radioactivity has thus far been limited to levels to which people are normally exposed. There are, in addition, places in the world where the levels of natural radiation exposure are abnormally high.

THE MINERAL SPRINGS

It has long been known that many mineral springs contain relatively high concentrations of radium and radon, and in many places in the world the radioactivity of the springs has been exploited for its alleged curative powers.[1] Spas in South America, Europe, Japan, and elsewhere have commercialized the high radioactive content of local waters, and in some places research laboratories are operated in which the physiological basis for the alleged curative effects is studied. Visitors are encouraged not only to drink and bathe in the radioactive waters but also to sit in "emanatoria," where they can breathe radon emanating from surrounding rock, as shown in Fig. 7–11 (Pohl-Ruling and Scheminzky, 1954).

Published values of ^{226}Ra in mineral waters ranged to more than 100 pCi/liter (United Nations, 1958), which is several thousand times greater than the values normally reported for public water supplies. However, to illustrate the fact that the values reported for spring waters are not typical of the drinking water of the region, the ^{226}Ra concentration in tap water of Bad Gastein is reported to be 0.62 pCi/liter, compared to 100 pCi/ml for some of the local springs (Muth *et al.*, 1957).

About 5 million liters of water that contain high concentrations of radon are discharged daily from hot springs in Bad Gastein. The water is conveyed to hotels and bathhouses, where 58 Ci/year are released to the atmosphere. Uzinov *et al.* (1981) have described the manner in which radon emanations from this water result in exposure to the city residents, tourists who visit the spas, and employees of the many bathhouses. They estimate that exposure

---

[1]Mineral waters were perceived as being beneficial to health in Roman times, many centuries before any knowledge of radioactivity existed. When the phenomenon of radioactivity was discovered, high levels were found to exist in the mineral waters and may have resulted in the trilogy: the waters are good for health, the waters are radioactive, hence radioactivity is good for health.

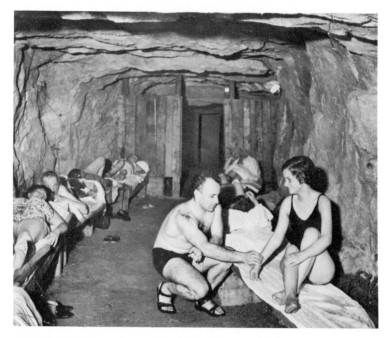

Fig. 7–11. Radon gallery at Bad Gastein, Austria. (COSY Verlag, Salzburg.)

of some of the attendants is as high as 40 WLM/year. A cytogenetic survey of persons subject to a gradient of radon exposure, including residents of Bad Gastein and several groups of employees of the spas, demonstrated dose-dependent increases in the frequency of chromosomal aberrations in the white blood cells (Pohl-Ruling and Fischer, 1979).

One of the most popular mineral springs localities in the United States is at Saratoga, New York, where reports of the medicinal value of the water go back to the early 18th century. It is known that many of the springs contain ^{226}Ra in amounts that exceed the EPA limit of 5 pCi/liter by a factor of 100. A survey has been made of persons using one spring, Hathorn No. 1, which attracts many visitors and contains an average of about 200 pCi/liter (Aulenbach and Davis, 1976). Twenty-seven long-time users of the water from this spring who were selected for study stated that they consumed the water for 5 to 65 years in amounts that ranged from about 200 to 3000 ml/day. Based on the information obtained in this way, it was calculated that the ^{226}Ra body burdens of the 27 individuals questioned ranged from slightly above normal to a high of 0.11 µCi, which is slightly above the limit of 0.1 µCi for industrial workers. More comprehensive surveys are needed to evaluate the public health significance of mineral springs such as these. There are few studies of the extent to which the radioactivity contained in mineral spring waters has been absorbed by human beings.

Monazite Sands and Other Radioactive Mineral Deposits

Major anomalies in the concentrations of radioactive minerals in soil have been reported in two countries, Brazil and India.

In Brazil, the radioactive deposits are of two distinct types: the monazite sand deposits along certain beaches in the states of Espirito Santos and Rio de Janeiro, and the regions of alkaline intrusives in the state of Minas Gerais (Roser *et al.*, 1964; Brazilian Academy of Sciences, 1977).

Monazite is a highly insoluble rare earth mineral that occurs in some beach sand together with the mineral ilmenite, which gives those sands a characteristic black color. The black sands are advertised for their radioactivity and are much sought by some Brazilian tourists for their perceived benefits to health. The external radiation levels on these black sands range up to 5 mR/hr, and people come from long distances to relax on the sands and enjoy the many hotels that have been constructed to care for their needs (Fig. 7–12). The most active of these Brazilian vacation towns is Guarapari, which has a stable population of about 7000 people and an annual influx of 10,000 vacationers. Some of the major streets of Guarapari have radiation levels as high as 0.13 mR/h, which is more than 10 times the normal background. Similar radiation levels are found inside some of the buildings in Guarapari, many of which, in parts of the village that are not built on monazite sand, are nevertheless elevated in radioactivity because the beach sands were incorporated into the building materials. Roser and Cullen (1964) undertook extensive external radiation measurements throughout the black sands region and concluded that almost all the approximately 60,000 inhabitants of the region were exposed to abnormally high radiation levels, but that only a small number (about 6600) were exposed to more than 0.5 rem/year. The population is too small to permit meaningful epidemiologic investigations to be conducted, but cytogenetic studies have shown a higher than normal frequency of abnormalities in blood chromosomes (Barcinski *et al.*, 1975).

The principal radionuclides in monazite are from the ^{232}Th series, but there is also some uranium present and, therefore, some opportunity for ^{226}Ra uptake. However, very little food is grown in the monazite areas of Brazil, and the diets of the local inhabitants are derived principally from outside sources. The exposures in the monazite areas are due primarily to external radiation, and the internal dose is not believed to be significant (Eisenbud *et al.*, 1964; Penna Franca *et al.*, 1970).

Fig. 7–12. Beach of the black sands at Guarapari, state of Espirito Santos, Brazil. The dark areas contain monazite sands over which the exposure rate is as high as 5 mR/hr. Tourists travel to these beaches because of local beliefs about the beneficial effects of radioactivity.

In the state of Kerala, on the southwest coast of India, the monazite deposits are more extensive than those in Brazil, and about 100,000 persons inhabit the area. The dose from external radiation is, on average, similar to the exposures reported in Brazil (500–600 mR/year), but individual exposures up to 3260 mR/year have been reported (Sunta *et al.*, 1982). The epidemiologic studies that might be possible with a population of this size have not as yet been undertaken, but some reports have been published. A higher than normal incidence of mongolism (Down's syndrome) has been reported in the high-background area (Kochupillai *et al.*, 1976), but this has been refuted (Sundaram, 1977). In contrast to Brazil, there have as yet been no reports of cytogenetic studies in humans. Mistry *et al.* (1970) reported that uptake of ^{228}Ra by food grown in the monazite area is greater than the uptake reported from Brazil.

A distinctly different source of exposure to natural radioactivity exists near the city of Araxa in the Brazilian state of Minas Gerais. The soil is generally poor in this area except in patches where it contains apatite, a phosphate mineral associated with a number of radioactive minerals that contain both uranium and thorium. The food grown in this area contains relatively high amounts of ^{228}Ra and ^{226}Ra, but the number of exposed persons is small. Penna Franca *et al.* (1970) have surveyed the dietary habits of the indigenous population of this region and have undertaken

Fig. 7–13. The Morro do Ferro (Mountain of Iron) in Minais Gerais, Brazil.

radiochemical analysis of the foods. Of the 1670 persons who live in the area, nearly 200 ingest radium in amounts that are 10–100 times greater than normal. The ratio ^{228}Ra : ^{226}Ra is about 6 to 1 in the diet.

A unique anomaly located near Poços de Caldas, also in the state of Minas Gerais, is the Morro do Ferro, a hill that rises about 250 m above the surrounding plateau (Fig. 7–13). Near the summit of the hill is a near-surface ore body that contains about 30,000 tonnes of thorium and an estimated 100,000 tonnes of rare earth elements. The ambient gamma-radiation levels near the summit of the hill range from 1 to 2 mR/hr over an area of about 30,000 m^2 (Eisenbud *et al.*, 1984). The flora from this hill have absorbed so much ^{228}Ra that they can readily be autoradiographed, as shown in Fig. 7–14.

Studies have been undertaken of the exposures of rats living underground on the Morro do Ferro. Of particular interest is the dose to these rodents due to inhalation of ^{220}Rn, which was found by Drew and Eisenbud (1966) to be present in the rat burrows in concentrations up to 100 pCi/ml. The dose to basal cells of the rat bronchial epithelium was estimated to be in the range of 3000 to 30,000 rem/year. Of 14 rats trapped and sacrificed for pathological study, none was observed to show any radiation effects. This is of little significance, since the Morro do Ferro is a relatively

Fig. 7–14. Autoradiograph of species of *Adiantum* from the Morro do Ferro in the state of Minas Gerais, Brazil. (Courtesy of Dr. Eduardo Penna Franca.)

small area, and if rats were affected by this exposure they could be replenished rapidly from the surrounding normal areas. By using thermoluminescent dosimeters implanted in trapped rats that were released and later recaptured, the external radiation dose to the rats was estimated to be between 1.3 and 6.7 rad/year.

Other studies being undertaken at the Morro do Ferro will be discussed in Chapter 11.

THE CHINESE MONAZITE AREA

About 73,000 persons live in an area of monazitic soils in Guangdong Province, China, where the external radiation exposure is about 330 mR/year (Wei *et al.*, 1985). The radiation levels are between three and four times normal, and Wei and associates have undertaken a comprehensive long-term study of the area, beginning in 1972 (Wei, 1980). The ^{226}Ra body burdens in the high-background area were reported to be 281 pCi, about three times higher than in the control area. If continued for many years, this study could result in statistically meaningful data on the effects of low-level radiation.

THE NATURAL REACTOR AT OKLO

Much to the surprise of the scientific world, it was discovered in 1972 that the site of an open-pit uranium mine called Oklo in the West African republic of Gabon was the fossil remains of a 2-billion-year-old natural reactor. Although contemporary uranium contains only 0.72% of the fissionable isotope ^{235}U, this nuclide has a half-life of only 700 million years, compared to the 4.5-billion-year half-life of ^{238}U. The percentage of ^{235}U present in the ore was thus very much greater 2 billion years ago, and the

TABLE 7–18

ESTIMATED ANNUAL EFFECTIVE DOSE EQUIVALENT RATE FROM NATURAL SOURCES IN NORMAL REGIONS[a]

Source	Effect dose equivalent rate (mrem/year)		
	External	Internal	Total
Cosmic (including neutrons)	30		30
Cosmogenic nuclides		1.5	1.5
Primordial nuclides			
^{40}K	12	18	30
^{87}Rb		0.6	0.6
^{238}U series			
^{238}U $\rightarrow$ ^{234}U	9	1.0	10
^{230}Th		0.7	0.7
^{226}Ra		0.7	0.7
^{222}Rn $\rightarrow$ ^{214}Pb		80	80
^{210}Pb $\rightarrow$ ^{210}Po		13	13
^{232}Th series			
^{232}Th	14	0.3	14.3
^{228}Ra $\rightarrow$ ^{224}Ra		1.3	1.3
^{220}Rn $\rightarrow$ ^{208}Th		17	17
Total (rounded)	65	134	200

[a]From UNSCEAR (1982).

TABLE 7–19

SUMMARY OF AVERAGE DOSE EQUIVALENT RATES (mrem/YR) FROM VARIOUS SOURCES OF
NATURAL BACKGROUND RADIATION IN THE UNITED STATES[a]

Source	Gonads	Lung	Bone Surfaces	Bone Marrow	Gastrointestinal tract
Cosmic radiation	28	28	28	28	28
Cosmogenic radionuclides	0.7	0.7	0.8	0.7	0.7
External terrestrial	26	26	26	26	26
Inhaled radionuclides	—	2800 [b]	—	—	—
Radionuclides in the body	27	24	60	24	24
Rounded totals	80	180	120	80	80

[a]From NCRP (1975c) unless stated otherwise.
[b]From NCRP (1984b).

conditions for initiating fission reactions apparently existed. Criticality was sustained for more than 1 billion years, during which time an estimated 15,000 MW-years of energy was released by the consumption of 6 tons of ^{235}U. Studies of the migration of the fission products and transuranic elements produced during this period have shown that, with few exceptions, migration was minimal. The implications of this finding for radioactive waste management are discussed in Chapter 11 (Cowan, 1976; IAEA, 1975).

Summary of Human Exposures to Natural Ionizing Radiation

The annual effective dose equivalent received by persons living in most parts of the world, where the natural radioactivity is within normal limits, is given in Table 7–18. Radon-222 and its short-lived daughters contribute about 40% of the total effective dose equivalent, a term which, as explained in Chapter 2, takes into consideration the volume of tissue irradiated by the nuclides, the dose equivalent to the tissue, and the relative radiosensitivity of the tissue. Radon-222 and its short-lived daughter products irradiate the basal cells of the bronchial epithelium, which has a mass of only a few grams.

The doses delivered to the various organs of the body that are most affected by natural sources of radiation can also be given in units of dose equivalent (as distinguished from *effective* dose equivalent). This is done in Table 7–19.

Chapter 8

Production and Reprocessing of Nuclear Fuels

The production of nuclear energy is based mainly on the fission of ^{235}U, which is present in natural uranium to the extent of 0.7%. Uranium-238, which is the most abundant isotope of uranium, is not readily fissionable but does transmute by neutron capture to ^{239}Pu, which is a fissile isotope. Thus, the source materials of atomic energy may be either fissile, like ^{235}U, or fertile, like ^{238}U.

Thorium has thus far been a minor source of nuclear energy. It occurs in nature almost entirely as the isotope ^{232}Th, which is a fertile nuclide that can be transmuted to the fissile ^{233}U. If full use is to be made of the world's potential resources in nuclear energy, thorium may ultimately be incorporated into reactor systems for the purpose of breeding ^{233}U, but comparatively little use will be made of thorium in this way for the foreseeable future in the United States.

Uranium

When interest in uranium first developed in World War II, commercially exploitable deposits were thought to be comparatively rare and to occur only in a few districts rich in the mineral pitchblende. The sources of uranium then known to exist were deposits located in the Belgian Congo, the Great Bear Lake region of Canada, and Czechoslovakia. It was known that much lower grade deposits existed on the Colorado Plateau, and these deposits had for some years been used as a source of radium. The full extent of the uranium resources of the southwest United States was not then appreciated. However, following an intensive exploratory program conducted by the U.S. government, the uranium industry grew rapidly during the post-World War II period.

The U.S. government encouraged exploration and mining in other coun- tries as well, because domestic production was not sufficient for its needs during the 1950s. In 1959, U.S. production of U_3O_8 reached nearly 35,000 tons, of which about one-half was produced from imports (U.S. DOE, 1984d, 1985a). The U.S. demand for uranium for military purposes slack- ened in the 1960s, at a time when commercial nuclear power began to develop. However, the large civilian market for uranium that was then forecast did not materialize because of the slackening of the requirements for nuclear power that began in the mid-1970s.

The present-day reserves of uranium in the United States are in the form of sandstone ores that have a relatively low uranium content.[1] The eco- nomically feasible uranium content expressed as percent U_3O_8 is expected to average about 0.10% until the end of the century. The U_3O_8 content of ores mined during and after World War II ranged from 20 to 60%, but the total amount of uranium available was not nearly as great as now. The Department of Energy estimated that the 1984 "reasonably assured" ura- nium resource was between 260 and 370 short tons of U_3O_8 (U.S. DOE, 1985a). Most of the ore is located in the Wyoming Basin and the Colorado Plateau.

About 11 million tons of ore containing 13,000 tons of U_3O_8 were mined in the United States in 1982. This raw material is mined in both open pits and underground workings and is then shipped to mills that produce ura- nium concentrate containing between 70 and 92% U_3O_8. The output of these mills is transported to uranium refineries, where the concentrates are converted into uranium compounds having a high degree of chemical purity. From there the uranium goes to isotopic enrichment plants, where the ^{235}U content is increased. Until now, enrichment has been accom- plished in huge gaseous diffusion plants located in Ohio and in the Ten- nessee Valley. Newly developed processes are expected to replace gas-

---

[1]Significant uranium reserves are also present in deposits of phosphates and could make major contributions to uranium production in the future. The radiological health conse- quences of phosphate mining are important whether or not uranium is recovered and will be discussed in Chapter 10.

eous diffusion in time. If not destined for enrichment, the uranium may be shipped as the oxide or metal to a number of privately or government-owned installations where natural uranium reactor fuel elements may be fabricated. A large fraction of the metallic natural uranium has been shipped to plutonium production facilities located at Hanford and Savannah River. The enriched uranium from the gaseous diffusion plants may also be shipped to fuel fabrication facilities.

URANIUM MINING

Uranium mining takes place mainly in underground workings and open pits. The mines vary in size from relatively small workings emplying one or two people using hand tools to thoroughly mechanized enterprises employing more than 100. Uranium mining by *in situ* underground leaching (solution mining) has also been demonstrated to be feasible and may become more important in the future (Hunkin, 1980). Solution mining involves solubilizing uranium by pumping water, to which oxidizing and complexing agents have been added, into an ore body and then pumping the solution to the surface for processing (Brown and Smith, 1980). The method reduces the risks to the miners and has minimal environmental impact.

The concentration of radon and its daughter products in the air of underground uranium mines can be hazardous to the workers unless proper ventilation is employed. As discussed in Chapter 2, it has been known since early in this century that the high incidence of lung cancer among miners in Bohemia and Saxony was due to their exposure to radon. Despite that experience, adequate protective measures against radon were not adopted when uranium mining began in the United States. The Atomic Energy Commission did not assume responsibility for the health of the uranium miners because of a technicality in the Atomic Energy Act.[2] This left the matter to the mining states, where enforcement of the rules of safe practice was inadequate, as a result of which many radon-induced cases of lung cancer have since developed among the miners. The federal government did not enter the picture insofar as enforcement is concerned until the late 1960s (Holaday *et al.*, 1968), when radon standards were promulgated.

---

[2]The individual states normally have responsibility for ensuring the safety of industrial activities within their borders. However, because the risks of the atomic energy program were not then fully known, and the states did not then have the required expertise, the Atomic Energy Act of 1946 gave the responsibility for health and safety to the AEC, *but only after the ore was removed from the ground*. Thus, the AEC authority and responsibility did not preempt that of the states insofar as mining was concerned.

Radon and its daughter products are the only significant radioactive contaminants discharged from the mines to the general environment. Required ventilation rates for the mines vary from 1000 to over 200,000 cubic feet per minute (cfm), and the discharged air contains radon in concentrations that range from 0.5 to 20 μCi/min per 1000 ft^3 of air (Holaday, 1959), which is no greater than the normal radon flux from about 1 km^2 of the earth's surface. The risks to the general public from the radioactive emissions from uranium mines is insignificant (Blanchard *et al.*, 1982).

THE URANIUM MILLS

The uranium ore is transported from the mines to mills, where the uranium concentrate U_3O_8 (yellowcake) is produced. Milling begins with a grinding process that reduces the ore to the consistency of fine sand, following which the uranium is separated by acid or alkaline leaching and concentrated either by ion exchange or solvent extraction. Although more than 95% of the uranium is removed from the ore by this process, almost all of the radioactive daughter products in the uranium series remain with the tailings contained in slurries that are discharged into holding areas. Because the U.S. mills are located in the arid regions of the Southwest, the impounded slurries dry rapidly and in the course of time become growing mounds of the residue (Clegg and Foley, 1958; Goldsmith, 1976; U.S. NRC, 1980a). It has been estimated that by 1983 approximately 1.75×10^8 metric tons of uranium mill tailings had been accumulated at 52 active and inactive sites, all but one of which (in Canonsburg, Pennsylvania) were located in the western states (U.S. DOE, 1985a).

Uranium milling in the United States reached a peak between 1960 and 1962, when there were 25 mills in operation, but this number has been greatly reduced by lessened demand for uranium. By 1984 there were only six operating mills in the United States. The piles of mill tailings are potential sources of environmental problems because of (1) emanation of ^{222}Rn, (2) dispersion of the tailings by wind and water, and (3) their use in building construction.

Radon Emanation

The ^{226}Ra concentrations of mill tailings piles have been found to vary from 50 to 1000 pCi/g of dry tailings (U.S. EPA, 1982). Radon production within the tailings piles will continue for hundreds of thousands of years because the piles contain almost all of the ^{230}Th originally present in the ore, and this long-lived nuclide ($T_{1/2} = 77,000$ years) is the parent of ^{226}Ra, which in turn decays to ^{222}Rn. Most models for the production of radon in tailings piles assume that radon enters the interstitial gases within a dry tailings pile by alpha recoil. The radon then diffuses to the surface of the pile, where it enters the atmosphere at highly variable emission rates (Tanner, 1980). The DOE (U.S. DOE, 1981) gathered data for 24 inactive mill tailings sites, where the measured emission rates showed great variability not only between sites but also in serial measurements at individual sites. There was a poor correlation between the predicted and measured emission rates, and no correlation was found between the radon flux and the tailings radium content or cover. Such correlations must surely exist and will no doubt be found when account is taken of such parameters as soil moisture, wind velocity, barometric pressure, and other factors that would be expected to influence the emission rate and radon concentration.

A generic environmental impact statement on uranium milling prepared by the Nuclear Regulatory Commission assumed the flux from dry tailings (U.S. NRC, 1980a) to be 1.0 pCi ^{222}Rn/m^2 sec pCi ^{226}Ra g dry tailings. Schiager (1986) notes that this emission rate is comparable with the rates of radon emission from normal soils. The increased radon flux from the tailings piles is due to the higher ^{226}Ra content of the tailings, which may contain several hundred picocuries per gram compared to 1–2 pCi/g for normal soils. Since the area covered by the tailings piles, usually on the order of 10–100 acres, is small compared to the areas not covered by tailings piles, the piles themselves do not make a significant contribution to the concentration of Rn in the general environment. Their influence is localized and can be detected only within about 1 mile (Healy, 1981; Shearer and Sill, 1969; Schiager, 1974).

The emanation rate has been reported to be sensitive to atmospheric pressure. It was reported that a change of only 1% in barometric pressure can result in emission rate variations of 50–100% (Clements and Wilkening, 1974). The relationship between atmospheric pressure and diffusion rate has been explained by some investigators as being due to more rapid release of radon during periods of low atmospheric pressure and dilution of interstitial soil radon by air pumped into the ground during periods of high pressure (Kraner *et al.*, 1964; Clements and Wilkening, 1974).

It is important to note that the emission rate applies only to radon, the parent of the radon series, since the daughter products are not gaseous and do not diffuse from the ground. Daughter product ingrowth occurs after release to the atmosphere, when the radon is being diluted by turbulence. Thus, for some distance downwind, the fraction of the daughter products associated with the radon will be much smaller than the fraction usually found in environments in which radon is present.

Based on the various studies performed, the radon emissions of themselves are a minor source of risk to individuals. However, the small risk applies to a large number of persons and persists for a long time. The NRC (U.S. NRC, 1980a) estimated that the radon generated by the tailings piles, if all mills were to be in full operation until the year 2000, would result in about six premature deaths (from cancer) per year during the period 1979–2000 in the United States, Mexico, and Canada. If we assume that the populations of these countries will reach an equilibrium of 600 million persons during this period, the risk that any one individual will die prematurely because of cancer produced by radon exposure is about 3×10^{-10}. The risk to individuals located within about 3 km of the piles would be much higher, about 4×10^{-4} per lifetime, but only a few such individuals would be at risk.

Dispersion by Wind and Water

Materials from the tailings piles can be dispersed into the environment by impoundment failure, erosion by flowing water, or wind action. Impoundment failure can be precluded during the foreseeable future by application of available engineering methods, and if cover is provided for the piles, erosion by wind or water can be avoided in the semiarid regions in which the tailings piles are located (Webb and Voorhees, 1984). However, considering the long half-life of ^{230}Th, the parent of ^{226}Ra, there can be no assurance that these protective features will be effective for the hundreds of thousands of years during which radium production will be sustained.

One cannot rule out the possibility that future changes in climate or population density may result in increased risks. There are currently only six operating mills at which the tailings piles are receiving continuing attention, but there are 25 abandoned tailings piles for which stabilization would be desirable. Public Law 92-314 gave the Department of Energy the responsibility for stabilization of the tailings at the 25 inactive uranium processing sites, according to the standards established by the Environmental Protection Agency (U.S. EPA, 1983).

Erosion by wind and water can be reduced by contouring the pile, providing cover, stabilizing the surface with rock or other material, and constructing dikes to divert floodwaters. Erosion of new tailings sites can be minimized by burying the tailings in shallow pits or by locating them away from sites subject to flooding. Radon inhibition can be accomplished by covering the pile with compacted earth or less permeable materials such as asphalt, clay, or cement. The reduction in radon emission that results from a given thickness of soil cover depends on the moisture content. The thickness of the required soil cover is reduced for soils having higher moisture contents.

Ground water contamination can be reduced by provision of underlying plastic ro clay barriers. These arc, of course, feasible only for new tailings

sites. Should this be done in areas of high precipitation and low evaporation, it would be necessary to seal the surface of the pile. Otherwise, the
underlying linear may act as a bowl that can cause contaminants to accumulate and move toward the surface.

The EPA has established an emission limit of 20 pCi ^{222}Rn/m^2 sec and
requires that the disposal method be designed to provide "reasonable
assurance" that radon emissions will not exceed that limit, averaged over
the disposal area for 1000 years. This, of course, leaves open the possibility
that the standards would be exceeded beyond 1000 years. The EPA estimated that the cost of complying with its standards will be \$310–540
million by the year 2000 (U.S. EPA, 1983).

In past years, liquid tailings were allowed to seep into nearby streams. It
was shown (Gahr, 1959) that the ^{226}Ra content of water in the Colorado
River below Grand Junction was 30 pCi/liter, compared to 0.3 pCi/liter
upstream. The San Miguel River below Uravan, Colrado, where a mill was
located, contained 86 pCi/liter, compared to 4.9 pCi/liter in water immediately upstream of the mill.

The Animas River in southwestern Colorado serves as a public water
supply for the cities of Aztec and Farmington, and the water is also used
for irrigation. In 1955 the radium concentration below Durango, where a
mill was located, was found to be 3.3 pCi/liter, compared to 0.2 pCi/liter
upstream. That considerable concentration was taking place in the stream
biota was shown by the fact that plants below Durango contained 660
pCi/g, compared to 6 pCi/g above Durango. Stream fauna below Durango
contained 360 pCi/g compared to 6 pCi/g above the mill.

It was found that using the then-existing ICRP recommendations as a
guide, consumers of untreated river water received about three times the
maximum permissible daily intake of radium (Tsivoglou, 1959, 1960a,b).
This included food grown on land irrigated with river water, as well as the
water consumed. About 61% was due to the radium content of food resulting from the contaminated irrigation water.

Studies by Shearer and Lee (1964) concluded that the radium was leach-
able from both the tailings piles and the river sediments. Thus, radium was
entering the rivers not only in untreated liquid wastes but in surface runoff
during rainfall as well. Steps were taken by the mill operators to correct
this problem, and by 1963 the radium content of the Animas River sediments had been reduced to three times background, compared to several
hundred times background several years earlier. The Animas River experience illustrates the potential danger, if precautions are not taken, of contamination of potable water by radium and other radionuclides and toxic
chemicals present in tailings piles.

It has been estimated that there are approximately 1.75×10^8 metric tons of tailings located at sites in the United States. Assuming an average of 280 pCi ^{226}Ra/g of dry tailings (U.S. NR,C 1980a), the ^{226}Ra content of the tailings piles is about 50,000 Ci, being produced by ^{230}Th decay at a rate of about 20 Ci/year. The tailings also include substantial quantitites of the nuclides of the ^{226}Ra decay series, which are in equilibrium with ^{226}Ra except to the extent that radon is lost by diffusion from the piles. Lead-210, ^{210}Po, and ^{210}Bi are assumed to be present in a concentration of 250 pCi/g, compared to 280 pCi/g for ^{230}Th and ^{226}Ra.

Breslin and Glauberman (1970) measured the airborne dust downwind from unstabilized tailings piles and demonstrated clear relationships between the distance from the tailings piles and the concentrations of uranium and ^{210}Pb. For the three tailings piles sampled, the air concentrations were well below permissible levels in two cases, but approached the upper limits recommended in 10CFR20 at a distance of about 1000 ft in the third case.

Use of Mill Tailings for Construction

The exposure of many Grand Junction residents to both gamma radiation and radon was increased by the use of mill tailings as a material of construction of homes and public buildings (U.S. DOE, 1980b; Committee on Armed Services, 1981). That practice was not discountinued until 1966. Under Public Law 92-314, enacted in 1978, Congress provided financial assistance to the state of Colorado to limit the radiation exposures that existed because of the use of tailings for construction purposes. The cleanup in Grand Junction was performed in accordance with guidelines issued by the Office of the Surgeon General in 1970, whose recommendations were applicable to radiation exposure levels in dwellings constructed on or with uranium mill tailings and are summarized in Table 8–1 (U.S. AEC,

TABLE 8–1

1970 REMEDIAL ACTION RECOMMENDATIONS OF THE SURGEON GENERAL FOR EXPOSURES TO MILL TAILINGS USED IN BUILDING CONSTRUCTION[a]

Recommendation	External gamma radiation (mr/hr)	Radon (WL)
Remedial action indicated	>0.1	>0.05
Remedial action may be suggested	0.05–0.1	0.01–0.05
No action required	<0.01	<0.01

[a]From U.S. Atomic Energy Commission (1972b).

1972b). The federal government provided 75% of the cost of remedial action, with the state providing the balance. Tailings were removed from about 740 buildings where external gamma radiation or indoor radon daughter exposures exceeded the Surgeon General's guidelines (Ramsey, 1981).

REFINING

The mill concentrates in this country are sent to any of several locations in which the uranium is converted to either the metal or some intermediate uranium compounds, such as orange oxide (UO_3) or green salt (UF_4). The principal steps in converting the concentrates to a form that is of acceptable chemical purity are shown in Table 8–2. These processes involve potential exposure of the employees to alpha-emitting dusts and, in the case of high-grade fuels, to radon and γ radiation (Eisenbud and Quigley, 1956).

The refining operations involve the mechanical processing of dry powders of uranium compounds, which can result in the discharge of uranium dust to the environment. Present-day plants are equipped with filtration equipment that effectively removes the uranium dust, and the monetary value of uranium is such as to preclude the possibility of its being discharged to the atmosphere in large quantities for sustained periods of time. However, the plants hastily constructed during World War II had insufficient control over dusts contained in exhaust air, and relatively large amounts of uranium were discharged to the outside atmosphere. Nevertheless, uranium is so abundant in the environment that the element was undetectable above the natural background at moderate distances from the plants (Klevin *et al.*, 1956).

The kinds of wastes produced by the refineries depend on the type of feed that is processed. During World War II and for a few years thereafter, when high-grade ores were processed that contained as much as 100 mCi ^{226}Ra per ton of ore, some of the sludges contained as much as 1 Ci ^{226}Ra per ton. However, the uranium industry has been operating with ores of much lower grade for several decades, and the uranium is separated at the mills, thus sparing the refineries the problem of disposing of waste products containing large amounts of radium. The limiting factor in the discharge of wastes from uranium refineries is apt to be the chemical wastes rather than their radioactive constituents.

There were 31 plants, laboratories, and storage sites involved with production of uranium from ore during the World War II era. Operation of those plants was discontinued by the mid-1950s, and they were cleaned up according to the then-existing understanding of what constituted adequate

TABLE 8–2

Principal Steps in the Refining and Conversion of Uranium[a]

Feed

Miscellaneous uranium concentrates
(approximately 75% by weight U_3O_8 or equivalent

↓

Production of orange oxide (UO_3)

1. Digestion of the uranium concentrates in nitric acid
2. Solvent extraction to remove impurities and reextraction into water
3. Boil-down of the uranyl nitrate solution from (2) to a molten uranyl nitrate hexahydrate
4. Denitration of the molten salt by calcination to produce orange oxide powder

↓

Conversion to green salt (UF_4)

1. Reduction of the orange oxide to brwon oxide (UO_2) by contacting with hydrogen
2. Conversion of the brown oxide to green salt by contracting with anhydrous hydrofluoric acid

↓

Reduction to metal

1. Reduction of the green salt to massive uranium metal (derbies) by a thermite-type reaction using magnesium as the reducing agent
2. Vacuum casting of several uranium "derbies" from (1) to produce a uranium ingot

↓

Product

High purity uranium metal in ingot form

---

[a] Adapted from U.S. Atomic Energy Commission (1957b).

decontamination. As time went on, more conservative criteria were developed, and many of the sites were found to be above the limits considered suitable for unrestricted access. In 1974 the Atomic Energy Commission, predecessor of the Department of Energy, initiated a program of decontamination known as the Formerly Utilized Sites Remedial Action Program (FUSRAP), which is currently under way and which will involve the removal of nearly 1 million yards of soil and building rubble. The estimated cost of the program is \$463 million in 1985 dollars (U.S. DOE, 1985b).

ISOTOPIC ENRICHMENT

The green salt (UF_4) from the refineries is converted to uranium hexafluoride (UF_6), which up to the present time has been shipped for isotopic enrichment to the large gaseous diffusion plants located in Portsmouth, Ohio; Paducah, Kentucky; and Oak Ridge, Tennessee. The UF_6 is pumped through cascades of porous barriers. Each stage of diffusion results in a slight enrichment in ^{235}U, and the number of stages is determined by the degree of enrichment required.

The gaseous diffusion plants are now considered to be obsolete. Centrifuges and other aerodynamic methods of separation are already in use in several countries. Electromagnetic separation from plasmas and chemical separation of isotopes have also been proposed, as well as the use of lasers (Tait, 1983). It is believed that enrichment by lasers will be the future method of choice in the United States. Bombardment of UF_6 or U vapor with multiple laser photons can result in up to 50% enrichment in a single pass (Knief, 1981).

Even more so than in the refineries, the economic value of the enriched uranium precludes the likelihood of widespread environmental contamination from these plants. In addition, as the uranium progresses through the diffusion plant, it becomes of increasing importance as a source of material for weapons and must be subject to strict accountability.

The enriched uranium from the diffusion plants is destined for assembly into either weapons or fuel elements for reactors. The depleted uranium is being used for shielding and for armor-piercing projectiles.

FUEL-ELEMENT MANUFACTURE

Fuel-element manufacture is currently carried on at a great many government and private facilities. Again, the relatively high cost of the uranium and the requirements for strict accountability make it unlikely that significant environmental contamination can occur from these plants, but the possibility of accidents cannot be discounted entirely. Uranium chips are pyrophoric, and if fires or explosions occur, more than normal amounts of activity may be released.

REPROCESSING SPENT REACTOR FUEL

When a reactor core has reached the end of its useful life, only a small percentage of the ^{235}U will have been consumed in fission and an addi-

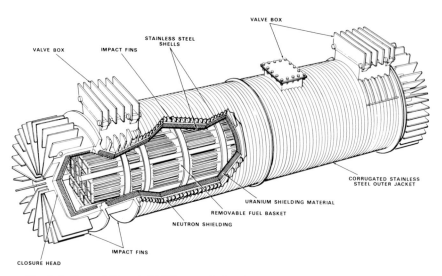

Fig. 8–1. General Electric IF 300 spent fuel cask, designed for transportation of all types of light-water-moderated fuels. During normal operation the cask is filled with water, which provides heat transmission to the walls by natural circulation. The fins are intended for impact protection. The outer surface of the cask is cooled by air blowers (not shown). (Courtesy of General Electric Company.)

tional small fraction of the ^{238}U will have been transmuted to ^{239}Pu and other transuranic elements. If the core is to be sent to a reprocessing plant for recovery of U and Pu, it must first be removed from the reactor and stored under water in fuel storage pools associated with the reactor, following which it can be transported (Fig. 8–1) to a fuel-reprocessing plant, in which the spent fuel is chemically treated to (1) convert the fission products into a form suitable for long-term storage (see Chapter 10) and (2) recover the remaining ^{235}U and the transuranic elements. The presence of large amounts of fission products and transuranic actinide elements in the irradiated fuel greatly complicates the processing procedures and makes it necessary to adopt elaborate measures to protect the operating personnel and to avoid environmental contamination.

History of Fuel Reprocessing in the United States

From World War II until 1966, all fuel reprocessing in the United States was performed at four government-owned centers of atomic energy development and production. Two of the plants, at Hanford and Oak Ridge, were built during World War II; the other two were built in the 1950s at Savannah River, South Carolina, and at the National Reactor Testing Station at Idaho Falls, Idaho. These plants were constructed to meet military needs and were not intended for the processing of civilian fuel.

There was an apparent need for commercial fuel reprocessing, but the initial attempts by private industry to meet the needs of the projected nuclear power development were failures for a combination of technical, economic, and political reasons (Colby, 1976). The first and only privately owned nuclear fuel-reprocessing plant in the United States, Nuclear Fuel Services Inc. (NFS), went into operation in 1966 in West Valley, New York, with a daily processing capacity of 1 metric ton of low-enriched uranium oxide fuel. Other privately owned plants were constructed by General Electric Company in Morris County, Illinois, and the Allied Chemical Corporation at Barnwell, South Carolina (Unger *et al.*, 1971; U.S. AEC, 1970a), but these plants never operated. The General Electric plant never went into operation because it proved to be technically and economically deficient, and the Allied Chemical plant in South Carolina encountered serious licensing difficulties. The plant in West Valley encountered licensing and economic problems that resulted in its shutdown in 1972 and subsequent abandonment of the venture by its owners.

Any incentive to correct the deficiencies of the three plants for the purpose of starting or restarting fuel reprocessing was eliminated in 1977 when President Carter announced a U.S. policy of prohibiting fuel reprocessing as a means of preventing the diversion of plutonium and the possible uncontrolled proliferation of nuclear weapons (Gilinsky, 1978–1979). The question of whether this was a justifiable policy became moot subsequently because lessened requirements for nuclear power eliminated the economic incentive to recycle the fuel.

Fuel-reprocessing methods vary, depending on the materials from which the fuel is fabricated, but all fuel-reprocessing plants use some form of the Purex process, a solvent extraction system using tributyl phosphate (TBP) diluted with kerosene (Etherington, 1958; Stoller and Richards, 1961). The fuel elements are prepared for reprocessing by being sheared into small lengths, after which the cladding is removed chemically. The fuel is then dissolved in nitric acid, following which a series of successive TBP extraction and stripping steps, under controlled chemical conditions, results in separation of the original solution into the required fractions, transuranic elements, uranium, plutonium, and fission products. The waste products are in the latter fraction and are produced in highly radioactive form. After being concentrated by evaporation, the wastes may be stored in underground tanks until the decay heat has subsided; they are then prepared for ultimate disposal, as will be discussed in Chapter 11.

Zirconium cladding may be removed by dissolution in ammonium fluoride solutions in a process that produces about 4600 gal/ton of fuel.

Stainless steel, type 304, which was once a popular cladding for uranium oxide fuels, is removed by sulfuric acid, producing about 3000 gal of solution per ton of fuel. The solutions that result from removal of the cladding contain of the order of 10^{-4} of the radioactivity in the fuel (Bruce, 1960).

In the Purex process, about 1200 gal of acidic wastes are produced initially per ton of fuel processed, but the volume may be reduced to about 55 gal/ton by evaporation followed by sodium hydroxide neutralization.

The intensely radioactive Purex concentrates cause radiolytic decomposition of H_2O and $NaNO_3$, which produces HNO_3 and has resulted in corrosion of the liners of the waste storage tanks. It has been estimated that the Hanford liners are corroding at a rate of 10^{-4} to 10^{-5} in./year (Bruce, 1960; Schneider *et al.*, 1971). Storage of high-level wastes in liquid form is only an interim procedure in the United States, and the wastes must eventually be converted to solid form, which facilitates long-term storage (see Chapter 11).

The large fission-product and transuranic actinide inventories, together with the very nature of the fuel-reprocessing methods, present opportunities for major environmental contamination unless strict procedures are followed to avoid release of radioactive substances to the vicinity of the plant. Fortunately, the potential hazards of these operations were anticipated even before the first chemical-reprocessing plants were built during World War II, and techniques have been used from the very start that have proved effective in safeguarding the operation of the government-owned plants.

Sources of Radioactive Emissions

As in other components of the nuclear industry, the opportunities for environmental contamination from fuel reprocessing can be divided into those that occur in normal plant operation and those that result from accidents so severe as to overwhelm the defenses against uncontrolled releases. The nature of the process is such that chemical explosions are conceivable, and since fissionable materials are processed, it is possible for critical masses to be assembled accidentally. However, as in the case of reactors, the potential hazards have been offset by conservative processes designed to minimize the probability of a serious accident. As in the case of reactor design, however, having achieved this objective, it is nevertheless assumed that an accident will happen, and compensating safeguards are provided to minimize releases to the environment. The safety of commercial fuel-reprocessing plants is regulated by the Nuclear Regulatory Commission with a system as rigorous as that used to regulate reactor design and operation.

Accidental criticality is prevented by limiting the amount of fissionable material being processed. The geometric design of process vessels and storage containers is carefully controlled and is a basic method by which the opportunity for assembling a critical mass is prevented. In some cases, neutron-absorbing components have been placed in tanks in the form of rings or parallel plates (Unger *et al.*, 1971).

Dispersal of airborne material in the event of an accident is controlled by constructing the process building as a series of shells having gradations of negative atmospheric pressure. The physical design of the processing cells is such that they can withstand the blast effects from the maximum credible explosion (Unger *et al.*, 1971).

The reprocessing process results in enormous quantities of fission products, the management of which will be discussed in Chapter 11. In this chapter, we will limit our discussion to the low-level wastes that are discharged to the atmosphere or nearby surface waters.

The primary sources of gaseous wastes are the fuel-element chopping and dissolution processes (Logsdon and Chissler, 1970). Owing to the relatively long storage time before the fuel is processed, most of the ^{131}I (half-life 8.2 days) has decayed, but enough remains to require special treatment of the gases.

A potential long-range problem exists because of ^{129}I, a nuclide with a half-life of 1.7×10^7 years. Because of its long half-life, this nuclide has received little attention. However, it is produced in fission with a yield of 1% and is, therefore, present in reprocessing wastes in relatively large quantities. Its long half-life means that it will accumulate in the environment, become part of the iodine pool, and deliver a thyroid dose to the general population that could increase in proportion to the rate of nuclear power production (NCRP, 1983). The radioiodines can be removed chemically with caustic scrubbers or other means, such as by reactions with mercury or silver.

Other radioactive gaseous releases from a nuclear fuel-reprocessing plant include ^{85}Kr and ^{3}H. The gaseous release of ^{3}H has been small compared to the releases in liquid form. The stack releases of ^{85}Kr are substantial but have not been a source of significant exposure in the vicinity of the processing plants. Doses to individuals from ^{85}Kr and ^{3}H emissions from reprocessing plants would be expected to be small, but they could accumulate in the general environment in a manner that would expose very large numbers of persons. The collective dose to be expected is, of course, sensitive to the assumptions made about the amount of nuclear power generation. If spent fuel reprocessing is resumed in the future, it will be necessary to develop policies to separate these nuclides from waste streams. The various available options are discussed by NCRP (1975c, 1979, 1983) and OECD (1980).

Thorium

Thorium (Cuthbert, 1958; Albert, 1966) is estimated to be three times more abundant in the earth's crust than uranium and may ultimately become an important source of nuclear energy as techniques are developed for converting ^{232}Th to ^{233}U in breeder reactors. The most important known occurrences of thorium minerals are in the monazite sands of Brazil and India. Although thorium has been used for many years in the manufacture of gas mantles and at present has a limited application in the atomic energy industry, the production capacity for thorium is so small that there is little potential for general environmental contamination with this material or its daughter products (Meyer et al., 1979).

Chapter 9

Power Reactors

The first nuclear reactor was operated briefly by Enrico Fermi and his associates in Chicago on December 2, 1942, less than 4 years after the discovery of nuclear fission. Under wartime pressure, reactor technology continued to develop rapidly, and only 1 year later a 3.8 thermal megawatt (MWt) research reactor began operation at Oak Ridge. Thereafter it remained in service for more than 20 years, during which time it served as the chief source of radioisotopes for research and industrial use in the United States and much of the Western world (Hewlett and Anderson, 1962; Tabor, 1963). Even more remarkable, the first of several reactors designed for plutonium production began operation in the state of Washington in 1944 at an initial power level of 250 MWt. These and additional units, at considerably higher power levels, also remained in service for more than 20 years. At present, about 1000 land-based reactors have been built and operated in various parts of the world. These include reactors built for research as well as production of power, radioisotopes, and plutonium. Nearly 200 vessels of the American and Russian navies are powered by nuclear reactors.

From 1945 until 1954 the reactor program of the United States was dominated entirely by the government and was closely identified with the military applications of nuclear energy. All reactors constructed in the United States prior to 1953 were located on government sites, but in that year a research reactor was placed in operation at North Carolina State University, the first reactor to go into operation outside an AEC facility.

Statutory changes that occurred in 1954 with passage of a revised Atomic Energy Act made it possible, for the first time, to disseminate information about reactor technology to private industry as well as to the world at large. This change coincided with the start of President Eisenhower's Atoms for Peace program, which expressed the desire of the United States to make available the civilian benefits of atomic energy on a global basis. Great impetus to the development of a civilian reactor industry came from the First United Nations International Conference on the Peaceful Uses of Atomic Energy, which took place in Geneva in the fall of 1955 (United Nations, 1956).

The 1955 conference was followed by the entry of many private companies into the business of designing and building reactors. It was generally recognized that it would take another decade for nuclear power to become economically viable, but meanwhile markets were developing for research reactors as well as for experimental "demonstration power reactors" that were financed in part by the U.S. government in cooperation with electrical utilities and manufacturers. In parallel with the civilian program, a joint AEC–U.S. Navy program to develop nuclear propulsion was also proceeding rapidly, and the first nuclear power submarine, the Nautilus, was launched in 1954. The use of nuclear power for naval propulsion was highly successful and accelerated the development of civilian nuclear power in many ways.

The use of electricity in the United States had been increasing at an average rate of about 6% per year for several decades prior to the 1970s, necessitating that the installed generation capacity be doubled every 10–12 years. The economics of nuclear power generation tended to favor construction of large plants, and as the plants increased in size (from 200 MWe in 1950 to 1000 MWe in 1970), nuclear power became more desirable from an economic point of view. In addition, national concern with increasing levels of air pollution in many communities during the 1960s and early 1970s favored installation of nuclear units, as did the increasing costs of fossil fuels and uncertainties in regard to the long-range availability of these fuels.

In 1963 the New Jersey Central Power Company announced that it was purchasing a General Electric Company 620 electrical megawatt (MWe) reactor. This was the first power reactor to be purchased in the United States purely on economic grounds and the decision heralded the birth of a new major industry. There followed a 15-year period of growth for the nuclear industry until the mid-1970s, when installation of additional nuclear capacity came to a sudden stop due to a number of interacting factors, of which only one was the accident at Three Mile Island Unit II, operated by the General Public Utilities Corporation near Harrisburg, Pennsylvania. (see Chapter 14). The accident resulted in about $1 billion in damage to the power plant and increased the already distrustful attitude of the public toward nuclear power. However, the accident occurred during a period of worldwide economic recession that had resulted in a reduction in the rate of growth of electrical demand, in both the United States and other countries. To complicate things further, interest rates in the United States rose precipitously during that period, which affected the economics of nuclear power plants because they are much more capital-intensive than power plants that use fossil fuels. The more stringent regulatory requirements imposed after the TMI accident also increased the cost of nuclear power. Finally, the fact of the TMI accident, the reduced demand for electricity, and the higher costs of nuclear power plant construction were all seized on by antinuclear individuals and organizations to further their opposition to nuclear power, causing further delays in licensing, which further increased the cost of the nuclear plants.

As of the beginning of 1986, no new nuclear plant has been ordered in the United States since the accident at TMI, and orders for about 100 plants were canceled, some of which were nearing completion. Whether there is a viable future for nuclear power in the United States remains to be seen (Manning, 1985). One must bear in mind that production of electricity by fossil fuels is also beset with problems, which, in contrast to those of the nuclear industry, may prove to be more difficult to solve. The political and economic problems associated with petroleum production and the environmental problems caused by fossil fuel combustion may be far less tractable in the long run. While renewable sources of energy such as solar and geothermal seem attractive, they do not seem to be viable options for the foreseeable future.

Despite its setbacks, nuclear power production is a major industry at the present time. By 1985 there were 528 nuclear power plants either in operation or under construction in the world (*Nuclear News*, 1985). A total of 3500 civilian power reactor years had been accumulated. Although the United States as a whole was generating only 13.5% of its electricity from nuclear fuel, nuclear power accounted for about 50% of production in New England and parts of the Middle West. France generated 58.7% of its electricity from nuclear fuel in 1984, and five countries (France, Belgium, Finland, Sweden, and Switzerland) generated more than 30% from nuclear fuel (IAEA, 1985).

The health physicist, physician, engineer, or health officer concerned with the effects of reactors on the environment should have a general understanding of those aspects of design that may affect the kind and quantity of radioactive effluents discharged to the environment under normal and abnormal conditions. It is also important to appreciate the basic methods used to prevent uncontrolled release of radioactivity to the environment and to understand the monitoring activities required to ensure compliance with the applicable limitations on discharges of radioactive wastes. Finally, in the event of an accidental release to the environment, the environmental specialist should be prepared to advise on the measures that can be taken to minimize the consequences of the release. The discussion that follows is intended to provide the general reader with some of the basic aspects of reactor design that affect safety of operation. The reader who wishes to pursue the subject more comprehensively is referred to several excellent texts on the subject (Marshall, 1983; Weinberg and Wigner, 1958; Thompson and Beckerly, 1964; Lamarsh, 1966, 1975; Glasstone and Sesonske, 1980; Okrent, 1981).

Some Physical Aspects of Reactor Design and Operation

Contemporary reactors, with only a few exceptions, use either natural uranium or uranium in which the amount of isotope 235 has been enriched. The amount of ^{235}U enrichment may vary from 0.7% (natural) to more than 90%, but the fuel in most civilian power reactors is enriched to about 3% ^{235}U. Plutonium can also be used as a fuel for reactors, as will be discussed later in this chapter.

Because of a number of difficulties with using uranium metal, the fuel material is uranium dioxide in most present-day reactors. Among these problems is the fact that the metal tends to swell after intense neutron bombardment. The metallic uranium is also very active chemically, which presents several problems, among which is its ability to react exothermically with both air and water. Uranium dioxide is a ceramic which melts at about 5000°F, but has the disadvantage of being a rather poor heat conductor, so the heat transfer requirements necessitate that the rods have rather small diameters, typically ½ to ¾ in.

The fuel may be fabricated as rods, pins, plates, or tubes, protected by a cladding whose function is to prevent the escape of fission products and protect the fuel from the eroding effect of the coolant. The cladding may be zirconium, stainless steel, or other special alloys. In most power reactors, the fuel is in the form of sintered UO_2 pellets less than 0.5 in. in diameter and about 1 in. long. The pellets are aligned within tubes of zircalloy or stainless steel tubes about 12 ft in length,[1] and the tubes are

---

[1] The UO_2 pellets are mounted within the tubes in such a way that helium-filled spaces exist. The helium is intended to improve heat transfer from the fuel to the cladding. These spaces serve as a plenum that accumulates the volatile fission products such as the halogens and noble gases that diffuse from the fuel.

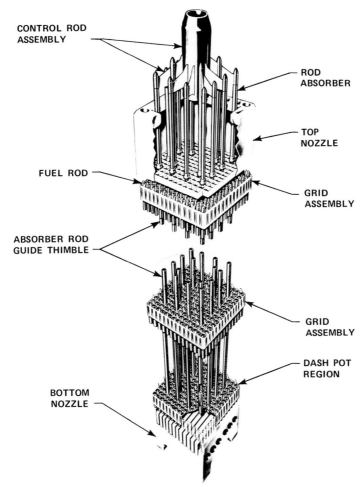

CONTROL ROD ASSEMBLY

ROD ABSORBER

TOP NOZZLE

FUEL ROD

GRID ASSEMBLY

ABSORBER ROD GUIDE THIMBLE

GRID ASSEMBLY

DASH POT REGION

BOTTOM NOZZLE

Fig. 9–1. Cutaway of a fuel and control rod assembly. (Courtesy of Westinghouse Electric Corporation.)

arranged in bundles, as shown in Fig. 9–1; within the bundles, at selected positions, control rod guide thimbles replace the fuel pins.

Fission results from the capture of a neutron by the nucleus of a fissionable atomic species. Because more than one neutron is released in the process, multiplication of neutrons may be achieved; this allows additional atoms of uranium to be split, which in turn yield additional neutrons to continue the multiplicative fission process.

Some of the neutrons produced by fission will escape from the reactor system and be lost. This can be minimized by surrounding the reactor core with a reflector which tends to scatter escaping neutrons back into the system. Other neutrons may be captured by the nuclei of nonfissionable materials in the reactor system; this can be minimized by selection of materials that have a low capture cross section[2] for neutrons and by careful control over high-cross-section impurities in the materials from which the system is constructed. The nuclear reaction becomes self-sustaining when, for every atom that fissions, one fission-producing neutron remains after allowing for escape by leakage or loss by nonfission capture. At this point the reactor is said to be in a "critical" condition.

The neutrons produced in fission are relatively energetic, and in most commercial power reactors the probability that they will produce fission is increased by reducing their energies to less than 0.1 eV. This is accomplished by distributing the fuel in discrete components between which is placed a moderator of low atomic number. The neutrons are reduced in energy as a result of elastic collision with moderator atoms. Most reactors are designed for thermal neutrons and are called "thermal reactors." A fast reactor has no moderator and depends for its operation on the production of fission by fast neutrons (i.e., energies greater than 100 keV). Almost all power reactors use water to reduce ("moderate") the neutron energies to the thermal region. A limited number of gas-cooled reactors that use graphite as a moderator have been built, principally in the United Kingdom.

The state of criticality of a reactor is expressed by k, the neutron multiplication factor, which is the ratio of the number of neutrons produced by fission in any one generation to the number of neutrons produced in the preceding generation. When criticality exists, k equals unity and $dn/dt = 0$, where n is the neutron flux, usually expressed as neutrons per square centimeter per second. The heat produced in a reactor, and hence its power level, is directly proportional to the neutron flux.

When k is less than unity, the reactor is subcritical, and a chain reaction cannot be sustained. When k is greater than unity, the reactor is supercritical, $dn/dt > 0$, and the power tends to increase with time. Unless the k is reduced to ≤ 1, the heat produced would increase to a level that would destroy the reactor. Conversely, when $k < 1$, the power will diminish and cannot be sustained at any given level without improving the neutron economy. This is usually done by adjusting the position of neutron-absorbing control rods.

A reactor is designed so that it is possible to maintain $k = 1$ for various

---

[2]The cross section is a quantitative expression of the probability of occurrence of a given reaction between a nucleus and an incident particle. The unit of cross section is the barn, equal to 10^{-24} cm^2/nucleus. The cross section of each nucleus for both fission and capture depends on the energy of the incident neutrons.

neutron densities corresponding to the desired power levels. To increase the power level, k is made slightly greater than unity for a brief period of time. An important parameter that describes the neutron kinetics at any given time is the reactivity, ρ, defined as

$$\rho = (k - 1)/k \qquad (9\text{--}1)$$

The control rods of a thermal reactor contain neutron-absorbing materials such as cadmium, indium, boron, or hafnium. Insertion of the rods into the core reduces the number of neutrons available for fission, thus reducing the value of ρ, which causes the power level to diminish. Conversely, the power level increases when the rods are withdrawn from the core.

The reactivity of a reactor is also affected by its temperature and the radiation history of the core. The effect of core temperature on reactivity is an important characteristic of a reactor and determines its ability to self-regulate should the power be increased inadvertently. The overall effect of temperature on reactivity is the result of a number of factors. For one, an increase in temperature results in increases in nonfission capture cross sections. Another effect is that the coolant density is decreased by a rise in temperature, and this affects reactivity by reducing the moderating ability of the coolant and increasing the number of neutrons that escape from the reactor core or are removed from the system by nonfission capture. Another way in which temperature may reduce reactivity is by causing vapor bubbles to form in the liquid coolant–moderator. A negative temperature coefficient of reactivity is a basic requirement for power reactor safety and serves to self-regulate the reactor. With a positive coefficient, dangerous instabilities would result from the fact that any power excursion that caused an increase in core temperature would cause a further increase in power and temperature because of increased reactivity and so on, until the destruction of the reactor (Lamarsh, 1975).

Reactivity is lost as the core ages owing to fuel burnup and because some of the accumulating fission fragments or their decay products have a high cross section for thermal neutrons and therefore increase the fraction of neutrons lost by capture. Xenon-135 and ^{149}Sm are particularly important in this regard. The concentration of ^{135}Xe increases for several hours after reactor shutdown and may reduce reactivity to such an extent as to prevent restart of the reactor for a day or more.

Thus, because of the effects of temperature and fission-product poisons, a hot, aged core is less reactive than the cold, fresh core, which requires that there should be a compensating loss in reactivity at start-up. In most power reactors, this is accomplished by the use of control rods. Another technique is the use of ^{10}B, which has a high capture cross section for thermal neutrons and which can be used to reduce reactivity when added in small amounts to the reactor core. The boron concentration in the cooling water can be controlled by the use of an ion exchanger. This technique is used in most pressurized-water reactors (PWRs), in which the boron concentration is generally lowered as the uranium in the fuel is used up. In the case of boiling-water reactors (BWRs), one cannot use dissolved boron because it leaves deposits on the fuel as the water boils. Instead, the poisons are fixed in solid form and are gradually depleted by neutron capture at just the right rate to correct for the loss of reactivity due to fuel burnup. This system is called the burnable poison method.

When the control rods are removed, the value of k increases from $k < 1$ to $k > 1$ and the power level begins to rise. At the desired flux density (power level) the rod positions are adjusted until $k = 1$, and the steady-state critical condition is achieved.

The rate at which the power will rise or fall depends on the amount by which k is greater or less than unity. If n neutrons are present per unit volume at the beginning of each generation and L is the generation time (or average neutron lifetime), one can roughly approximate the change in the number of neutrons as:

$$(dn/dt) = n \ (k - 1)/L \tag{9-2}$$

Integrating, we obtain $n = n_0 \exp t(k - 1/L)$, where n_0 is the initial neutron flux (and thus the initial power level) and n is the flux at any time t.[3]

If we call the ratio $L/(k - 1)$ the reactor period T, so that $n = n_0 \exp t/T$, then T is equal to the time it takes to increase the power by a factor of e, sometimes called the e-folding time.

Under most conditions, L is constant, being characteristic of the core. The period is dependent solely on $k - 1$, and reactor power will rise or fall in an exponential manner, depending on whether $k - 1$ is positive or negative.

The above relationships show that the reactor period is sensitive to changes in the generation time L. The average time that elapses between the production of a fission neutron and its ultimate capture is about 10^{-3} sec in a natural uranium reactor. Assuming a value of $k - 1 = 1.005$ and a value of $L = 10^{-3}$ sec, the power level in each second would increase by a factor of about 150. This would result in the reactor being uncontrollable.

Fortunately, production of a small fraction of the fission neutrons is delayed. They result from the beta decay of certain fission products. The size of this fraction depends on which fissionable material is involved and

---

[3]This brief description of basic reactor kinetics is deliberately oversimplified. The reader is referred to advanced texts on reactor theory for a more exact discussion (Lamarsh, 1966).

the energy of the fission-causing neutrons, and varies from 0.0024 for ^{233}U to about 0.007 for ^{235}U. The delayed fraction actually controls the value of L, which is about 0.1 sec for ^{235}U. The delay time itself varies—some neutrons are delayed by many tons of seconds, other only a few milliseconds. The mathematics for dealing with these delayed neutrons is too complex to develop here, but they have a controlling effect on the rate of power increase for cases where $1 < \rho < 1 + \beta$, where β is the fraction of fission neutrons delayed. In this range of values for ρ, criticality depends on the delayed neutrons, and they greatly slow down the rate of power increase from a doubling time of about once every 0.14 sec to a doubling time of many seconds, depending on the exact value of ρ. However, should ρ exceed $1 + \beta$, then the reactor will become critical without the delayed neutrons. In such a case the reactor is said to be prompt critical and the rate of increase is limited only by the time required for the fission neutrons to thermalize (10^{-3} to 10^{-4} sec). A basic objective of reactor design is that a prompt critical condition must not be possible.

The reactor designer must select materials that have the desired nuclear, thermal, and structural properties. Corrosion must be minimized, because corrosion products become radioactive in passing through the reactor and complicate its operation. The effects of intense radiation on the physical properties of materials must also be taken into consideration.

One of the objectives of design is to contain the fission products within the fuel elements by maintaining the integrity of the cladding under normal operating conditions. However, should cladding failure occur, there must be provision for operating safely, even though some fission products may escape from the fuel into the coolant.

The reactor must be equipped with sufficient instrumentation that the operator can know when the system is not functioning properly and take whatever remedial measures are required. When prompt response is essential, the information obtained by instruments is fed into electrical systems, which are designed to take corrective action automatically. The challenge to the designer is to provide a plant that will produce power efficiently and yet at all times be under complete and safe control.

The choice of materials for the core, coolant, and reactor structure must be such as to minimize the possibility of exothermic reactions. Many of the materials that are potentially useful in reactor construction react exothermally with each other under certain conditions. Examples include metallic uranium, sodium, and zirconium, all of which react exothermally with water. Some of the energy that becomes available during a serious reactor accident may be due to chemical reactions between these materials.

One of the most fundamental estimates which must be made in the course of design is the temperature of the hottest channel in the core.

Systematic peaking in both the radial and axial directions does occur and can be predicted by calculation and experiment. The product of the radial and axial power-peaking factors can be used to determine the amount of power that will be produced in the hottest part of the core, assuming that the peaking factors in both directions coincide. In addition to this type of peaking, which is due to design factors, it is necessary to consider variations from the average power density that will result from small manufacturing deviations in fuel-rod diameter, the amount of enrichment, fuel density, the extent to which fuel is out of round, and other similar factors. The most conservative approach is to assume that the least favorable of all such deviations coincide spatially. The product of all such factors gives a so-called hot channel factor, which becomes a fundamental limitation in thermal design. The amount of power that can be produced by the reactor is limited by the temperature of the hot channel and the requirement that failure of the cladding be avoided.

In a pressurized water reactor, boiling does not normally occur but may be present should the temperature rise sufficiently. A moderate amount of boiling is not deleterious and may, in fact, increase the amount of heat transfer from the fuel. However, if the boiling, instead of being nucleate, produces a vapor film along the surface of the fuel element, heat transfer can be greatly reduced, and the temperature of the fuel and its cladding may rise dangerously. The reactor is designed so that the temperature in the hot channel does not approach the temperature at which departure from nucleate boiling (DNB) occurs.

The severe thermal and radiation environment within a reactor can affect reactor materials by producing changes in such physical properties as thermal conductivity, resistivity, hardness, and elasticity. In addition, dimensional changes in some materials may occur, and in graphite these changes may be associated with the storage of relatively large amounts of energy, which under certain conditions can be released as heat. This phenomenon was responsible for an accident to an air-cooled, graphite-moderated reactor at Windscale in the United Kingdom in 1957 (see Chapter 14).

More than 80% of the energy produced within the core originates from the kinetic energy of fission fragments. The balance originates mainly from particulate radiations, the range of which is of the order of 10^{-3} cm. Thus, most of the energy is converted to heat near the point of fission within the fuel. About 90% of the energy produced by the fission is absorbed within the fuel. About 4% is absorbed within the moderator, some of the energy (mainly in the form of neutrinos) is lost to the system, and the rest is absorbed within the reactor structure.

Whether or not the changes produced by a given radioactive environment are deleterious to materials can be answered by subjecting test specimens to conditions simulating those expected in the reactor. The fuel, cladding, welding materials, and various pieces of hardware that may be included in fuel-element design are particularly important. For example, irradiation may affect the rate at which fission gases such as xenon and krypton are released from the fuel into the helium-filled gap that in light-water reactors separates the fuel and the cladding. This could affect the rate of heat transfer to the cladding. The effect of irradiation on the integrity of the cladding must also be studied. If the physical properties are changed so that the cladding erodes or otherwise becomes unable to withstand the rigors of the thermal and radioactive environment within the core, fission products may be released. These and other similar questions can best be answered by actual irradiation of test specimens.

During every phase of reactor design the engineer and physicist must examine each decision for its possible effect on the safety of the system. Decisions regarding materials, dimensions, equipment, operating temperatures, and other aspects of design must be carefully reviewed.

Types of Reactors

Reactors can be constructed to serve as sources of either radiation or heat. Included among those constructed as radiation sources are: (1) "production" reactors in which products of neutron irradiation such as plutonium or ^{60}Co are produced, (2) research reactors such as those located on university campuses and other research centers, and (3) industrial-type test reactors that are used to study the effects of radioactivity on materials of construction and equipment components. The reactors used as sources of heat are used primarily as sources of power for electric generators but may also be used to generate steam for space or process heat. The discussion here will be limited to power reactors.

LIGHT-WATER REACTORS

With but few exceptions, the power reactors constructed by U.S. power companies have been of the so-called light-water type, in which the fuel is enriched to about 3% and water is used as both moderator and coolant. Prior to about 1963, when the light-water reactors emerged as economically viable and operationally practical, a number of other systems were demonstrated, some in the laboratory and some as power generators. These included reactors in which the fuel and coolant were mixed homogeneously, reactors that were moderated and cooled by organic fluids, and other types, all of which gave way to the light-water reactors, which were

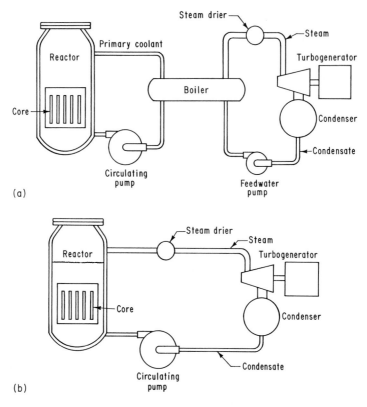

Fig. 9–2. (a) Schematic flow diagram of pressurized water reactor. (b) Schematic flow diagram of boiling-water reactor. (c) Schematic flow diagram for large, high-temperature gas reactor (Gulf General Atomic). (d) Schematic diagram of a liquid-metal fast breeder reactor.

adopted for use by the U.S. Navy. The LWRs include two types, the pressurized water reactor and the boiling water reactor.

In the PWR (Fig. 9–2a), the reactor core is enclosed in a 6- to 8-in. steel pressure vessel through which water is circulated, absorbing the heat being produced. This cooing loop is maintained under pressures in excess of 2000 psi and is called the primary system. The water does not boil because of the high pressure, but passes through the heat exchanger to produce steam in the secondary loop (Fig. 9–2a), which is maintained at a lower pressure to permit boiling. The steam thus produced is pumped through a steam dryer and then to a turbogenerator, from which the steam tailings are condensed and returned to the boiler (steam generator). Thus, in the PWR the steam used to drive the turbogenerator does not pass

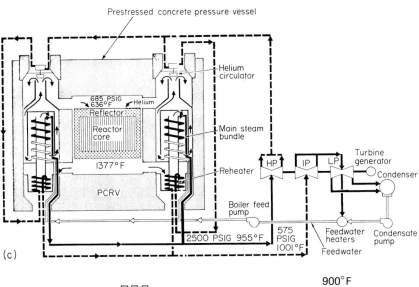

(c)

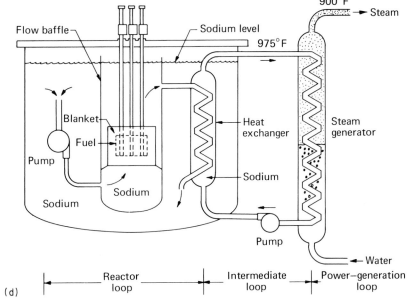

(d)

Fig. 9–2 (*continued*).

through the reactor but receives its heat via the steam generator. Radioactivity is retained in the primary system except to the extent that there is leakage to the secondary system.

In the boiling-water reactor (Fig. 9–2b) the water passing through the reactor core is allowed to boil, and the steam passes through a dryer directly to the turbogenerator, from which the steam tailings are condensed and returned to the reactor.

HIGH-TEMPERATURE GAS-COOLED REACTORS

The British (Kaplan, 1971; Goodjohn and Fortescue, 1971) have in the past placed considerable emphasis on the development of high-temperature gas-cooled reactors (HTGRs) and have had a number of such reactors in operation since the mid-1950s. This system has received relatively little attention in the United States, but one HTGR has been operated since 1979 by the Public Service Company of Colorado at Fort St. Vrain.

In the HTGR, the reactor core is constructed from hexagonally shaped graphite fuel elements within which the fuel is contained as rods of uranium or thorium carbide. The latter would serve as a fertile material to breed ^{233}U, which is fissionable. As shown in Fig. 9–2c, the reactor is cooled by helium maintained under pressure. One characteristic of the HTGR is that it operates at higher temperatures and pressures and is, therefore, more efficient thermodynamically and discharges less waste heat to the environment per unit of power generation.

Reactors can also be designed to operate without a moderator, using fast neutrons to initiate fission. Among the advantages of the fast neutron reactor is that no moderator is required, and hence it is possible to design cores that are small in comparison with thermal reactors. In addition, the capture cross sections of most substances for fast neutrons are relatively low, permitting a wider choice in the selection of construction materials. The fission products also have low capture cross sections for fast neutrons, and greater burnup of fissionable material is, therefore, possible because certain fission products do not act as poisons in fast reactors. Finally, because the nonfission capture cross sections of fast neutrons by the fertile material are relatively high, the fast reactor is inherently favorable for breeding fissionable ^{239}Pu from ^{238}U.

One of the disadvantages of the fast reactor is the fact that the selection of a coolant is restricted to those having no moderating effect. Liquid sodium, which is frequently used for cooling fast reactors, has obvious disadvantages because of its chemical reactivity with air and water, but this is offset by excellent heat transfer properties and the fact that high pressures are not required when sodium is used as a coolant. The small size of the liquid-cooled fast reactor reduces the surface areas available for heat transfer and, thus, further limits the choice of coolants to those having high conductivity and specific heat.

Because of the favorable neutron economy referred to earlier, it is possible to design fast reactors with relatively small amounts of excess reactivity, but an extensive core meltdown might result in a configuration having a higher reactivity than the original configuration. Should sufficient melting occur, it is possible that the molten fuel could rearrange itself in such a way as to produce a critical mass. Although this can probably be avoided by proper design, this factor has led to a great deal of conservatism in regard to the use of fast reactors for civilian power (Okrent, 1965). However, the French are about to put a 1200-MWe sodium-cooled power reactor into operation.

THE BREEDER

The breeder reactor is one in which some of the neutrons that are not used for fission are absorbed in "fertile" nuclei such as ^{238}U or ^{232}Th, which transmute to ^{239}Pu and ^{233}U, which are fissionable. This results in a fuel economy in which more fissionable material is produced than is consumed.

Light-water reactors are inherently inefficient and convert only 1–2% of the potentially available energy of the uranium mined into heat. In contrast, a breeder reactor can economically use up to about 75% of the energy contained in uranium, thereby achieving fuel efficiencies about 40 times greater than those of light-water reactors (Seaborg and Bloom, 1970).

Of the various alternative ways of designing a breeding reactor, the so-called liquid metal-cooled fast-breeder reactor (LMFBR), has received the most attention in the United States. The LMFBR is illustrated schematically in Fig. 9–2d. It is cooled with liquid sodium in loop A, from which the reactor-produced heat is passed to an intermediate loop B, which, in turn, produces steam in loop C, which drives a conventional turbogenerator. The intermediate loop is required because the sodium becomes highly radioactive passing through the reactor, and the ultimate consequences of a possible sodium–water reaction in the event of failure of the barriers between loops B and A or C would be somewhat mitigated by the fact that the radioactive sodium in loop A would not come in contact with the water of loop C. The high boiling point of the sodium is an attractive feature that enables the reactor to operate at high temperatures and low pressures.

A conventional 1000-MWe LWR that uses uranium and recycled plutonium will, during a 30-year lifetime, use about 4000 metric tons of uranium and 5300 kg of plutonium. A similar sized LMFBR would use 200 metric tons of depleted uranium and 2300 kg of plutonium, but in addition to the power generated, it would produce 50 metric tons of depleted uranium and 7700 kg of plutonium. Prior to the worldwide reduction in the requirements for electricity that began in the mid-1970s, it was commonly believed that limitations in the supply of uranium would require development of the breeder reactor. Most of the countries that use nuclear power, including the Soviet Union, the United Kingdom, and, most notably, France, have included the development of breeder reactors in their programs. Following two decades of research and development, the United States decided to construct an advanced LMFBR to produce 380 MWe at Clinch River, Tennessee. However, considerable public opposition to the breeder concept developed, and Congress withdrew support for the program.

Low-Level Discharges from Light-Water Reactors

The radionuclides that accumulate in reactors are primarily those produced by fission within the reactor core and secondarily the activation products formed when traces of corrosion products and other impurities contained in the coolant–moderator undergo neutron bombardment in passing through the core. In this way, radionuclides of such elements as chromium, cobalt, manganese, and iron are produced. The fission product inventory is very much larger than the inventory of corrosion products, but the nature of reactor operation is such that the corrosion products may be present in relatively greater quantities in aqueous wastes.

Uranium oxide has the desirable characteristic that fission products are trapped efficiently within its crytsal structure, thereby minimizing their escape to the coolant. As noted earlier, a helium-filled gap between the fuel and the cladding contains volatile fission products (iodines and noble gases) that are not trapped within the fuel.

The inventory of each fission product can be calculated during any period of reactor operation for any given power level history, and their decay can be calculated during periods of shutdown or reduced power. If a reactor operates sufficiently long that equilibrium with the short-lived nuclides has taken place, the distribution of the radionuclides as a function of time after reacter shutdown is as given in Fig. 9–3. The total amount of radioactivity as a function of time after shutdown for various radiation histories is given in Fig. 9–4, in which it is seen that a 1-MW reactor will accumulate about 21 MCi of radioactivity after 500 days at full power. The principal radionuclides contained in a reactor core that has

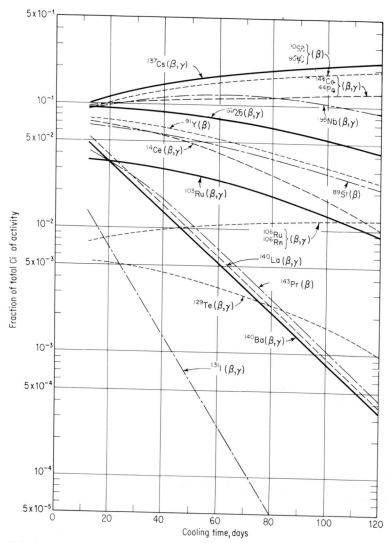

Fig. 9–3. Principal radionuclides in fission products at various times after reactor shutdown. It is assumed that the reactor has operated for an extended period and that an approximate equilibrium has been attained prior to shutdown. (From Glasstone, 1955.)

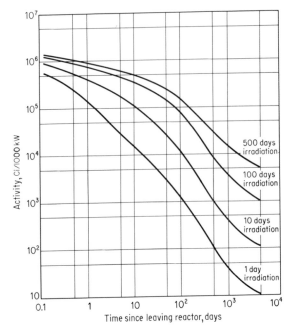

Fig. 9–4. Fission-product inventory in a reactor core after various periods of irradiation and shutdown. (From Parker and Healy, 1956.)

been shut down for 1 day subsequent to 2 years of continuous operation are given in Table 9–1.

EMISSIONS TO THE ENVIRONMENT

Light-water reactors during normal operation produce both gaseous and liquid wastes, some of which are discharged directly to the environment. These wastes originate from both the radionuclides produced by activation of elements in the primary coolant and the fission products. The fission products may originate from traces of uranium present on the surfaces of fuel elements or other reactor components, but the major source is leakage or diffusion through the fuel cladding. The radionuclides in the coolant system exist in gaseous form, as dissolved solids, and as suspended solids. Chemical-processing systems are provided that remove these radionuclides and concentrate them into a form for ultimate disposal (Fig. 9–5) (Blomeke and Harrington, 1968; Cottrell, 1974).

The wastes are accumulated in storage tanks and are then passed through a waste-gas stripper. Steam that is generated in the stripper scrubs incoming water free of gas as it passes downward through sections of

TABLE 9–1

INVENTORY OF SELECTED RADIONUCLIDES 1 DAY
AFTER 2 YEARS OF REACTOR OPERATION[a]

Selected isotopes	Half-life	Activity in fuel (kCi/ MWt)
^{3}H	12.3 years	0.0043
^{85}Kr	10.7 years	0.25
^{89}Sr	51 days	24
^{90}Sr	28.9 years	1.8
^{90}Y	64 hr	1.8
^{91}Y	58.8 days	32
^{99}Mo	66.6 hr	40
^{131}I	8.06 days	28
^{133}Xe	5.3 days	54
^{134}Cs	2.06 years	0.61
^{132}Te	78 hr	34
^{133}I	20.8 hr	22
^{136}Cs	13 days	0.74
^{137}Cs	30.2 years	2.4
^{140}Ba	13 days	46
^{140}La	40.2 hr	49
^{144}Ce	284.4 days	35

[a] National Academy of Engineering (1972).

porcelain saddles. The stripped gases, after passing through a condenser, are then routed to the waste-gas system. The effluent from the waste-gas stripper is passed to an evaporator, from which the vapors are condensed and passed to a demineralizer having cation and mixed-bed resins. The sludge from the evaporator bottom is passed to storage tanks, from which it goes to drumming stations. The effluent from the demineralizer passes to storage tanks, where it can be sampled. If it is of satisfactory quality, it can be discharged via the condenser discharge canal (Fig. 9–6), or it may enter the purification system for reuse in the primary system. In the event the quality is not satisfactory, the water may be returned for processing.

The induced activities that occur in the primary coolant depend on the materials of construction of the core, pressure vessel, pumps, piping, and other components in contact with the water. Induced activation of impurities ordinarily present in water is minimized by removal of the contaminants by water treatment before the coolant is introduced into the reactor.

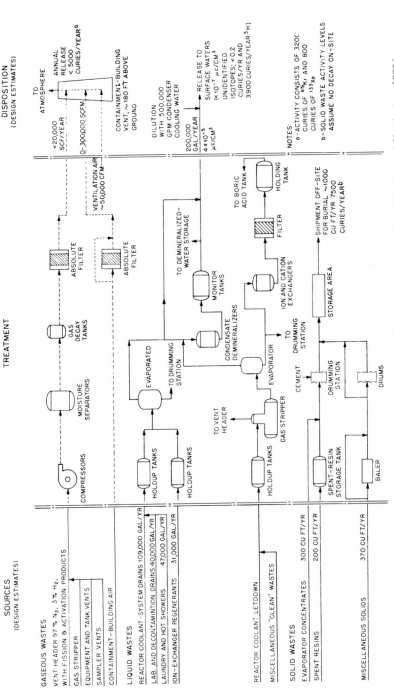

Fig. 9–5. Waste management flow diagram for a large pressurized-water reactor. (From Blomeke and Harrington, 1968.)

Fig. 9–6. Condenser discharge for a pressurized-water reactor. Low-level liquid wastes can be released into this tunnel, through which condenser coolant is flowing at a rate of 300,000 gal/min. The pipe at the left edge of the canal is a sampling manifold. Sampling intake pipes can also be seen. (Courtesy of Consolidated Edison Company.)

Stainless steel is an important material in reactor construction and results in formation of a number of nuclides that find their way into the coolant. These include ^{60}Co, ^{59}Fe, ^{51}Cr, ^{54}Mn, and ^{55}Fe, all of which have moderately long half-lives and ultimately present a waste-disposal problem in addition to being an operational problem because of their γ activity.

In-pile tests of uranium oxide fuel have shown that the various radioac-

TABLE 9–2

ESCAPE-RATE COEFFICIENTS[a]

Elements	Escape-rate coefficient (sec^{-1})
Cs, I, Xe, Kr, Rb, Br	1.3×10^{-8}
Sr, Ba	1.0×10^{-11}
Zr, Ce, and rare earths	1.6×10^{-12}
Te	1.0×10^{-9}
Mo	2.0×10^{-9}

[a] U.S. Atomic Energy Commission (1960a).

tive elements diffuse from the fuel at different rates, depending on their relative mobility. Table 9–2 summarizes the escape-rate coefficients measured in pressurized-water reactor fuel. Elements such as cesium, iodine, xenon, krypton, rubidium, and bromine have escape-rate coefficients that are a factor of 10^3 greater than those of strontium and barium.

LIQUID WASTES

The fission and activation products that diffuse from the fuel into the circulating water can be removed continuously by means of a purification system that is designed to have sufficient capacity to allow continued reactor operation in the event of minor failures in 1% of the fuel rods.

Tritium is produced in light-water reactors in quantities that are relatively copious compared to other radionuclides. Fortunately, tritium is a pure β emitter of very low energy, which ordinarily enters the environment in the form of water. It does not concentrate significantly in biological systems and has a relatively rapid turnover rate. The maximum permissible concentration (MPC) in drinking water is relatively high, and large amounts of tritium can be discharged to the environment without exceeding the permissible levels of human exposure (Moghissi and Carter, 1973; NCRP, 1979).

The tritium originates in two ways. Albenesius (1959) first demonstrated that tritium is produced in fission at a rate of about 1 atom per 10,000 fissions. Tritium is also produced by spallation following neutron irradiation of ^{10}B, a major source of tritium in reactors that use ^{10}B as a burnable poison.

Differences in the diffusion rates of the various radionuclides from intact fuel cladding cause the coolant radioactivity in most reactors to be

TABLE 9–3

ANNUAL AVERAGE RADIONUCLIDE COMPOSITION
RELATIVE TO THE CONCENTRATION OF ^{137}Cs
FROM A TYPICAL PRESSURIZED-WATER
REACTOR

Nuclide	Relative concentration[a]
^{3}H	5.3×10^2
^{51}Cr	$<1.7 \times 10^{-2}$
^{54}Mn	2.8×10^{-1}
^{55}Fe	8.1×10^{-2}
^{59}Fe	$<2.4 \times 10^{-3}$
^{58}Co	2.1×10^{-2}
^{60}Co	1.1×10^{-1}
^{65}Zn	$<3.1 \times 10^{-3}$
^{90}Sr	6.6×10^{-4}
^{91}Y	2.6×10^{-3}
^{95}Zr-Nb	$<2.8 \times 10^{-2}$
^{103}Ru	$<3.1 \times 10^{-2}$
^{106}Ru	$<8.6 \times 10^{-2}$
^{131}I	$<5.2 \times 10^{-3}$
^{134}Cs	4.8×10^{-1}
^{137}Cs	1.0×10^0
^{144}Ce	$<3.8 \times 10^{-2}$

[a] ^{137}Cs $= 1.0$.

relatively depleted in less labile fission products such as ^{90}Sr or ^{89}Sr. However, in the event of fuel failure, direct exposure of the uranium oxide fuel to the coolant can increase the amounts of radiostrontium in the coolant. Nevertheless, the radioactivity will be dominated by the more soluble fission products such as Cs and I.

The exact composition of the liquid wastes from light-water reactors will vary from reactor to reactor, depending on the materials of construction and the condition of the fuel. Table 9–3 lists the relative radionuclide composition of the primary coolant of a pressurized-water reactor. The concentrations listed are relative to ^{137}Cs, which is the most labile of the nonvolatile long-lived fission products. The effect of the relatively low diffusivity of the strontium nuclides is shown by their low concentration relative to ^{137}Cs.

The light-water coolant can be purified continuously by drawing off a fraction of the coolant for treatment. Following filtration to remove suspended radionuclides, the water then passes through cation- and anion-exchange resin beds. If necessary, the coolant can be passed through a gas stripper in which the water is percolated over plates across which a countercurrent stream of steam is passed, which serves to remove dissolved gases such as air, fission gases, and hydrogen (Coplan and Baron, 1978; Moghissi et al., 1978a).

Apart from the coolant purification process, there are many other ways in which radioactive liquid wastes are produced. Leaks of coolant from valves, flanges, and pumps result in contamination of sump water. Components, which are removed for repair, must sometimes be decontaminated, and this will result in contaminated water, as will the operations of washing casks, sluicing resin beds, laundering contaminated clothes, and washing contaminated laboratory ware. Provision must be made for containment of these waste liquids and their treatment and ultimate disposal. The system for treating such wastes at a PWR where low-level contaminated water is accumulated for periodic processing is shown in Fig. 9–5.

GASEOUS WASTES

The methods of managing gaseous wastes from a PWR differ from the methods used for a BWR. In a PWR, the waste fission-product gases that are stripped from liquid wastes can be passed through a condenser, which removes the condensible portion, passing the remainder to holdup tanks and then to catalytic recombiners, which remove the radiolytic hydrogen. The residual gases can then be pumped to holdup tanks or can be stored on activated charcoal to permit decay, following which the gases can be passed to the exhaust stack through a high-efficiency filter. After 3 months of holdup the radioiodine has decayed and the remaining radioactivity is due mainly to ^{85}Kr, which can be released to the atmosphere or, in special circumstances, condensed cryogenically or removed for off-site disposal.

Other sources of PWR gaseous wastes are leaks from the primary system. Small quantities of gases from the primary coolant can leak directly to containment and be vented to the atmosphere when the containment building is purged to permit access by personnel. Small leaks can also develop in the heat exchanger, causing quantities of primary coolant to pass to the secondary loop. When this happens, radioactive gases can pass to the atmosphere without treatment via the boiler blowdown.

In a BWR, the gaseous fission products, mainly the noble gases, boil off with the steam, pass through the turbine, and then enter the condenser, from which they are removed by the air ejector. Leakage of air into the condenser occurs because it is normally operated under vacuum. Small quantities of air that leak into the condenser mix with radiolytic hydrogen and oxygen as well as the gaseous fission products. When the first BWRs began operation in the late 1960s, the air ejector vented to the environs through a stack that allowed a holdup time of only a few minutes, too short to permit decay of the short-lived noble gases. The amount of radioactivity in the gaseous discharges from BWRs have been greatly reduced by catalytic recombination of the radiolytic gases, which account for about 80% of the air ejector exhaust. Recombination of the oxygen and hydrogen thus result in a fivefold reduction in the gas volume, which slows the passage time through the exhaust system and provides time for additional noble gas decay.

Two short-lived radionuclides, ^{89}Kr ($T_{1/2}$ 3.2 min) and ^{137}Xe ($T_{1/2}$ 3.8 min), decay to ^{89}Sr and ^{137}Cs, which are trapped on high-efficiency filters. The gases in modern BWRs are then passed through refrigerated activated-charcoal beds, which remove the remaining radioiodines and noble gases, thereby allowing additional time for radioactive decay. Another option is to compress the gases for storage in tanks to allow sufficient time for almost complete decay of all but ^{85}Kr, which has an 10.8-year half-life. The gaseous discharges from a typical BWR for various levels of waste gas treatment is given in Table 9–4 (Collins *et al.*, 1978).

The noble gases are the main source of gaseous radioactivity from BWRs. Stigall *et al.* (1971) found that the inherent barriers to ^{131}I transport through a BWR system function with a high degree of efficiency and that ^{131}I is not a significant nuclide in the gaseous releases from this type of reactor.

RADIATION EXPOSURE OF THE PUBLIC FROM REACTOR EMISSIONS

The radioactive emissions from operating power reactors in the United States are routinely reported by the NRC (see, e.g., Baker and Peloquin, 1985; Tichler and Benkovitz, 1984). The emissions have resulted in insignificant doses to the general population and in many cases are less than the dose from natural radioactivity in the production of power plants fueled with coal (Chapter 7).

The doses received by members of the public from nuclear power reactor emissions are routinely a fraction of that permitted by ICRP and NCRP recommendations. The NRC regulations for emissions from power reactors were the first formal application of the ALARA principle by a regulatory agency, and are contained in Appendix I of Part 50 of Title 20 of the Code of Federal Regulations. The limits are based on what was found to be practicable in actual nuclear power plant operations; they require that the calculated dose to individuals beyond the site perimeter be less than 5 mrem/year from liquid discharges and that the dose from gamma radiation from gaseous emissions be less than 10 mrem/year. The limit from beta

TABLE 9–4

TYPICAL BWR ANNUAL RELEASES FOR 30-MIN HOLDUP, TREATMENT BY CHARCOAL DELAY, AND
TREATMENT BY CRYOGENIC DISTILLATION[a]

		Ci/yr per 3400-MWt reactor		
Radionuclide	Half-life	Base case (30-min holdup)	Charcoal delay	With cryogenic distillation
83mKr	1.9 hr	44,000	[b]	12
85mKr	4.5 hr	84,000	80	22
85Kr	10.8 year	290	290	280
87Kr	76 min	240,000	[b]	72
88Kr	2.8 hr	280,000	5	76
89Kr	3.2 min	2,800	[b]	60
131mXe	12 day	220	18	
133mXe	2.2 day	4,300	[b]	[b]
133Xe	5.3 day	120,000	460	13
135mXe	15 min	11,000	[b]	3
135Xe	9 hr	330,000	[b]	34
137Xe	3.8 min	9,700	[b]	39
138Xe	14 min	390,000	[b]	90
131I	8.0 day	5.0	[b]	0.0026
133I	21 hr	2.1	[b]	0.011

[a]From Collins et al. (1978).
[b]Less than 1 Ci/yr for noble gases, less than 10^{-4} Ci/yr for radioiodine.

radiation in the gaseous emissions is 20 mrem/year (U.S. Code of Federal Regulations).

The NRC issues annual reports that summarize the dose received by people who live between 2 and 80 km from nuclear power reactors. In 1981, when 71 nuclear power plants were operating at 48 sites in the United States, the mean individual dose commitment from all pathways ranged from a low of 1×10^{-5} mrem to a high of 0.05 mrem (Baker and Peloquin, 1985). All told, 98 million persons lived within the 2–80-km annuli, and their collective dose commitment was 160 person-rem from emissions during 1982 (Tichler and Benkovitz, 1984).

No estimates are given for people who live less than 2 km from the reactor because there are very few such individuals, and since the NRC annual limit of 10 mrem is readily met, their collective dose would be insignificant.

Reports that summarize emissions from the operation and maintenance

TABLE 9–5

COMPARISON OF DISCHARGES OF THE PRINCIPAL RADIONUCLIDES FROM INDIAN POINT I AND FALLOUT FROM WEAPONS TESTS (MEASURED IN CURIES)[a]

	^{90}Sr	^{54}Mn	^{137}Cs	^{3}H (tritium)
Annual discharge from Indian Point I (1968)	0.008	5.8	2.3	810
Fallout from weapons				
a. On Hudson River watershed (35,000 km²)	825	1236	1320	205,000
b. On Hudson River surface	3.7	5.5	5.9	920
c. On mixing zone of river, 16 km above and below plant	0.58	0.58	0.93	144

[a] For purposes of comparison, the year of heaviest fallout (1963) is compared to the year of maximum reactor discharge (1968). These data have been assembled from several sources.

of reactors on U.S. naval vessels are also issued periodically (Rice *et al.*, 1982).

There has been a gradual reduction in both liquid and gaseous emissions from the power reactors due to consistent improvement in fuel quality for all LWRs. Additional reduction in BWR emissions have resulted from installation of stack gas treatment systems such as those described earlier, which allow longer decay times before emission to the atmosphere. Despite the fact that large-scale testing of nuclear weapons had been terminated in 1963 (see Chapter 12), the man-made radioactivity of the Hudson River estuary was dominated through 1970 by radionuclides introduced from fallout, as can be seen from Table 9–5, which compares the annual discharges from the Consolidated Edison reactors at Indian Point to fallout from weapons tests in past years. The pressence of ^{137}Cs from such fallout is so predominant that the reactor contribution could only be detected by using the isotope ^{134}Cs as a tag. This nuclide is not present in bomb fallout but is formed in reactors by neutron capture in ^{133}Cs, the daughter of the fission product ^{133}Xe.

Systems are now available by which the liquid wastes can be so thoroughly purified that the decontaminated water can be returned via storage tanks for reuse in the primary system. Of course, this does not apply to tritium, which cannot be separated by any practical means and is returned with the treated wastewater to tanks, where it can be stored awaiting reuse in the reactor.

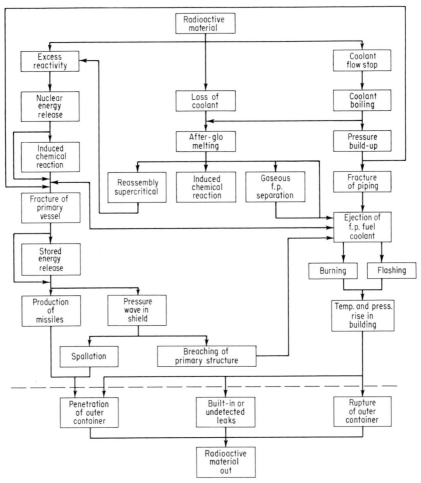

Fig. 9–7. The various factors involved in accidental releases of fission products from reactors. Not all factors will be involved in any one type of accident. (From Brittan and Heap, 1958.)

Reactor Accidents

Thus far the discussion of releases from light-water reactors has been limited to normal operating conditions. Malfunctions can, of course, develop, not all of which are abnormal any more than it is abnormal for an incandescent bulb to burn out or an electrical appliance switch to fail in the home. It is considered normal for the cladding of a small percentage of fuel elements to develop minor imperfections and for small boiler leaks to develop in PWRs. There are many other possible defects that can develop and must be anticipated and dealt with in design. Although they have a high probability of occurrence during the lifetime of the reactor, their consequences can be dealt with in routine fashion.

Accidents that might result from relatively common failures of relays or valves can be avoided by redundancy in design, high standards of quality assurance during manufacture of the components, or stringent inspection procedures during plant operation. However, certain catastrophic failures can be postulated that could have severe consequences, but for which it is not possible to provide absolute assurance that the radioactivity can be contained. In general, it has been shown that the probability that a given accident will occur is inversely related to the severity of its consequences.

TYPES OF REACTOR MISHAPS

A sequence of events by which one can analyze possible reactor accidents is given in Fig. 9–7, which illustrates three important types of accidents: excessive reactivity, loss of coolant, and coolant-flow stoppage. Although various combinations of the blocks can be assembled to describe many accidents resulting from malfunctions of the reactor, not all the blocks will apply in any one mishap.

The more common types of accidents that must be considered in the course of risk assessment include stuck control rods, loss of coolant, loss of station pumping power, and the sudden addition of cold water to a core. Each of these and other contingencies must be studied and its effect on the premanence of the reactor evaluated. The loss-of-coolant accident may result in very serious consequences and will be discussed in detail.

The principal accidents that are known to have occurred at nuclear reactors have been summarized by Bertini (1980).

RELATIVE HAZARDS FROM VARIOUS NUCLIDES

The fission fragments produced during reactor operation vary in mass number from 72 to 160 and include more than 80 isotopes produced in the frequency distribution shown in Fig. 9–8. It is seen that the yields of the mass numbers, which are plotted on a logarithmic scale, range from about $10^{-5}\%$ to nearly 10% (Walton, 1961).

The inventory of fission products, activation products, and transuranic elements in the reactor core can be estimated by computer programs for any specified operating or shutdown history (Steinberg and Glendenin, 1956; Bell, 1973). Although nearly 800 nuclides are produced, not all them are radioactive and others have such short half-lives that they are not

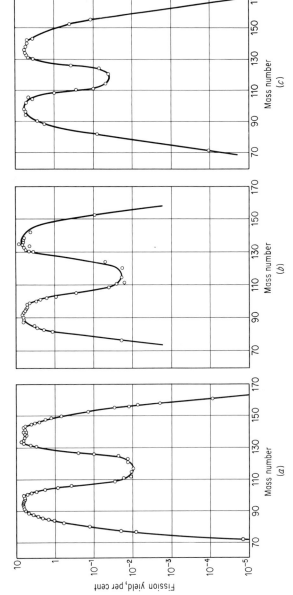

Fig. 9–8. Mass yields for slow neutron fission of (a) ^{235}U, (b) ^{233}U, and (c) ^{239}Pu. (From Steinberg and Glendenin, 1956.)

significant in risk assessment. The initial yields of the fission products are given in Fig. 9–8. If stable nuclides and those with half-lives less than about 26 min are excluded, the list reduces to the 54 nuclides shown in Table 9–6.

In the event of a meltdown, the relative contribution of each nuclide to the dose received by the population at various distances from the plant depends on the quantity present in the core, its volatility, its chemical and biological behavior once it enters the general environment, and the effectiveness of containment.

It has been widely accepted that the radioiodines (and the nuclides of tellurium, which decay to the radioiodines) are responsible for most of the dose received by persons downwind from the plant. It has been estimated that the nuclides of these two elements would be responsible for 83% of the dose received during the first day, as shown in Table 9–7. However, if the dose in integrated over periods of several years or longer, then ^{137}Cs dominates in many of the postulated accidents.

POTENTIAL CONSEQUENCES

There have been many investigations of the consequences of the "maximum credible reactor accident." Some of these studies were made early in the history of the World War II atomic energy program and led to great conservatism in the sites selected for the location of the early reactors. However, it must be remembered that when the first reactors were designed their performance could only be predicted from theoretical considerations, and time was required for operating experience that would demonstrate the validity of the underlying safety principles.

The early studies (Parker and Healy, 1956; Marley and Fry, 1956; U.S. AEC, 1957a) did, however, make it clear that the enormous inventory of radioactive fission products in a reactor core could cause catastrophic consequences if they were disseminated to the environment in a mishap. The most comprehensive early study was that undertaken by the AEC in 1957, "Theoretical Possibilities and Consequences of Major Accidents in Large Nuclear Power Plants" (U.S. AEC, 1957a). This study analyzed the consequences of destruction and volatilization of the core of a 500-MWt reactor in which the fission product inventory would be 4×10^8 Ci, measured 24 hr after an accident. The reactor was assumed to be located about 3 miles from a major city.

The study was performed at a time when power reactor technology was in its infancy and the first commercially owned nuclear generating station was still several years away. By defining the enormous potential consequences of dispersing a reactor core of this size, if no mitigating features

TABLE 9–6

INITIAL ACTIVITY OF RADIONUCLIDES IN THE CORE
OF A 3200-MWe NUCLEAR REACTOR[a]

No.	Radionuclide	Radioactive inventory	
		Source (Ci $\times$ 10^{-8})	Half-life (days)
1	Cobalt-58	0.0078	71.0
2	Cobalt-60	0.0029	1,920
3	Krypton-85	0.0056	3,950
4	Krypton-85m	0.24	0.183
5	Krypton-85	0.47	0.0528
6	Krypton-88	0.68	0.117
7	Rubidium-86	0.00026	18.7
8	Strontium-89	0.94	52.1
9	Stronitum-90	0.037	11,030
10	Strontium-91	1.1	0.403
11	Yttrium-90	0.039	2.67
12	Yttrium-91	1.2	59.0
13	Zirconium-95	1.5	65.2
14	Zirconium-97	1.5	0.71
15	Niobium-95	1.5	35.0
16	Molybdenum-99	1.6	2.8
17	Technetium-99m	1.4	0.25
18	Ruthenium-103	1.1	39.5
19	Ruthenium-105	0.72	0.185
20	Ruthenium-106	0.25	366
21	Rhodium-105	0.49	1.50
22	Tellurium-127	0.059	0.391
23	Tellurium-127m	0.011	109
24	Tellurium-129	0.31	0.048
25	Tellurium-129m	0.053	0.340
26	Tellurium-131m	0.13	1.25
27	Tellurium-132	1.2	3.25
28	Antimony-127	0.061	3.88
29	Antimony-129	0.33	0.179
30	Iodine-131	0.85	8.05
31	Iodine-132	1.2	0.0958
32	Iodine-133	1.7	0.875
33	Iodine-134	1.9	0.0366
34	Iodine-135	1.5	0.280
35	Xenon-133	1.7	5.28
36	Xenon-135	0.34	0.384
37	Cesium-134	0.075	750
38	Cesium-136	0.030	13.0
39	Cesium-137	0.047	11,000
40	Barium-140	1.6	12.8

TABLE 9–6 (*Continued*)

No.	Radionuclide	Radioactive inventory Source (Ci × 10⁻⁸)	Half-life (days)
41	Lanthanum-140	1.6	1.67
42	Cerium-141	1.5	32.3
43	Cerium-143	1.3	1.38
44	Cerium-144	0.85	284
45	Praseodymium-143	1.3	13.7
46	Neodymium-147	0.60	11.1
47	Neptunium-239	16.4	2.35
48	Plutonium-238	0.00057	32,500
49	Plutonium-239	0.00021	8.9×10^6
50	Plutonium-240	0.00021	2.4×10^6
51	Plutonium-241	0.034	5,350
52	Americium-241	0.00017	1.5×10^5
53	Curium-242	0.0050	163
54	Curium-244	0.00023	6,630

[a]From U.S. Nuclear Regulatory Commission (1975), Appendix VI.

TABLE 9–7

RELATIVE CONTRIBUTIONS OF THE MAJOR RADIONUCLIDES TO FIRST DAY DOSES AT 0.5 MILE

Radionuclide group	Curies (3000-MWt reactor)	Relative dose[a]
Noble gases	3.44×10^8	0.8
Iodines	7.15×10^8	54.8
Telluriums	1.76×10^8	28.8
Cesiums	0.152×10^8	1.0
Ceriums	3.65×10^8	6.2
Rutheniums	2.07×10^8	1.0
Others	33.25×10^8	7.4
Total relative dose		100.0

[a]American Nuclear Society (1984). Adapted from U.S. Nuclear Regulatory Commission (1975).

were incorporated into its design, the report served a useful purpose by pinpointing the kind of research that would be needed to determine which radionuclides would be disseminated in the event of an accident and the manner in which the releases could be contained.

In the 1957 study there were essentially two boundary conditions: (1) the partially volatilized reactor core is fully contained and no contamination of the environs occurs, and (2) 50% of the core is assumed to be volatilized without containment, resulting in restrictions in the use of 150,000 square miles of land on which almost 4 million people live. For the latter case, the cost estimates ranged up to $7.2 billion.

The assumption that 50% of all nuclides in the reactor core would be volatilized proved to be unrealistic, and as a result of laboratory investigations and studies of accidents that occurred, it was subsequently assumed that in the event of a loss-of-coolant accident in which no mitigating features were effective, volatilization of the core contents would be limited to 100% of the noble gases, 50% of the iodines (of which one-half would be released from the reactor building), and 1% of the remaining fission products. These are the assumptions that have been used for many years in assessing the consequences of power reactor loss-of-coolant accidents (10CFR100), but it will be seen later that even these assumptions appear to be unrealistically high.

A major contribution toward understanding the likelihood and consequences of a major reactor accident was the publication in 1975 of the Reactor Safety Study, directed by Norman Rasmussen of the Massachusetts Institute of Technology and sponsored by the Nuclear Regulatory Commission (U.S. NRC, 1975). The study was a probabilistic assessment of the risks associated with reactor accidents, and compared the probabilities and consequences of accidents of varying degrees of severity with the probabilities and consequences associated with other natural and manmade sources of risk. The findings of the study are shown in Figs. 9–9 and 9–10, where it is seen that at all levels of risk the probability of causing a given number of fatalities is orders of magnitude lower than for other manmade hazards such as death from aircraft accidents, dam failures, fires, and explosions. The likelihood of a given number of fatalities due to natural hazards is similarly shown to be orders of magnitude greater than the likelihood of death from a nuclear power plant accident for all sources except meteorites. The risk of dying because of a reactor accident, assuming 100 operating power reactors, was estimated to be 5×10^{-10} per year, compared, for example, to a risk of about 10^{-3} per year of dying as the result of other accidents. The conclusions of the Rasmussen report were at once hailed by the proponents of nuclear power as convincing proof of the safety of nuclear power reactors and regarded with suspicion by the oppo-

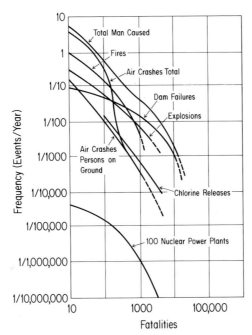

Fig. 9–9. Frequency of accidents of varying severity due to man-caused events. (From U.S. Nuclear Regulatory Commission, 1975.)

nents of nuclear energy. The NRC then constituted an expert panel under the chairmanship of H. W. Lewis of the University of California at Santa Barbara (Lewis *et al.*, 1979), with the charge to review the Rasmussen report and "clarify the achievements and limitations" of the study. The Lewis committee lauded the Rasmussen report as a substantial advance over previous studies and for its sophisticated use of probabilistic methods of analyzing the consequences of reactor accidents. However, the committee concluded that the error bounds of the probabilities given in the report were probably larger than stated, although one could not determine whether the probabilities assigned in the Rasmussen study were overstated or understated.

The Lewis report, like its predecessor, was also both praised and condemned: from the pronuclear point of view it was thought to be supportive of the Rasmussen conclusion, rather than a repudiation as claimed by others. It is of interest that a study, similar in purpose to the Rasmussen report, conducted by the German government came to conclusions that were quantitatively similar to those reported by Rasmussen (German Federal Minister of Research and Technology, 1979).

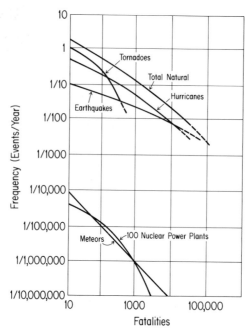

Fig. 9–10. Frequency of natural events that cause varying numbers of fatalities. (From U.S. Nuclear Regulatory Commission, 1975.)

The severity of a reactor accident in which core damage occurs is dependent on the extent to which radioactivity, mainly as ^{131}I and ^{137}Cs, is released to the environment. The 14 reactor accidents in which core damage occurred through the 45-year-old nuclear energy program are tabulated in Table 9–8, in which it is seen that significant quantities of ^{131}I were released only in the 1957 accident at Windscale in the United Kingdom and in the 1986 accident at Chernobyl. Some of these accidents will be discussed in Chapter 14.

The Windscale accident was different from the others in that it was a plutonium production reactor that was air-cooled, graphite-moderated, and fueled with metallic natural uranium. Part of the core was consumed in a fire, and 20,000 Ci of ^{131}I was released to the environs via the cooling air.

The Chernobyl reactor was also graphite-moderated, and had the disadvantage of having been built without some of the protective features that are routinely adopted in Western countries.

Following the 1979 accident at Three Mile Island, which will also be discussed in Chapter 14, a number of investigators began to question the

TABLE 9–8

REACTOR ACCIDENTS THAT INVOLVED CORE DAMAGE

Year	Location	Name of reactor	Type	Extent of contamination
1952	Canada	NRX	Experimental	None
1955	Idaho	EBR-1	Experimental	Trace
1957	United Kingdom	Windscale	Military production reactor	20,000 Ci ^{131}I
1957	Idaho	HTRE-3	Experimental	Slight
1958	Canada	NRU	Research reactor	None
1959	California	SRE	Experimental	Slight
1960	Pennsylvania	WTR	Research	None measured
1961	Idaho	SL-1	Experimental	10 Ci ^{131}I
1963	Tennessee	Orr	Research	Trace
1966	Detroit	Fermi	Experimental power	No release outside plant
1969	France	St. Laurent	Power	Little, if any
1969	Switzerland	Lucens	Experimental	None
1979	Pennsylvania	TMI-II	Power	Slight
1986	U.S.S.R.	Chernobyl-4	Power	Extensive

assumptions made concerning the release of iodine nuclides from a damaged core. It was noted that the fraction of ^{131}I released from the reactor building was not 0.25, which is the assumption required by the NRC, but about 3×10^{-9}. The releases of ^{131}I during other accidents were also known to be far less than assumed, and further laboratory experiments have in fact also suggested that the assumed release of ^{131}I from a damaged core was much too high.

A major inquiry into the fractional releases of radionuclides from damaged cores has been completed by a Special Committee of the American Nuclear Society (ANS, 1984). It was concluded that the releases of iodine and certain other elements have been overestimated by one or more orders or magnitude for many of the postulated accidents. A major reason for this discrepancy has been a failure to appreciate that reducing conditions exist within a reactor system during an accident, and this favors formation of CsI, a highly soluble nonvolatile compound. This greatly reduces the release of radioiodine in vapor form, with important effects on the dose received by surrounding populations. The ANS report also concluded that the tellurium release was overestimated because reactions with zirconium and stainless steel were not previously considered. If the results of the ANS study are accepted, the impact on regulatory requirements will be considerable, particularly in regard to siting restrictions and emergency evacuation planning.

ENGINEERED SAFEGUARDS

A number of engineered safeguards are provided by the reactor designer to prevent release of the radioiodines and other radionuclides to the general environment. As noted earlier, every effort is made to design the reactor system so that the probability of accidents that can result in core damage are minimized. However, having designed the reactor to achieve this objective, the designer then makes the assumption that core damage will take place nevertheless and that exposure to nearby populations must be controlled by engineered safeguards. Among the most important of these safeguards are those that are provided to avoid a massive release of fission products in the event of sudden structural failures of the primary coolant systems. This is the loss-of-coolant accident (LOCA).

A fundamental question is whether a massive failure of the piping or other components of the primary system is, in fact, credible (U.S. AEC, 1970b). There have been no such accidents in more than 40 years of experience with nuclear systems and, more important, massive failures are unknown in high-pressure central steam boilers, with which there is considerably more experience than is available in the nuclear industry. A study of some 500 boiler steam drums, designed for pressures over 600 psi and represetning 4000 boiler-years of operating experience, showed no failures of the steam drums themselves. Failures did occur in other parts of the high-pressure steam system, but they were not the massive type of failure that would cause a sudden release of coolant from a water reactor (Miller, 1966).

Should such a massive failure occur, the coolant would flash to steam and the core would quickly become subcritical from void formation due to boiling. In the absence of proper cooling, the fuel would overheat. At a temperature of about 2500°F, the zirconium cladding would react with steam and produce H_2, O_2, and heat. Without intervention, the fuel itself could melt. To limit overheating of the core, an emergency core cooling system is provided which would flood the core in the event of a LOCA and keep the temperature below that at which the cladding would fail.

An essential additional safeguard is the containment building designed to confine the steam and any entrained radioactive substances. The containment structure must be designed to withstand a variety of mechanical stresses, including the release of steam from the reactor system and the missile impact that could conceivably be produced (Gwaltney, 1969). Containment structures have become more complex since the reactors have become larger and have been built closer to centers of population. The first containment vessels were simple spheres, about 125 ft in diameter and fabricated from 1-in. welded steel plate. Subsequent containment vessels have been constructed of concrete, with linings fabricated from steel plate.

Means can be provided to condense the steam rapidly so that the pressure is greatly reduced, making it possible to use a less massive containment building. In one such system (Fig. 9–11), designed by General Electric Co., the steam would be vented into a water-filled pool, which would serve as a heat sink by condensing the steam. Another type of vapor suppression system, not widely used, employs an ice condenser in which the steam is condensed by venting through an ice-filled structure (Weems *et al.*, 1970).

Additional protection is provided by a system of sprays that can wash the radioiodine and other fission products from the containment atmosphere into a sump. Addition of sodium thiosulfate to the spray adds to the efficiency of the radioiodine scavenging process (ANS, 1971; Parsly, 1971).

Finally, the containment structure itself must be designed so that leakage is minimized. It is conventionally assumed that the containment atmosphere leaks at a rate of 0.01%/day, and tests must be performed regularly to prove the leak rate is smaller than this. Methods are available for testing the leak rates (Zapp, 1969).

Analysis of the loss-of-coolant accident and several other types of accidents and the design of safeguards intended to mitigate the effects of such an accident occupy much of the vast amounts of material that are required to be submitted to the NRC by an applicant for a power reactor license. During the past decade the technical information derived from laboratory study, pilot scale experiments, and theoretical analysis has been adapted to computerized methods of predicting the pressure and temperature transients involved in the various mishaps that might occur. This work has verified that the original assumptions are very conservative in almost every case. In a few cases in which the early practices were not conservative, new regulations were issued.

The wisdom of the methods used by the United States and other Western countries to protect against the effects of nuclear reactor accidents has been demonstrated by the contrast in the consequences of the accidents at Three Mile Island and Chernobyl. Both of these accidents are discussed in Chapter 14.

SITE SELECTION CRITERIA

Selection of a site for a power plant, whether it is to be nuclear or fossil, is a complex procedure on which many constraints are imposed (Gifford, 1974; Okrent, 1981). These include economic factors, availability of rights-

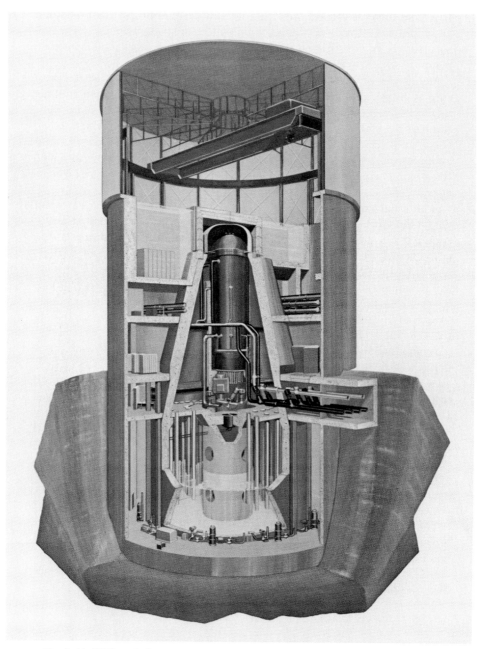

Fig. 9–11. BWR conical concrete pressure suppression containment. (Courtesy of General Electric Company.)

of-way for transmission lines, availability of a natural source of cooling water (or the feasibility of installing cooling ponds or towers), the geological and seismological characteristics of the region, the distance to the load center, and ecological or public health implications of atmospheric and liquid effluents from the plant. Until 1969, nuclear power plants were regulated exclusively by the AEC, whose authority was limited to control over the radiological hazards, and who determined whether the plant could be built without jeopardizing the public health and safety only insofar as radiological hazards are concerned. Other environmental factors were not then subject to federal regulations. However, with the passage of the National Environmental Policy Act of 1969 the picture changed drastically, and a federal agency with the authority to issue a license, of any kind, was henceforth required to examine the full spectrum of environmental consequences of the proposed action. A comprehensive analysis (environmental impact statement) must be submitted by the applicant to the licensing agency and reviewed by other agencies of the government concerned with the various relevant environmental effects. This chapter will review that part of the licensing process which is concerned with radiological effects.

The detailed reactor siting criteria used by the AEC are contained in Part 100 of Title 10 of the Code of Federal Regulations. The rationale for these criteria is given in technical document TID-14844, "Calculation of Distance Factors for Power and Test Reactor Sites," by DiNunno *et al.* (1962), the purpose of which is to provide a uniform approach to site evaluation. A more detailed discussion of the evolution of reactor siting practices is given by Okrent (1981).

According to Part 100, three zones surround the reactor:

1.   An "exclusion area" surrounding the reactor in which the reactor licensee will have the authority to determine all activities including exclusion or removal of personnel and property form the area.

2.   A "low population zone," which is the area immediate surrounding the exclusion area and which contains residents, the total number and density of whom are such that there is a reasonable probability that appropriate protection measures could be taken in their behalf in the event of a serious accident. The guide does not specify a permissible population density or total population within this zone because the situation may vary from case to case.

3.   A "population center distance," which is the distance from the reactor to the nearest boundary of a densely populated center containing more than about 25,000 residents.

Using these definitions, methods are proposed (DiNunno *et al.*, 1962) for calculating the size of the required exclusion area, low population zone, and population center distance for a contemplated reactor design. The distances should be selected so that the following criteria can be met in the event of an accident:

1. The exclusion area should be of such size that an individual located at any point on its boundary for 2 hr immediately following onset of a fission product release would not receive a total radiation dose to the whole body in excess of 25 rem or a total radiation dose in excess of 300 rem to thyroid from iodine exposure.

2. The low population zone should be of such size that an individual located at any point on its outer boundary who is exposed to the radioactive cloud resulting from the release would not receive a total radiation dose to the whole body in excess of 25 rem or a total radiation dose in excess of 300 rem to the thyroid from iodine exposure.

3. The population center should be located at least 1⅓ times the distance from the reactor to the outer boundary of the low population zone.

The NRC is careful to note that the doses of 25 rem for whole-body exposure and 300 rem for thyroid exposure are not to be construed as being permissible in the event of accidents. "Rather, this 25 rem whole-body value and 300 rem thyroid value have been set forth in these guides as reference values, which can be used in the evaluation of reactor sites with respect to potential reactor accidents of exceedingly low probability of occurrence, and low risk of public exposure to radiation."

Document TID-14844 defines the basic ground rules for estimating the amount of radioiodine available for leakage to the environment and the emergency doses to the surrounding population that should serve as the design criteria. Ever since publication of these criteria, reactor designers have been devising safeguards, some of which have been previously described, that have the effect of greatly reducing the doses permitted by the NRC regulations. An example of the progress that has been made is shown in Table 9–9, in which are summarized the calculated off-site doses under a variety of circumstances following loss of coolant in a large PWR. Whereas for design purposes 10CFR100 suggests a thyroid dose limit of 300 rem in the low population zone following a loss-of-coolant accident, the calculated thyroid dose would be less than 1 rem if a reasonable combination of safeguards were in operation.

Table 9–9 gives the dose estimates based on (1) the assumption that the radioiodine and noble gases released to the environment are contained in the gap between the fuel and the cladding and (2) the more pessimistic assumption of TID-14844, that 25% of the core inventory of radioiodine is released to the environment. Gap release would result from cladding failure and the TID-14844 assumption is based on failure of both the cladding and fuel.

TABLE 9-9

SUMMARY OF OFF-SITE EXPOSURE CALCULATIONS FOR LOSS-OF-COOLANT ACCIDENT[a]

Dose	2-hr Exposure at 520 m (minimum exclusion radius)	Total exposure at 1100 m (minimum low population zone radius)
I. Thyroid dose (based on zero to 5% of airborne as CH_3I)		
Containment leakage terminated in 1 min by isolation valve seal water system	0.7 rem	0.36 rem
Gap release[b]—continuous leakage with 2 spray pumps and 5 fan filters operating	0.8–1.45 rem	0.42–0.68 rem
Gap release[b]—continuous leakage with one spray pump and 3 fan filters operating	1.7–2.7 rem	0.85–1.4 rem
Gap release[b]—continuous leakage with 5 fan filters operating	8.8 rem	4.8 rem
Gap release[b]—continuous leakage with 2 spray pumps	0.95–6.7 rem	0.55–13.9 rem
10CFR100 suggested limit	300 rem	300 rem
II. Whole-body dose		
Containment leakage terminated in one minute by isolation valve seal water system	<1 mrem	<1 mrem
Gap release—continuous leakage	18 mrem	68 mrem
TID-14844 release—continuous leakage	3.8 rem	4.9 rem
10CFR100 suggested limit	25 rem	25 rem

[a] USAEC Docket 50, Exhibit B.
[b] TID-14844 initial iodine leakage inventory of 25% of core equilibrium quantity will result in thyroid dose 10 times value shown.

LICENSING PROCEDURES

Whether a given reactor design is suitable for a given site is evaluated under the rules promulgated in Parts 50 and 100 of Title 10 of the Code of Federal Regulations. The first formal application to the NRC must be accompanied by (1) a preliminary safety analysis report (PSAR) and (2) the environmental impact statement.

The PSAR is a comprehensive document of several volumes that includes a complete description and safety assessment of the site, whose hydrology, geology, meteorology, and ecology must be adequately documented. The analysis must include a preliminary design of the reactor with an analysis and evaluation of the design and performance of structures, systems, and components of the facility. A description of the quality assurance program to be applied in design, fabrication, construction, and testing must also be included, as well as a preliminary plan for the applicant's proposed organization and personnel training program. To the extent that full technical information is not yet available, the PSAR must identify the research and development programs that must be completed before the design can be finalized. Finally, the PSAR must include a preliminary plan for dealing with emergencies.

The PSAR is reviewed by the NRC staff and by a statutory committee of outside experts known as the Advisory Committee on Reactor Safeguards (ACRS) (Okrent, 1981). The staff and ACRS review is accompanied by many conferences with the applicant and a massive amount of highly technical and voluminous correspondence. If the reviews are favorable, and depending on whether or not the public participates in the proceeding, the commission can elect to hold a public hearing by an Atmoic Safety and Licensing Board (ASLB). In a few cases, public intervention at these hearings has been intensive and has delayed the licensing procedures.

If the ASLB rules in favor of the application, then a license to construct the reactor is normally issued by the NRC.

Several years later, as the reactor is nearing completion, essentially the same procedure is followed for the operating license. The applicant submits a Final Safety Analysis Report (FSAR), which goes through the same reviews leading up to public hearings and a final decision as to whether the license will be issued. The licensing procedures have become overly complex and time-consuming, and modifications are badly needed that can streamline the procedures without sacrificing their effectiveness.

Chapter 10

Various Other Sources of Exposure

Other chapters of this book are concerned with environmental radioactivity from natural sources, the processing of uranium, production of nuclear power, and the use and testing of nuclear weapons. There are a number of other ways in which exposure to radioactive substances can occur. These include their use in radioluminescent paints, as coloring agents in glass or ceramics, in smoke detectors and power generators, and as tracers and radiation sources in the biomedical sciences. Finally, transportation of radioactive substances must be considered.

In some publications (NCRP, 1977c; Moghissi *et al.*, 1978b), products such as tobacco and phosphate fertilizers that contain natural radioactivity are included among consumer products such as smoke detectors, ceramic glazes, and other manufactured products in which radionuclides are used. However, the radioactivity of tobacco and fertilizers is of natural origin and is inadvertently present in products that find their way to the marketplace. These were discussed in Chapter 7. This chapter will discuss products such as ceramic glazes and smoke detectors, in which radioactive substances are deliberately used to achieve a desired objective.

Before proceeding to a discussion of these and other uses of radioactive materials, the special case of radium will be reviewed because of the unique place this element occupies in the history of radioactivity and its relationship to luminous-dial painting, which was the first use of a radioactive substance in manufactured products.

233

The Early History of Radium

Radium was discovered by the Curies in December 1898, but they did not produce the first 100 mg until nearly 4 years later. It is reported that about 1300 Ci of radium was sold in the United States between 1912 and 1961, when the use of artificially produced radionuclides replaced radium to such an extent that new production had ceased (Stevens, 1963; U.S. PHS, 1971). Approximately 60 known deaths resulted from the use of radium in luminizing compounds. Luminous paints that contained radium were used not only in timepieces, but also for compasses, religious articles, aircraft instruments, and even luminous rings attached to the knobs of chamber pot covers, which could thus be found more easily in the dark of night (Holm, 1978)!

Before the dangers of radium came to be appreciated, an unknown fraction of the total production was also used in quack medicine, as discussed in Chapter 1. This practice resulted in additional cases of radium-induced bone cancer.

Of the 1300 g known to have been produced in the United States, only about 480 g were accounted for in 1971. The balance, about 820 g, may have been used largely for luminous compounds, static eliminators, or other ionization sources that were discarded as refuse when they served their purpose (U.S. PHS, 1971). It is possible that some of the radium so far unaccounted for its unknowingly stored in safe deposit boxes or attics by survivors of early radiologists (Villforth, 1964; Villforth et al., 1969).

From the time of its discovery and continuing up to the present, radium has been a source of many problems. In addition to the tragic misuses of radium in luminous-dial painting and quack medicine, in which the radium was handled in loose form without precaution, problems developed because hydrogen and oxygen produced by radiolysis from the water of crystallization of the radium compound caused pressure to develop within the capsules into which radium was sealed for use in radiography. This resulted in occasional ruptures of sources, necessitating expensive building decontamination. In one case, rupture of a single 50-mg radium sulfate capsule used for instrument calibration caused such extensive contamination that abandonment of a building was necessary (Gallaghar and Saenger, 1957).

Another problem associated with radium sources, as well as sources using artificially produced γ emitters such as ^{137}Cs, is that they are small and are frequently lost in hospitals or in industry, where they are used for industrial radiography. Every large city has had instances in which frantic searches of hospital incinerators, plumbing systems, municipal sewage treatment plants, or landfills were undertaken to find lost radium. Villforth et al. (1969) analyzed 415 mishaps involving radium up to 1969.

Exposure from Radioluminescent Paints

It has been estimated that 3 million radium-bearing timepieces were sold in the United States annually at one time (U.S. PHS, 1971) and that an additional 30,000 miscellaneous devices containing radium were sold annually. These included static eliminators, fire alarms, electron tubes, gauges, and educational products.

The radium content of men's wristwatches ranged from about 0.01 to 0.36 μCi (Seelentag and Schmier, 1963). In Europe, a number of studies have resulted in estimates that the per capita gonadal dose from the use of radium in timepieces ranged from 0.5 to 3.3 mrem/year (Robinson, 1968). Others have estimated the per capita gonadal dose from wristwatches containing 0.1 μCi ^{226}Ra to be 6 mrem/year and the annual absorbed dose to the skin under the face of the watch to have been as much as 165 rem/year. Although clocks contain more radium than wristwatches, people are normally located at a greater distance from them. The whole-body dose equivalent from clocks containing about 0.5 μCi ^{226}Ra has been estimated to average about 7–9 mrem/year among 10 million people who lived or worked in proximity to them (NCRP, 1977c).

Radium has been largely replaced by less expensive, artificially produced radionuclides such as tritium and ^{147}Pm. Although no wristwatches containing ^{226}Ra have been sold in the United States since about 1968, it was estimated in 1977 that 10 million such watches were still in use (NCRP, 1977c). Both tritium and ^{147}Pm are regulated by the NRC, which limits the ^{147}Pm content per watch to 100 μCi and the tritium content to 20 mCi. The minimum thickness of a watch case is sufficient to absorb the beta particles from both tritium and ^{147}Pm, but radiation exposure may result from inhalation or skin absorption of tritium that was volatilized.

The wearer of a wristwatch containing ^{147}Pm receives an annual gonadal dose equivalent of about 0.25 mrem/year from penetrating radiation (McDowell-Boyer and O'Donnell, 1978).

Uranium and Thorium in Ceramics and Glass

Uranium has been a popular coloring agent in ceramic glazes in past years, producing colors that ranged from orange-red to lemon yellow. The dose to the hands in contact with these glazes ranges from 0.5 to 20 mrad/hr for glazes produced prior to 1944, but there is evidence that the dose from ceramics produced since 1944 may be less by a factor of about 5 (Menczer, 1965). In 1961 the Atomic Energy Commission (AEC) ruled that a license would not be required for the use of uranium in ceramic glazes if the uranium content of the glaze was less than 20% for flatware and 10% for glassware (Simpson and Shuman, 1978).

Uranium has also been used at concentrations of a few hundred parts per million to enhance the appearance of porcelain teeth (Thompson, 1978). The principal dose is from irradiation of tissues within the oral cavity, estimated to be 1.4 rem/year to a small volume of epithelial tissue. The presence of ^{40}K in the porcelain raises the dose to about 1.6 rem/year. A limit of 37 ppm has been proposed as a standard, which would result in a dose of about 1.0 rem/year. A program to replace the uraniferous teeth has not been considered warranted because the risk is small. About 20 million bridges and dentures would require replacement, at a cost of many billions of dollars.

Thorium is incorporated into optical glass used in specialized instruments, but this ordinarily involves no risk to the user because the lenses are enclosed. It has been reported that thorium has been found to be present in eyepieces, but the extent of this practice is not known (NCRP, 1977c).

A more general source of exposure is the presence of thorium and uranium in ophthalmic lenses (Goldman and Yaniv, 1978). These elements enter the glass-making process because of their natural association with rare earth elements used to tint the glass. In 1975 the Optical Manufacturers Association adopted a voluntary standard to limit the radioactivity of opthalmic glass to less than 0.45 alpha particle/cm^2 min from the lens surface (OMA, 1975). Goldman and Yaniv analyzed the dosimetric implications of the radioactivity in eyeglasses and concluded that the critical tissue was the germinal layer of the cornea, estimated to be at a tissue depth of about 50 μm. On the assumption that an individual might wear spectacles 16 hr a day for a total of about 6000 hr per year, the maximum annual dose to the germinal layer of the cornea has been estimated to be about 500 mrem.

Thorium in Gas Mantles

Thorium has long been used for producing luminescence in gas mantles, of which an estimated 25 million are used annually, mostly by campers. Between 250 and 400 mg Th is used in each mantle (O'Donnell, 1978). The dose to campers and other consumers is estimated to be less than 0.1 mrem/year, which is sufficiently low to permit these devices to be distributed to the public without regulation. However, the dose received may vary, depending on the age of the mantle and its duty cycle. The thorium daughter products are separated in the manufacturing process and are thus not initially incorporated in the mantle, but they build in over a period of a few days. Radium-224, ^{212}Bi, and ^{212}Pb are volatilized in the process of burning but will again build in if the mantle is not used for a few days (Mohammadi and Mehdizadeh, 1983; Luetzelschwab and Googins, 1984).

Use of Specific Radionuclides ("Isotopes") in Research and Industry

Production of radionuclides for use in medicine, research, and industry has become an important by-product of reactor operation and particle accelerators during the past 40 years. In general, these nuclides, which have come to be known loosely as "isotopes," may be employed for a wide variety of uses, depending on their chemical and physical properties. The total quantitites of the various radionuclides produced each year, the fraction of each that decays, the fraction that remains in use, and the amounts that go to waste disposal sites are difficult to estimate.

A few radioactive isotopes that exist in nature among the heavy elements have been used as tracers, and early in this century it was shown that radium D (^{210}Pb) could be used as an indicator of lead in studies of the solubilities of lead compounds (von Hevesey, 1966). Subsequently, with the aid of the first particle accelerators such as the cyclotron and the Van de Graaff generator, it became possible to produce many radionuclides that do not exist naturally in significant quantities. These included ^{14}C,^{32}P, ^{24}Na, and ^{131}I, elements which are important in many biological processes. The potential importance of these artificially produced radionuclides as tracers was recognized in the 1930s, but the amounts then available were small and the instrumentation by which applications could be developed existed in very few laboratories.

The situation changed dramatically with the discovery of fission. In addition to the copious amounts of many new radionuclides that became available as by-products of the fission process, the high neutron fluxes available in reactors made it possible to produce new isotopic species by neutron irradiation. Transuranic nuclides such as ^{241}Am and ^{238}Pu could also be produced for application in smoke detectors and thermoelectric power generation. Officals of the wartime nuclear energy program were quick to recognize the potential value of these isotopes and took steps early in 1946 to make them available to science and industry. When the Atomic Energy Commission was formed in 1947, the utilization of isotopes for peaceful purposes became a major program objective.

The importance of radionuclides as tracers and radiation sources is too well known to require elaboration. "Isotopes" are now standard tools of research workers, and equipment such as scintillation counters, geiger counters, and scalers are part of the normal laboratory scenery in university, government, and industrial laboratories everywhere. The variety of radionuclides is more than matched by the number of uses to which they are put.

The sensitivity of the instrumentation for tracer studies permits many isotopes to be used in such small quantitites that their use does not require Nuclear Regulatory Commission licenses. The NRC regulations that pertain to the use of the by-product materials are contained in Part 30 of Title 10 of the Code of Federal Regulations (CFR).

There are also many clinical procedures in which radionuclides are used for either diagnostic or therapeutic reasons. It was estimated in 1980 that 10–12 million doses of radiopharmaceuticals were administered in the United States. Sales of radiopharmaceuticals in the United States increased at a rate of 25% per year during the period 1965–1975, but have since slowed to a rate of increase of 10–15% per year. Between 1966 and 1975, the annual frequency of radiopharmaceutical use in the United States increased from 3.7 to 37 per 1000 inhabitants.

Americium-241 in Smoke Detectors

The first smoke detectors, introduced in 1951, contained about 20 µCi of radium, but the use of that material has been discontinued in favor of ^{241}Am (Johnson, 1978), which has made it possible to market inexpensive, efficient devices that are easily installed and can now be found in many homes and most hotel rooms, offices, and factories. Each unit contains a small ionization chamber in which the air between two electrodes is ionized by the presence of a small foil containing the radioactive source. Ionization produced by the radioactive source allows an electric current to flow across the gap between the electrodes. The current flow is reduced by smoke particles, thereby making it possible to actuate an electronic circuit and to sound an alarm. The americium is in the form of the dioxide, which is fabricated between a backing of silver and a front cover of gold or gold/palladium alloy that is sealed by hot forging (Wrenn and Cohen, 1979). Each detector now contains from 0.5 to 1 µCi ^{241}Am. However, smoke detectors of earlier design (circa 1970) contained an average of 79 µCi, which was gradually reduced by improved design. Twelve million units per year were being sold by the mid-1980s, containing a total of about 8.5 Ci ^{241}Am (Harris, 1986).

Americium-241 has a half-life longer than 400 years and emits both alpha and photon radiation. Because of the widespread use of these devices, a great deal of attention has been given to the exposure and risk to occupants of households in which they are used. The highest dose is from external radiation and is reported by Wrenn and Cohen (1979) to be 0.014 mrem/year to an individual sleeping 6 ft from the detector for 8 hr per day.

Exposure of the general population to internal emitters disseminated by incineration of these devices or their disposal in landfills has been shown to be less than that from external radiation.

The exposures from smoke detectors involve minuscule risk in exchange for the great benefits that have been obtained. Nearly 9000 persons per year lose their lives from fires that occur in buildings, and 90% of these fatalities occur in private homes. Wrenn and Cohen (1979) concluded, from an analysis of reports by others, that 45% of these lives would have been saved had smoke detectors been available.

Radionuclides as Sources of Power

The decay heat of radionuclides has been used effectively to supply the electrical energy required for scientific instruments and communications equipment aboard space satellites. The power units are known by the acronym SNAP for satellite nuclear auxiliary power. Between 1961 and the end of 1982, 34 radioisotope generators were employed in 19 space systems. The first of these devices were developed in the late 1950s, with the advent of miniaturized transistorized electronic circuits with low power requirements. They proved to be useful for supplying power to instruments and telemetering equipment in weather stations located in remote areas (Morse, 1963). In 1961 the first SNAP device was launched into space to provide 2.6 $W_{(e)}$ of power to a satellite used in the TRANSIT navigational system. The design life of the unit was 5 years, but the power plant was still operational 15 years later, when the satellite was decommissioned (Bennett *et al.*, 1984). All of the SNAP devices used in outer space have depended on the decay heat of ^{238}Pu, an alpha emitter with a half-life of 86.4 years. Thermal efficiencies of the systems have been gradually improved from 5.1% for the first TRANSIT satellite to 6.6% for the unit planned for the Galileo mission, planned to be launched by the space shuttle in 1986 to begin a 2-year trip to Jupiter, where it will begin an investigation of that planet and its satellites.[1] The Galileo mission will be equipped with two radioisotope thermal generators, with a total initial power of nearly 600 $W_{(e)}$, compared to 2.6 $W_{(e)}$ for the generator used on the first TRANSIT mission.

When the SNAP program began in the late 1950s, the massive programs of testing nuclear weapons in the atmosphere that were then under way influenced the SNAP safety policies that were developed. The devices

---

[1]The Galileo mission, and Ulysses, a second deep probe designed to investigate the sun and also scheduled for 1986, will be delayed as a consequence of the Shuttle accident of January 1986.

were then designed so that in the event of accidents during the launching procedures, or during the postlaunch period when the vehicle remained subject to the control of the range safety officer, or until the moment the vehicle achieved a satisfactory orbit, the integrity of the isotopic power unit would be maintained to prevent dispersion of the fuel. However, a conflicting requirement was also placed on the designers of the early SNAP units: if the generator reentered the earth's atmosphere after attaining orbit, it was required that the heat of reentry would result in total dispersion of the fuel in a fine particulate form.

The requirement that the fuel remain contained within its metallic canister in the event of an abort during the early stages of a satelite launch was based on the desire to avoid massive exposure to ^{238}Pu in case of an accident in the immediate environs of the launch site, and the need to avoid the partial disintegration of the canister that might occur if low-grade burnup occurred prior to entering orbit. Since the early flights contained on the order of 10^4 Ci ^{238}Pu, a total burnup in the upper atmosphere, with subsequent dispersion and gradual fallout, would have resulted in a small increase above the fallout of ^{239}Pu and other radionuclides that had been occurring for many years as a result of nuclear weapons testing. This philosophy was subsequently changed to require that the capsule retain its integrity under all circumstances, or at least until the ^{238}Pu decayed. However, while the original policy was in effect, as will be discussed in Chapter 14, a SNAP device containing 17,000 Ci of ^{238}Pu was volatilized by the heat of reentry about 150,000 ft above the Indian Ocean and resulted in a 50-year dose commitment to the respiratory lymph nodes of the world's population of about 36 mrem (Shleien et al., 1970). In contrast, a NASA spacecraft (Nimbus-B1) which was launched from California in 1968 was destroyed by the range safety officer at an altitude of 30 km because of a guidance error, and the radioisotope generators fell into the Santa Barbara channel, from which they were recovered intact (Bennett, 1981).

In 1970 the spacecraft Apollo 13 was damaged on its way to the moon and the lunar module with its attached SNAP generator reentered the atmosphere over the South Pacific, where the generator landed in the 6-km Tonga Trench. The package was never recovered, but subsequent surveys could find no evidence of radioactivity in the atmosphere or the Pacific Ocean, leading to the conclusion that the containment remained intact.

Both the United States and the Soviet Union have developed reactors for use in outer space, but only one such device has been launched into space by the United States. It was known as the SNAP-10A and was launched in 1965. It began operation after being placed in a 4000-year orbit, but shut down as a result of an electrical malfunction after only 43 days. The reactor remains in orbit. Its initial radioactive inventory contained 2×10^5

Ci of fission products, but it has decayed by now to less than 100 Ci and will diminish to less than 1 Ci after 100 years (Bennett, 1981).

A satellite reactor placed in space by the Soviet Union as part of its Cosmos series reentered the atmosphere and disintegrated over Canada in 1978. This incident is described in Chapter 14. Shortly after the first Cosmos accident, the United Nations established a Working Group on the Use of Nuclear Power Sources in Outer Space (U.N., 1981).

A second Cosmos satellite that contained an enriched uranium reactor was reported to have reentered the atmosphere in February 1983. The reactor was designed to disintegrate on reentry. Balloon samples of dust recovered from the stratosphere over New Mexico 1 year later showed the presence of enriched uranium that was attributed to the Russian reactor (Leifer and Juzdan, 1984).

The initial power output of the SNAP units has increased from 2.7 $W_{(e)}$ in the TRANSIT satellites launched in 1961 to 584 $W_{(e)}$ for the Galileo mission that was scheduled to begin its voyage to Jupiter in mid-1986. The Galileo mission has been designed to use isotope generators that contain 2.8×10^5 Ci in the two heat sources.[2] This is a huge quantity of ^{238}Pu, which requires maximum assurance that containment of the fuel will be achieved in the event of any of the many types of accidents that can be hypothesized. These include launch vehicle explosions, which might lead to fuel release by the shock wave of the explosion, the thermal effects of the fireball, projection of the generators at high velocity, and their subsequent impact on hard surfaces in the immediate vicinity of the launch pad. If the launch is successful, there remains the possibility of explosion of the rocket fuel prior to entry of the spacecraft into orbit or, having attained orbit, reentry into the earth's atmosphere as the result of a misfunction (General Electric Co., 1985). The methods of probabilistic risk assessment referred to in Chapter 9, which have played an important role in the safety program of the nuclear power industry, were originally pioneered by the space program to provide estimates of the probabilities and consequences of mission failures at each step from prelaunch opertations to post-reentry recovery. The need for sophisticated methods of estimating the probabilities of mission failures was only related in part to the fact that some of the spacecraft would be equipped with radioisotope generators. These analytical techniques are required by the inherent characteristics of the space program: huge quantities of potentially explosive propellants are used and many of the missions are manned. The risks to populations in many parts of the world must be considered. The methods used to ensure

[2]As a result of various improvements in design, the Galileo power systems will require 0.5 kCi/$W_{(e)}$, compared to 3.4 kCi/$W_{(e)}$ for the early TRANSIT satellites.

safety in the U.S. space program proved to be remarkably effective during the first two decades of its existence, but the explosion of the shuttle Challenger in January 1986 demonstrated the difficulty of identifying and quantifying all possible contingencies.

Transportation of Radioactive Substances

Currently about 500 billion shipments of all kinds are made in the United States each year, of which about 100 million (0.02%) involve hazardous materials that are either flammable, explosive, toxic, or radioactive. The radioactive shipments number about 2.8 million per year, and contain about 9 million curies, not including spent fuel (Wolff, 1984; U.S. NRC, 1977c).

The rules and regulations that govern transportation of radioactive material are complex owing to the varied types of shipments to which they are applicable (U.S. GPO, 1983). In the United States, the Department of Transportation (DOT) has primary responsibility for regulating such shipments, most of which involve small quantities of radionuclides intended for use in research laboratories or medical facilities.

The first regulations for transportation of radioactive materials were drafted by the U.S. Postal Service as a result of the finding that photographic film was being fogged because of comingling with radium shipments (Pelletieri and Welles, 1985). The problems associated with shipments of radioactive materials became more complex as a result of developments during World War II, following which federal agencies concerned with hazardous shipments were concerned mainly with shipments of uranium and thorium ores, process residues, and radionuclides being shipped to and from research and medical institutions. There was as yet no nuclear power industry, and significant shipments of potentially fissionable materials were preempted by the newly formed Atomic Energy Commission. In 1961 the International Atomic Energy Agency (IAEA) issued proposed regulations based on the U.S. rules, in which it was recommended that member states adopt those rules for the sake of uniformity. The IAEA-suggested regulations were modified in 1973, and the U.S. Department of Transportation revised its rules to be in essential conformity with IAEA. The DOT regulations are contained in Title 49 of the Code of Federal Regulations and, since they are subject to change, should be consulted directly by persons in need of detailed current requirements. Additional regulations that augment those of the DOT are published by the Nuclear Regulatory Commission (10CFR71) and the U.S. Postal Service (39CFR124).

International regulations have been promulgated by IAEA (1973), the International Civil Aviation Organization (1983), the International Maritime Organization, and the International Air Transport Association (1982).

Under DOT regulations, a radioactive material is defined as one that has a specific activity in excess of 0.002 μCi/g of material. This is a very conservative definition, as can be seen from the fact that it is only about 2.5 times the specific activity of elemental potassium.

The exact regulatory requirements depend on the kinds and amounts of nuclides involved and the type of vehicle being utilized. Most shipments involve relatively innocuous materials that can be contained in "type A" containers, which are fiberboard or wooden boxes, or steel drums, designed to withstand moderately rough handling conditions. The requirements for type A shipments vary, depending on whether the material is in a form that is capable of being disseminated in the environment. The quantities that can be shipped in type A containers vary from 0.04 Ci ^{226}Ra (in a potentially dispersible form) to 1000 Ci ^{14}C (in a nondispersible form). All but 3.5% of the shipments are classified as type A or are exempt from shipping requirements.

Type B shipments are required to be shipped in containers that have been tested to withstand more rigorous stresses than type A containers. These include a 9-m drop to a hard surface, a fall of more than 1 m, landing on the upraised tip of a 15-cm-diameter steel bar, and 30-min exposure to a temperature of 1475°F for fissionable materials. A water immersion test requires that the package be submerged for not less than 8 hr under at least 1 m of water. Type B packages account for only 3.5% of all radioactive shipments, but 90% of the curies.

TABLE 10–1

ANNUAL UNCLASSIFIED SHIPMENTS OF RADIOACTIVE MATERIAL[a]

End use	No. of shipments	Percent of shipments	Percent of Ci
Medical	1,730,000	62.2	34.3
Other	519,000	18.6	0.2
Industrial	213,000	7.6	63.1
Power	114,000	4.1	0.7
Waste	181,000	6.5	1.5
Research and teaching	17,100	0.6	0.1
Unspecified	7,550	0.3	0.1

[a]From Wolff (1984).

The applicable regulations are highly detailed as to methods of packaging and labeling and should be consulted directly for guidance on the current requirements (Wolff, 1984).

The shipments can include a wide variety of radioactive materials, from a few microcuries of a relatively innocuous nuclide to high-level wastes or spent fuel. The types of shipments made annually in recent years are summarized in Table 10–1. Medical uses of radionuclides account for 62% of the radioactivity shipped, followed by industrial uses, mainly in the form of radioactive sources. Table 10–1 does not include many classified shipments by the defense agencies.

The many millions of shipments of radioactive materials have been involved in many accidents. Most of the accidents have occurred during handling, rather than during transport. There have been no failures of type B packages. Type A packages failed 13 times during the 10-year period 1971–1981, but the consequences were minor, having been limited by the small amounts of radioactive material that can be shipped in this way. There were no failures of type B packages during this period, although 45 such packages were involved in accidents (Emerson and McClure, 1985; Wolff, 1984).

Thus far the discussion has been limited to shipments of relatively small quantities of radioactive material. Millions of such shipments have been made, and a solid actuarial base has been established that permits one to conclude that the precautions which have been taken are adequate to protect the public from either direct radiation from the packages or escape of the material to the environment. Shipments of high-level wastes, spent fuel, or large amounts of transuranic wastes occur with less frequency, but involve huge quantities of radioactivity and therefore require correspondingly greater degrees of protection.

The casks in which such shipments are made are huge in size but can be accommodated either on flatbed trailers or railroad cars. Special features of design are needed to remove the heat that is generated during shipment. The required shielding and the need for structural strength add to the mass of the shipment, which may weigh as much as 100 tons. A shipping cask for

---

Fig. 10–1. The safety of a spent fuel shipping cask was put to a severe test when a rocket-propelled locomotive was crashed into it at more than 80 miles per hour. The 28-ton cask received minor surface dents during the test, but did not crack open or leak any contents. The cask at impact was knocked into the air and bounced twice on the ground before coming to rest between the rails of the track. The front half of the locomotive was totally crushed by the impact, and the trailor on which the cask was mounted was bent around the locomotive in a U shape. These results confirmed predictions made earlier by project engineers on the basis of scale model tests. (From U.S. Department of Energy.)

spent fuel is illustrated in Fig. 8–1. The casks are subjected to severe tests before their designs are accepted by the NRC. This is illustrated in Fig. 10–1, which shows a cask struck by a rocket-propelled railroad car driven at 120 km/hr. The 28-ton cask was hardly damaged despite the fact that the trailer on which the cask was mounted was bent around the locomotive in a U shape and the front of the locomotive was badly damaged.

Chapter 11

Radioactive Waste Management

Radioactive waste management is a subject concerning which the technical aspects are perhaps less complex than the sociopolitical ramifications. The subject of radioactive waste storage is fearful to the general public, and this has been reflected in a "not in my backyard" (NIMBY) syndrome; that is, there is general agreement that there is a need to store radioactive wastes, but each locality wants them stored somewhere else. Some of the sociopolitical ramifications of the subject will be discussed in Chapter 15.

There are several kinds of radioactive wastes, classified according to their physical and chemical properties as well as the source from which the wastes originate. Thus, for technical, legal, and political reasons, wastes that originate from the military programs must be handled separately from wastes from the civilian program (U.S. DOE, 1983).

Among the physical properties that influence the manner in which a radioactive waste should be managed are the half-life of the nuclide and the chemical form in which it exists. In this chapter, the following categories of wastes will be discussed:

1. Low-level wastes that consist of residues from laboratory research: slightly contaminated paper and other laboratory debris, biological materials, scrap metal, and building materials. There are two other categories of

low-level wastes which accumulate in huge volumes and are deserving of special attention. These are: (a) uranium mill tailings and (b) wastes generated in the cleanup of uranium, radium, and thorium processing plants—the so-called remedial action program of the Department of Energy (DOE), and the decommissioning of reactors. These two special classes of wastes were discussed in Chapter 8.

2.   High-level wastes, which can be considered in three subcategories: (a) spent fuel from civilian nuclear power reactors, (b) liquid and solid residues from the reprocessing of civilian spent fuel, and (c) liquid and solid wastes from the reprocessing of fuel used for military purposes.

3.   Transuranic wastes, which are mainly alpha-emitting residues from military manufacturing.

Low-Level Wastes

As this book goes to press, the United States is facing a crisis in the management of low-level wastes, not because the wastes have proved to be so hazardous, but because of widespread public concern about the potential risks of shallow land burial, which is the disposal option for low-level wastes most favored by technical specialists.

As far back as 1958 a panel of marine scientists of the National Research Council considered the impact of low-level radioactive waste disposal into the Atlantic and Gulf coastal waters and concluded (NAS–NRC, 1959) that relatively large quantities of radioactive wastes could be deposited safely in shallow coastal waters. Twenty-eight possible locations were selected that could be used for this purpose without limiting the areas for other uses. The total quantity of radioactivity that could be deposited in *any one* disposal area in any one year was estimated to be about 250 Ci of ^{90}Sr or the biological equivalent of other isotopes. Compared to actual practice, this was a rather liberal recommendation. The calculated quantities of selected radioisotopes equivalent in hazard to 250 Ci of ^{90}Sr are given in Table 11–1. These data are presented to emphasize the relatively large quantities of some of the more common radionuclides that could be dumped safely in coastal waters. The practicality of disposing of low-level wastes in Pacific coastal waters was also examined by a committee of the National Research Council with similar conclusions (NAS–NRC, 1962).

The National Research Council study seemed to justify practices that had already been used by the Atomic Energy Commission (AEC) and its contractors. The Atlantic and Pacific Oceans and the Gulf of Mexico were used beginning in 1946 for disposal of packaged low-level wastes originating mostly from research and development facilities. Mixtures of low-level

TABLE 11–1

QUANTITIES OF SELECTED RADIOISOTOPES EQUIVALENT TO 250 Ci OF ^{90}Sr, SHOWING THE INITIAL QUANTITIES THAT WILL DECAY TO 250 EQUIVALENT Ci ALLOWING 1 MONTH AND 1 YEAR CONTAINMENT[a,b]

	Curies		
Isotope	No containment	1 Month containment	1 Year containment
^{24}Na	5.0×10^7	10^{24}	10^{183}
^{32}P	15.5	68.6	1.1×10^9
^{35}S	3.1×10^6	3.9×10^6	5.6×10^7
^{42}K	3.1×10^6	10^{14}	10^{226}
^{45}Ca	1.6×10^5	1.8×10^5	7.5×10^5
^{59}Fe	1.2×10^3	1.9×10^3	3.3×10^5
^{60}Co	6.2×10^3	6.3×10^3	7.0×10^3
^{64}Cu	5.0×10^4	10^{21}	10^{201}
^{65}Zn	1.4×10^4	1.5×10^4	3.8×10^4
^{90}Sr	250	250	250
^{131}I	9.3×10^2	1.2×10^4	10^{16}
^{137}Cs	9.3×10^4	9.3×10^4	9.3×10^4

[a] NAS-NRC (1959).
[b] Equivalence based upon ratios of permissible sea-water concentrations.

wastes and cement, which was permitted to harden before disposal in the ocean, were contained in 55-gal, 18-gauge steel drums. When this practice was discontinued in 1970, the total quantity deposited in the oceans by the United States totaled 94,6000 Ci, much of which consisted of short-lived or relatively innocuous nuclides used in tracer applications and clinical practice. European countries have deposited about 700,000 Ci in waters of the Atlantic Ocean (Holcomb, 1982). These quantities are insignificant compared to the amounts listed in Table 11–1. The practice of ocean dumping was discontinued in the United States mainly because of mounting pressure against sea disposal of any kind, and in the absence of any evidence that the practice was potentially hazardous.

One cannot help but note that during the same period when this relatively small amount of low-level waste was being introduced into the oceans, far greater quantities were introduced as a result of nuclear weapons testing. From Table 11–2 we see that megacurie quantities of ^{90}Sr, ^{137}Cs, ^{14}C, and tritium were introduced into the Pacific Ocean alone, with only minimal impact on food consumed by humans (Eisenbud, 1981a).

TABLE 11–2

ESTIMATED DEPOSITION OF NUCLIDES IN THE PACIFIC OCEAN[a]

Nuclide	Curies produced by weapons tests, 1945–1972 (MCi)	Estimated deposition in Pacific Ocean (MCi)
^{3}H	4500	1140
^{14}C	5.8	1.9
^{90}Sr	17	7.1
^{137}Cs	27	11.3
^{239}Pu	0.4	0.17

[a]From Eisenbud (1981a). (Reproduced from *Health Physics*, **40,** 435, by permission of Health Physics Society.)

Despite the fact that the oceans have been the recipients of such enormous quantities of radioactivity from fallout, marine sources of food have not contributed significantly to the dose received from fallout in those countries of the world for which data are available. The UNSCEAR emphasis has been on the terrestrial food chains because most foods are derived from land sources. For example, in San Francisco, where representative diets have been monitored for ^{90}Sr for many years, fish and shellfish account for no more than about 0.2% of a *per capita* ^{90}Sr intake that has ranged between about 1000 and 3000 pCi/year. Similar findings have been reported from diet studies in New York (U.S. DOE).

In a special report to the President and the Congress, the National Advisory Committee on Oceans and Atmosphere (1984) recommended that Congress and the administration should revise the present policy of excluding use of the ocean for low-level radioactive waste disposal, but that ocean disposal should not be initiated until it has been established that the fate and effects of such disposal will not result in adverse effects.

By far the largest volume of low-level radioactive waste has been disposed of by near-surface land burial. Unitl 1963, industry arranged to send such waste to AEC sites for burial, but in that year AEC withdrew from offering that service and began to license private companies for the operation of waste burial grounds. By 1971 there were six commercially operated licensed shallow land burial sites, as shown in Table 11–3 (Holcomb, 1980), at which a total of about 600,000 m^3 of low-level radioactive waste was stored. Although it has not been shown that any of the six burial sites created a public health problem, intense local opposition to operation of the sites began to develop, and three of the sites have been closed.

TABLE 11–3

<small>Commercial Shallow Land Burial Sites[a]</small>

Location	Year first licensed	Site operator	Licensing authority	Year closed
Maxey Flats, Kentucky	1962	Nuclear Engineering Company	State	1977
Beatty, Nevada	1962	Nuclear Engineering Company	State	
Sheffield, Illinois	1967	Nuclear Engineering Company	NRC	1978
Richland, Washington	1965	Nuclear Engineering Company	State and NRC	
Barnwell, South Carolina	1971	Chem-Nuclear Systems, Inc.	State and NRC	
West Valley, New York	1963	Nuclear Fuel Services	State and NRC	1975

[a]From Holcomb (1980).

In 1975 the burial grounds at West Valley, New York, were closed despite the fact that studies conducted by state and federal agencies concluded that continued operation would have no significant adverse effect on public health and safety (Giardina *et al.*, 1977).

Low-level contamination of ground water and milk supplies near a second burial site at Maxey Flats caused the state of Kentucky to place a temporary ban on further operations, and in the following year the lease for the land on which the burial grounds were located was canceled. Studies by the Environmental Protection Agency (EPA) did find migration of tritium, ^{90}Sr, and a few other nuclides, but only in very small amounts (Montgomery *et al.*, 1977). The tritium content of milk within about 3 km of the site ranged from 300 to 6500 pCi/liter. An individual consuming 1 liter/day of milk containing the highest tritium level would receive an annual total body dose of about 0.4 mrem. The highest level of tritium-contaminated well water would deliver a dose of 0.1 mrem/year.

A third site, at Sheffield, Illinois, closed in 1978 because it reached its licensed burial capacity and the state was opposed to any expansion.

There are only three low-level burial sites in the United States as of this writing, located in Beatty, Nevada; Barnwell, South Carolina; and Hanford, Washington. In late 1979 the governors of two of these states, Nevada and Washington, closed their sites temporarily because of poor packaging practices on the part of the shippers. It was necessary to suspend the license of the Nevada burial site in March 1979, when it was found that employees at the site were removing contaminated hand tools, electric motors, and other items. As the result of an intensive effort, an estimated 25 pickup truckloads of radioactively contaminated equipment, as well as several loads of large items, were recovered and returned to the burial site (Wenslawski and North, 1979).

The only commercial burial site in the eastern half of the country is at Barnwell, South Carolina, which receives more than half of the low-level wastes generated in the United States.

In addition to the burial sites that are operated commercially, there are five major shallow land burial facilities located at the major production and research centers operated by the Department of Energy. These facilities serve the Department of Energy and other government agencies.

Although no damage to public health has been shown to have resulted from burial site operations, it is evident that there have been flaws in the management methods. These included poor packaging, water infiltration, and insufficient initial compaction, which resulted in creation of a "bathtub" due to subsidence. All of the examples of mismanagement were of a type that could have been corrected. However, the various mishaps were so widely publicized that widespread public opposition developed which has jeopardized the nation's ability to manage its low-level radioactive wastes. All of this has resulted in considerable frustration on the part of users of radioactive materials, as exemplified by a New York State study group, which expressed concern that low-level waste disposal services were likely to be severely disrupted unless the state acted to provide an in-state alternative to out-of-state shipments. At risk was the human value of the more than 19 million annual medical procedures using radiopharmaceuticals, the billion-dollar economic value of these procedures, the work of more than 5000 persons who perform research within the state of New York, and the operation of more than 300 industrial facilities within the state (N.Y. State Low-Level Waste Group, 1983).

About 30% of all waste shipped to the burial sites comes from medical institutions. The problems of disposing of these biomedical wastes illustrate how bizarre the subject has become. About half of these wastes, which are shipped at great expense to various parts of the United States, contain small amounts of ^{3}H and ^{14}C generated in the course of liquid scintillation counting, a procedure that is in widespread use in biomedical laboratories. The wastes are contained in 20-ml plastic vials, of which 84 million are used in the United States each year (Roche-Farmer, 1980). These vials contain a total of about 500,000 liters of organic solvent, frequently toluene, which contain a total of only 8 Ci of ^{3}H and ^{14}C. The disposal problems are complicated by the flammable nature of the solvents used, and fires have occurred during transportation of these wastes and at the burial sites. The total amount of radioactivity is so insignificant that the vials could be incinerated at the biomedical institutions without risk to the public. Any incinerator capable of properly combusting animal carcasses and other biomedical wastes would be more than adequate to receive the scintillation vials (Eisenbud, 1980; Philip *et al.*, 1984).

In 1981 the Nuclear Regulatory Commission issued revised standards that would permit incineration of the scintillation vials (U.S. NRC, 1981). However, resistance to incineration of "radioactive waste" was so strong in, for example, New York City that the City Council prohibited implementation of the NRC regulation, despite an urgent resolution by the Committee on Public Health of the New York Academy of Medicine (NYAM, 1983) that emphasized the safety of the procedure.

NEED FOR VOLUME REDUCTION

The national inventory of low-level radioactive waste is growing at such a rapid rate (about 10^5 m^3/year) that methods of volume reduction are needed. The Department of Energy has established a number of research projects concerned with the development of incineration techniques. Substantial reductions in volume are possible, since about 40% of the newly generated wastes are combustible (Borduin and Taboas, 1981). However, volume reduction up to the present has been accomplished largely by compaction.

THE LOW-LEVEL RADIOACTIVE WASTE POLICY ACT

By 1980 it was becoming apparent that the management of low-level wastes required intervention by the federal government, and in December of that year Congress passed the Low-Level Radioactive Waste Policy Act. The act assigned to the individual states responsibility for assuring proper disposal for low-level wastes generated within their borders, and suggested that the required facilities could best be provided through regional interstate compacts, which must be approved by Congress and reauthorized by it at 5-year intervals. The act further specified that any state or regional compact could exclude wastes originating outside the region after January 1, 1986.

NUCLEAR REGULATORY COMMISSION REQUIREMENTS

The Nuclear Regulatory Commission licensing requirements for near-surface burial of radioactive wastes are given in Part 61 of the Nuclear Regulatory Commission regulations (Title 10, Part 61, of the Code of

Federal Regulations). The regulations define low-level wastes by what they are not! They are not high-level wastes, nor thorium or uranium mill tailings, nor spent nuclear fuel, nor transuranic waste. In practice, low-level wastes are generally those that derive from the use of radionuclides ("by-product material") and contaminated equipment and material that originate from the operation of nuclear power plants or other industrial facilities. Surprisingly, definitions of high- and low-level wastes are so vague that one cannot always be certain whether the waste is suitable for near-surface burial or must be handled as high-level waste (Jacobs et al., 1985). The Nuclear Waste Policy Act of 1982 defines high-level radioactive waste as "the highly radioactive material resulting from the reprocessing of spent nuclear fuel, including liquid waste produced directly in reprocessing and any solid material derived from such liquid waste that contains fission products in sufficient concentrations." The act then adds that the definition can be extended to "other highly radioactive material that . . . requires permanent isolation." Over the years the definition of high-level waste has apparently hinged on the source of the material rather than the amount of radioactivity (Sinclair, 1984).

High-Level Wastes

Any discussion of high-level wastes must differentiate between "defense wastes," which have been accumulated during the past 40 years from the production and use of plutonium at government installations, and "commercial wastes," which originate from the production of civilian nuclear power.

Defense Wastes

The principal accumulations of defense wastes are located at Hanford, Washington, and Savannah River, South Carolina, with lesser amounts at Idaho Falls, Idaho. The wastes originated from the processing of irradiated uranium for the purpose of plutonium separation. Additional defense wastes originate from the reprocessing of fuel from nuclear-powered naval vessels. The waste liquid from these processes has been stored in concrete-encased steel tanks a few meters below the surface of the ground, and by 1982 the total volume of the defense wastes was about 80 million gallons. The original highly acid liquids were usually neutralized to reduce corrosion, causing a precipitate to settle to the bottom of the tanks, where in some cases it has attained a nearly rocklike hardness that makes it very difficult to transfer the waste to other containers when necessary.

It is known that the tanks develop leaks (U.S. AEC, 1973; Catlin, 1980) and it is generally recognized that an alternative to tank storage is required

(NAS–NRC, 1978, 1985a; U.S. DOE, 1978). Suggestions that have been made include immobilization of the waste in the present tanks, mixing the waste with concrete or other solidifying agents for injection underground as grout, and conversion to stable solids for burial. Whatever decision is made, the costs will be enormous. At the Savannah River plant, it has been estimated that the total cost of producing a slurry for disposal in bedrock will be $205 million (1978 dollars) and that conversion to solid form for on-site underground storage would cost nearly $2 billion (1978 dollars) (U.S. DOE, 1978).

TRANSURANIC WASTES

A special category of radioactive defense wastes are those contaminated with relatively minor amounts of transuranic (TRU) radionuclides (Moghissi, 1983). These originate mainly in the production and fabrication of plutonium for military purposes, and both ^{238}Pu and ^{239}Pu are the most prevalent contaminants.

It had been a DOE requirement since 1970 to separate TRU-contaminated wastes according to whether they contained more or less than 10 nCi of TRU per gram of waste. Wastes that contained less than 10 nCi/g could be disposed of by shallow burial, whereas wastes containing greater amounts were placed in retrievable storage for eventual transfer to a permanent repository. In 1982, the National Council on Radiation Protection and Measurements examined the TRU disposal practices and concluded that there was no basis for the 10 nCi/g criterion, that higher concentrations could be disposed of by near-surface land burial, and that what was required were site-specific limits based on geochemical and ecological conditions existing at individual sites (NCRP, 1982). Following publication of this report, the Nuclear Regulatory Commission changed the criterion from 10 to 100 nCi/g. The economic consequences of this change are illustrated by the fact that by 1984, two years after establishment of the new criterion, the U.S. government had already saved $250 million by diverting material being held for expensive geologic isolation to shallow land burial (Sinclair, 1984).

A waste isolation pilot plant (WIPP) intended to provide for permanent isolation of Defense Department TRU wastes is being constructed in a deep mined cavity in bedded salt in Eddy County, New Mexico (NAS–NRC, 1985a).

CIVILIAN HIGH-LEVEL WASTES

The volume of waste from civilian nuclear power plants likely to become available in the foreseeable future has been greatly reduced by the slowdown in the development of nuclear power. In 1970 it was assumed that there would be 735 MWe of installed nuclear capacity in the United States by 2000 (ORNL, 1970), but 10 years later the estimate had been reduced to 180 MWe (ORNL, 1980). In 1970 it was estimated that the total inventory of spent uranium would be about 90,000 metric tons, but by 1984 the estimate had been reduced to about 45,000 metric tons, a figure that serves as the basis for national planning at the present time (U.S. DOE, 1984a).

The spent fuel elements from a commercial reactor, which consist of zirconium-clad rods of uranium oxide, are placed immediately after removal in a large basin of water (the spent fuel storage pool) located adjacent to a reactor. It was originally intended that the fuel rods would remain in these cooling basins for only about 6 months—to allow for reduction in their radioactivity and temperature—before being shipped to a reprocessing center (Chapter 8). However, as mentioned in Chapter 8, the reprocessing of commercial spent fuel was halted in 1977 during the administration of President Carter because of concern that the separated plutonium might be diverted for military or terrorist purposes. As it turned out, uranium economics changed drastically at about the same time because of the reduction in demand for nuclear power, and fuel reprocessing could no longer be justified on economic grounds. As a result, the spent fuel elements have been piling up on the nuclear power plant premises, and if reprocessing capacity is not provided, the spent fuel itself may become the waste material. However, should the economics of nuclear power change in the future, it may again become necessary to reprocess spent fuel.

OPTIONS FOR THE PERMANENT MANAGEMENT OF RADIOACTIVE WASTES

Several methods of isolating high-level wastes have been considered during the past 40 years, including on-site methods of solidification and disposal, use of the seabed and subseabed, injection as a grout in deep rock fissures, insertion in Greenland glaciers, and geologic isolation in deep mined cavities. Proposals have also been made that the wastes be lifted into outer space or into the sun by rocket, or be transmuted to more rapidly decaying elements in giant accelerators. Many of these options must be considered infeasible for the foreseeable future because of economics, safety, and the current state of technology (USAEC, 1974; U.S. DOE, 1980a). Of these various options, on-site solidification and disposal, use of the marine environment, and geologic isolation in mined cavities appear to be the most viable options.

On-Site Solidification and Disposal

At the Oak Ridge National Laboratory and in the Soviet Union, moderately radioactive wastes have been mixed with cement to form a grout, which is then injected through wells into rocks that have been fractured hydraulically (NAS–NRC, 1985a; Spitsyn and Balukova, 1979). Recent studies have suggested that this technique might be suitable for certain defense wastes at both Savannah River and Hanford (NAS–NRC, 1978, 1981), but it has also been shown that at these sites wastes having intermediate levels of activity can be hardened into concrete and buried in near-surface trenches.

The Marine Environment

The use of ocean waters for the disposal of low-level wastes has already been discussed in this chapter. The United Kingdom has for some years followed the practice of discharging fission products from a fuel-reprocessing plant at Sellafield (formerly known as Windscale) near the coast of West Cumbria. The Sellafield site has for many years been discharging fission products into the Irish Sea (Dunster, 1969; Howells, 1966). Experimental releases were first made in 1952, when about 10,000 Ci of effluent were discharged over a period of about 6 months. Based on studies of the ecological behavior of the releases during this and subsequent experiments, the quantities released were increased, beginning about 1969. The annual releases of ^{137}Cs from 1952 until 1984 are given in Fig. 11–1. The discharges of cesium, and other nuclides as well, peaked dramatically in 1976 and then decreased as a result of changes in operating practice (British Nuclear Fuels Ltd., 1984; Black, 1984).

By studying the dietary habits of the nearby populations in relation to radioecological factors, it was possible to derive working limits of permissible contamination of silt, sand, seaweed, and fish. The early studies identified contamination of local fish and a species of seaweed used in cooking as the critical pathway for human exposure (Dunster, 1958). Having defined the working limits of contamination of these vectors of human

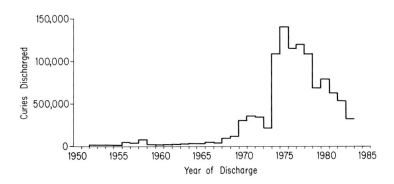

Fig. 11–1. Discharges to sea of cesium-137 from Sellafield. (From Black, 1984.)

exposure, it was possible to calculate the permissible rates of release of
the radioactive aqueous wastes. It was found that the ^{106}Ru content of
seaweed was limiting and that the permissible discharge rate of ^{106}Ru was
70,000 Ci/month. If it were not for the seaweed–food pathway, fish would
have been limiting, in which case the permissible release could have been
more than 1 million Ci/month.

A formal inquiry into the public health impact of these releases was
begun in 1983, in response to widespread concern that resulted from a
television program in which it was alleged that there was an increase in
the incidence of leukemia among young people in the vicinity of the plant.
The report from the National Radiological Protection Board (1984) concluded that young people who lived in the vicinity of the plant from 1950
through 1970 would have received a total dose of about 350 mrem during
that 20-year period, which is about 13% of the dose received from background sources in the area.

Subseabed Disposal

The sediments of the deep ocean floor are uniform over a considerable
area, show little evidence of mixing, have high ion sorptive capacity, and
there is virtually no movement of interstitial water (U.S. DOE, 1979b). The
pressure is so high that the water will not boil if the sediments are heated
by the decaying radionuclides. Because the subseabed has so many attractive features, about a dozen countries have formed a Seabed Working
Group to coordinate research on the possible use of the deep sea sediments as a repository for high-level nuclear wastes.[1] Several emplacement
concepts have been proposed, of which the most attractive is the pen-
etrometer, a free-falling ballistically shaped container that could be released from a vessel and penetrate as far as 30 m into the sediments at the
ocean bottom.

The subseabed concept is at a relatively early stage of development and,
if ever implemented, will require that present international restrictions on
"ocean dumping" be modified before it is put into effect.

DEEP GEOLOGICAL REPOSITORIES

The currently favored method of disposing of high-level wastes is in
deep underground mined cavities. After 25 years of political indecision as
to how to deal with the high-level waste problem, Congress enacted the
Nuclear Waste Policy Act in late 1982, and it was signed into law on

---

[1]The member nations of the Seabed Working Group are Belgium, Canada, Federal Republic
of Germany, France, Japan, Netherlands, Switzerland, United Kingdom, and United States.

TABLE 11–4

Sites Proposed for Nomination as
Suitable for Characterization[a]

State	Site	Geologic Medium
Mississippi	Richton Dome	Domal salt
Nevada	Yucca Mountain[b]	Tuff
Texas	Deaf Smith County[b]	Bedded salt
Utah	Davis Canyon	Bedded salt
Washington	DOE Hanford Site[b]	Basalt

[a]U.S. Department of Energy (1985c).
[b]Preliminary recommendation for detailed characterization.

January 7, 1983 by President Reagan. The law is highly specific as to how the U.S. program is to proceed. The process began with the issuance of guidelines for site selection, which were published in the summer of 1984 (U.S. DOE, 1984b) and enumerated the geological and other considerations that would qualify or disqualify a proposed site. The DOE is then required to nominate at least five sites as suitable for investigation and then select three sites for recommendation to the President.

The five preliminary sites that have been proposed for characterization by DOE are listed in Table 11–4 (U.S. DOE, 1985c). The act establishes January 31, 1998 as the date when the first repository will begin operation.

The concept of waste isolation in mined cavities depends on a multibarrier system that has the following component parts:

1. The solid form in which the wastes exist. For intact spent fuel, the waste form is the original clad fuel. For processed spent fuel, the radionuclides can be incorporated into solids such as glass or highly insoluble ceramics. The temperatures of the waste and host rock are important characteristics, which depend not only on the age of the waste but also the concentration of waste contained in the waste form, the spacing of the canisters, and the thermal properties of the rock in which the wastes are placed.

2. A canister that contains the waste and is designed to resist corrosion.

3. An "overpack" of highly adsorbent materials such as clay, which serves to reduce corrosion of the canister by retarding the migration of corrosive ions in ground water. The overpack will also help to retard the migration of radionuclides that may escape from the canister.

4. The geochemical properties of the repository environment, which govern the rate of migration of the radionuclides in ground water.

5. The complex physical and biological pathways by which the radionuclides pass from the immediate vicinity of the repository to the biosphere and eventually to humans.

CRITERIA FOR SITE SUITABILITY

General guidelines for determining the suitability of a repository site have been published by the Department of Energy in conformance with the requirements of the Nuclear Waste Policy Act (U.S. DOE, 1984b). An important requirement is that it should be possible to establish that the ground water time of travel from the repository to the "accessible environment" should be no less than 1000 years. The nature and rates of hydrologic processes during the past 2 million years should be such that, if continued into the future, the repository would not be affected deleteriously during the next 100,000 years. The guidelines then go into a great deal of detail concerning the required characteristics of the host rock, and specify that the repository should be placed at least 200 m below the ground surface.

An important criterion is that there should be no known natural resources at the site that might, in the foreseeable future, be of sufficient value to be commercially attractive.

Geochemical and hydrological conditions should be such that the annual solubilization rate for the radionuclide inventory in the repository, after 1000 years, should be no more than 10^{-5}.

Other required characteristics are tectonic stability, adequate rock mass, and suitable thermal conductivity, porosity, and permeability. An ideal rock mass would be large, homogeneous, dry, relatively free of fractures, and capable of sorbing or precipitating released radionuclides (NAS–NRC, 1982).

GEOLOGIC SETTINGS

Several types of rock are potentially suitable for repository sites, of which the following have received the most serious consideration.

Salt

Salt occurs in many localities throughout the world, either as extensive bedded deposits or in the form of domes. Salt has been considered a leading contender for repositories at least since 1957, when a National Academy of Sciences committee called attention to the many advantages of this type of rock (NAS–NRC, 1957a, 1970). A demonstration repository is being constructed by the Department of Energy near Carlsbad, New Mexico, but its use will be limited by law to TRU defense wastes, although there is provision for demonstration of the feasibility of storing high-level wastes. Among the advantages of salt beds are their age (greater than 200 million years), which gives assurance of geologic stability, and isolation from aquifers. Salt has high thermal conductivity, low permeability, and plastic characteristics that permit fractures to close. A disadvantage is that, because of its plasticity, the repository would become self-sealed after closure and wastes would thereafter not be readily retrievable. Another possible disadvantage is that the salt is frequently associated with deposits of potash or hydrocarbons and might result in human intrusion some time in the future (Klingsberg and Duguid, 1980). An obvious disadvantage is the solubility of salt, which requires that there be assurance that intrusion by ground water will not occur.

A curious characteristic of salt is that brine inclusions are frequently present, which, in a thermal gradient, tend to migrate toward a source of heat. This phenomenon is the result of differential rates of solution at the walls of the inclusion, caused by the temperature gradient. The brine inclusions can thus move toward the canister and cause undesirable corrosion. This need not be a significant issue if the thermal gradients are properly specified in the repository design (NAS–NRC, 1983a).

Basalt

Basalt originates from volcanic flows that occur extensively in eastern Washington, Oregon, and other localities in the United States. The basalt near Hanford, Washington, is one of the five candidate sites selected by the Department of Energy. One recent study has called attention to a number of critical geological problems related to basalt, including the effects of repository heating on the basalt, the abundance of faults and fractures, and uneven stresses in the basalt (NAS–NRC, 1983a).

Tuff

Tuff is a compacted volcanic ash that occurs throughout the western United States. The tuff of the Nevada Test Site near Las Vegas is one of the five candidate sites selected by the DOE as a possible location for a first repository. Tuff is characterized by relatively low permeability and porosity, but the negative aspects associated with tuff include its relatively high water content and the fact that the deposits are in areas of recent vulcanism and fault movement. A major advantage of tuff is that the deposits are above the water table.

Granitic Rocks

The granites and other crystalline rocks are a major class of rocks that have not been included among the candidates for the first repository, but have been selected as the preferred rock type in both Canada and Sweden. Positive aspects are high thermal conductivity and structural strength, low porosity and permeability, and, usually, low water content. The second round of repository site selection will concentrate on crystalline rocks. An important disadvantage is uncertainty about the degree of fracturing and its influence on the movement of ground water.

NEED FOR INTERIM STORAGE

Because spent fuel storage capacity at the reactor sites is limited, an acute need is developing for some form of "away from reactor" storage (AFR). As part of the Nuclear Waste Policy Act, the federal government is required to examine the need for construction of interim storage capacity and to make appropriate recommendations to Congress. The costs of the facilities, if needed, would be borne by the electric utilities. The act also provides that, pending development of a system for permanent isolation of high-level wastes, the government would construct monitored retrievable storage facilities. These facilities would be built by the DOE in accordance with Nuclear Regulatory Commission licensing requirements, and the cost would be borne by the users (U.S. DOE, 1985c).

METHODS OF RISK ASSESSMENT

There is no precedent in technology for the long periods of time for which risk assessments are required in radioactive waste management, and there is as yet no firm policy, either national or international, on the length of time for which safety must be assured or, for that matter, the amounts of radioactive materials that should be permitted to enter the biosphere in future millennia.

It is convenient to start a discussion of the methods of risk assessment with a description of the manner in which the radioactivity of high-level wastes diminishes with time. This is illustrated in Fig. 11–2, where the potential risk is plotted for 10 million years against "water dilution volume" (WDV), an overly simplified but useful measure of risk that simply considers the amount of water required to dilute the radioactive material to the concentration that is safe for drinking water (NAS–NRC, 1983a). The WDV does not consider the many ways in which the risk can be increased, as by biomagnification, or decreased, as by immobilization in

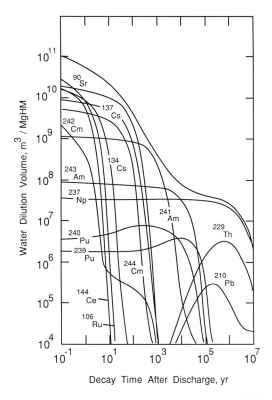

Fig. 11–2. Water dilution volume of pressurized-water reactor (PWR) high-level waste. (From National Academy of Sciences–National Research Council, 1983a.)

sediments. It is seen that after a few hundred years, the fission products are no longer a problem and the remaining nuclides are members of the actinide family of elements, except for ^{210}Pb, which grows in from ^{230}Th and peaks at about 10^5 years.

There have been a number of attempts to answer the question "How long is long enough?" by comparing the WDV for nuclear wastes with that of the uranium ore from which the fuel was originally derived. Figure 11–3 does this by comparing the WDV for the high-level waste and uranium ore on the same time scale shown in the earlier figure (NAS–NRC, 1983a). Two comparisons are shown, using WDVs calculated by both the NRC regulatory limits and the International Commission on Radiological Protection (ICRP) recommendations of allowable annual intake. Using the NRC regulatory limit, the WDV for high-level waste crosses that for the quantity of uranium ore from which the fuel was obtained in less than 1000 years, but

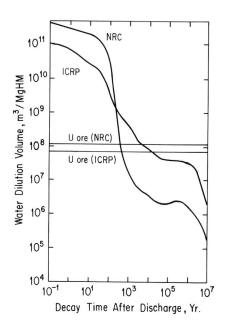

Fig. 11–3. Water dilution volumes of PWR high-level waste and its parent uranium ore. (From National Academy of Sciences–National Research Council, 1983a.)

the crossover according to the ICRP limits does not take place for more than 10^4 years. Figure 11–3 illustrates the sensitivity of such comparisons to changes in the underlying biophysical parameters used to calculate dose.

The ICRP and NRC values were the same for many years, but in 1980 the ICRP (ICRP, 1980b) changed the absorption factors for transfer of a number of radionuclides from the intestines to blood, the net effect of which was to greatly increase the long-range potential hazard from ^{237}Np. The difference between the two crossover times is mainly, but not entirely, due to that one change.

CRITERIA FOR OVERALL REPOSITORY PERFORMANCE

The design of high-level waste repositories has been handicapped by the fact that limits of permissible dose have been slow in evolving. There is no problem insofar as limits to the maximally exposed individual are concerned, since the repositories should have no trouble in meeting the dose limits prescribed for light-water reactors, i.e., 10–25 mR/year.[2]

The Department of Energy has undertaken risk assessments for the waste isolation pilot plant under construction in New Mexico (U.S. DOE, 1979b). It was concluded that a person living near the storage site would be subject to a 50-year bone dose commitment of 1.1×10^{-8} rem and a whole-body dose commitment of 8.7×10^{-11} rem. Even allowing for the uncertainties in dose modeling discussed in Chapter 4, it appears that isolation of high-level radioactive wastes in geological repositories can be achieved.

Establishing limits for collective dose is more complicated. The EPA regulations stipulate that the repository should be built so that it results in no more than 1000 deaths in 10,000 years, using present risk coefficients for development of fatal cancers. If these risks are applied to the world's population, and if one assumes that a new generation of people is born every 30 years and that the world's population will reach steady state at 10^{10} persons, it follows that 3×10^{12} persons will be born in the next 10,000 years and that the risk of dying because of radiation-induced cancer will be 3×10^{-9}. Although these risks are exceedingly small, it appears that the geological storage system is capable of meeting this limit (Smith *et al.*, 1982).

WHAT CAN WE LEARN FROM NATURE?

The models used to predict the performance of the repository are constructed by linking together the rates at which ground water will seep into the repository, the rates of corrosion of the canister, solubilization of the waste form, and migration of the radionuclides through the backfill and rocks. After a thousand years or more the nuclides reach the biosphere, and then begins the task of modeling the pathways by which the nuclides reach human beings. As noted earlier, Kocher *et al.* (1983) have estimated that the uncertainty in the dose estimates, based on transport only in the biosphere, may cover four or five orders of magnitude, and the uncertainty in the transport models used to describe the movement from the repository to the biosphere is indeterminate.

Most of the information on leaching rates, retardation factors, k_d, and other parameters of the dose models is obtained in laboratory measurements. A report by a panel of the National Academy of Sciences (NAS–NRC, 1982) noted that the leaching rates of solids under natural conditions

[2]There has been some concern over the possibility that someone in the far distant future might drill or dig their way into the repository proper, perhaps in the course of exploration or mining for natural resources. Whether or not this is a real problem depends on one's assumption about the durability of societal memory.

are very much lower than those obtained by laboratory measurement and that the differences are frequently two orders of magnitude or more.

A number of investigators have used natural analogues to infer the behavior of a deep geological repository. The best known of these studies involves a natural fossil fission reactor in the Republic of Gabon, West Africa (Cowan, 1976). The reactor existed in what is now known as the Oklo uranium mine about 1.8 billion years ago, when conditions in the uranium deposit were such as to sustain criticality for an estimated 10^5 years, during which time huge amounts of fission products and transuranic elements were produced. The radionuclides have long since decayed to stable nuclides at the end of the decay chain, and the presence of these nuclides, along with the stable nuclides produced in fission, has been detected in the environs of the fossil reactor. There is evidence that most of the nuclides of interest migrated for very short distances, of the order of meters, before they decayed. The fractional mobilization rates of the individual radionuclides from the reactor zones have been estimated to be on the order of 10^{-7} to 10^{-10} years. The mobilization rates were so low that most of the radionuclides decayed before they migrated more than a few meters from their source.

A second natural analogue from which similar conclusions have been drawn is a highly weathered deposit of thorium and rare earth elements located near the summit of a hill in the state of Minas Gerais, Brazil. The hill, which is known as the Morro do Ferro (Fig. 7–13), has been studied as an analogue for an ancient high-level waste repository that has been invaded by ground water and eroded to the surface. The ore body contains about 30,000 metric tons of thorium, which is being used as a chemical analogue for Pu^{4+}. Lanthanum, a rare earth element, is being used as an analogue for Cm^{3+} and Am^{3+} (Eisenbud et al., 1984). As at Oklo, the mobilization rates for the two analogues for the transuranic actinide elements have been shown to be on the order of 10^{-9} per year. Thus, even in this near-surface, highly weathered, wet ore body, the mobilization rates are so low as to ensure in situ decay of the transuranic elements plutonium, americium, curium, and, probably, neptunium (Krauskopf, 1986).

One feature that Oklo and Morro do Ferro have in common is the presence of abundant quantities of clay minerals. At Morro do Ferro, the primary minerals in which the analogue elements were originally contained have long since been destroyed by weathering, and the analogue elements are presently immobilized in amorphous form associated with clays and iron oxides.

There are surely conditions in nature that would result in more rapid dispersal than has been seen at Oklo and Morro do Ferro. On the other hand, the fact that some mineral deposits are stable in nature over geologic time is indisputable, and it would seem that long-term behavior of the wastes could best be predicted on the basis of the geophysical and geochemical conditions that tend to result in the natural stabilization of chemical elements.

The Special Problems of Gaseous or Highly Soluble Long-Lived Radionuclides

There are four radionuclides which, by reason of their long half-lives and either volatility or solubility, are accumulating in the general environment. These are tritium, krypton-85, carbon-14, and iodine-129, which are produced both in the generation of nuclear power and in the explosion of nuclear weapons.

Tritium

The annual production of fission product tritium by a 1000-MWe light-water reactor is in the range 15,000 to 25,000 Ci (Kouts and Long, 1973). When the tritium migrates from fuel during reactor operation, a fraction reacts with the zirconium cladding and becomes immobilized as zirconium hydride (NCRP, 1979). It has been found that about 13% of the total tritium contained in the fuel is in the form of zirconium hydride, and about 87% remains as ^3H in the fuel pellets. However, if the cladding has minor defects, the tritium can be released directly to the circulating water. Tritium production by neutron interaction with soluble boron can account for an additional 500–1000 Ci/year.

Tritium remaining in the fuel will be released if the fuel is reprocessed, as a result of which the tritium emissions from a fuel-reprocessing plant are far greater than from reactors. Not only is the bulk of the tritium contained in the fuel being dissolved, but the reprocessing plant has a far greater throughput of fuel.

Tritium from all sources is disseminated in the environment as water and enters the hydrological cycle. The impact of tritium produced in weapons testing has been far greater than that from all other sources, as can be seen from Fig. 11–4, where the contributions of the various sources to the tritium concentration of surface waters since 1960 are projected to the year 2000. Assuming there are no further major programs of atmospheric weapons testing, tritium from fallout will continue to diminish with the radiological half-life of 12.3 years, and toward the end of this century the contribution of nuclear power will be about equal to the fallout residual.

Although copious quantities of tritium have been and will continue to be produced, the dose to humans will be small. It has been estimated (NCRP,

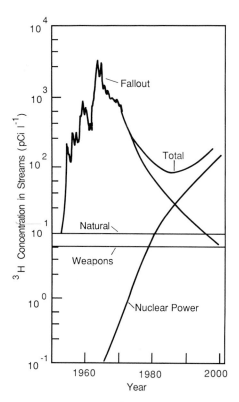

Fig. 11–4. Contributions to the concentration of tritium in streams and fresh water from all sources. (From National Council on Radiation Protection and Measurements, 1979.)

1979) that, as shown in Fig. 11–5, the dose to humans peaked in the mid-1960s at about 0.2 mrem/year.

KRYPTON-85

Krypton-85 is a noble gas nuclide that is produced copiously in fission. Krypton does not participate in metabolic processes, and the principal dose to an individual immersed in a cloud of this nuclide is to the skin, resulting from the 0.25-MeV beta emission.

Krypton-85 is produced in a 1000-MWe power reactor at an annual rate of about 500,000 Ci. Less than 1% of the krypton produced leaks through the fuel cladding during normal reactor operation, and for this reason the [85]Kr releases from normally operating reactors are insignificant compared to the releases from fuel reprocessing plants (NCRP, 1975b). About 50,000

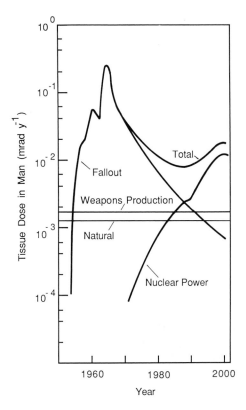

Fig. 11–5. Projected tissue dose rate in map. (From National Council on Radiation Protection and Measurements, 1979.)

Ci of ^{85}Kr was released as a result of the accident at Three Mile Island (NCRP, 1980b).

Because of its inertness, ^{85}Kr distributes uniformly throughout the earth's atmosphere within a few years after release. The ^{85}Kr concentration reached about 10 pCi/m^3 in 1970, mainly from nuclear weapons testing and plutonium production. It has been estimated that the skin dose to the world's population could reach about 2 mrem/year early in the next century, but this was based on the somewhat larger nuclear power growth projections of the early 1970s and also assumed that all fuel would be reprocessed without control over the ^{85}Kr. Krypton separation from exhaust gases by cryogenic and other methods is feasible. The EPA now requires that the krypton emissions be reduced from 500 kCi per 1000 MWe to 50 kCi per 1 MWe, a reduction of 90%, from all processing plants built after 1983. As noted earlier, the question of whether there will be reprocessing in the United States is uncertain. If the spent fuel is stored, the ^{85}Kr will decay with its half-time of 10.7 years. If the fuel is reprocessed, 90% of the ^{85}Kr must be removed for extended storage until the gas has decayed sufficiently for release.

CARBON-14

The three sources of ^{14}C in the environment are nature, nuclear weapons testing, and reactor operation. Carbon-14 from power reactors will accumulate in the atmosphere, but the dose to human beings will be insignificant for the foreseeable future. The contribution from fallout will continue to dominate the concentrations in the environment, about two orders of magnitude greater than that from reactor operation (NCRP, 1985b).

IODINE-129

Iodine-129 is one of the longest-lived nuclides produced in fission, with a half-life of 1.57×10^7 years. It is estimated that by the year 2000, about 2500 Ci of ^{129}I will have been produced by power reactors. Iodine is such a soluble element and the half-life of ^{129}I is so long that the ^{129}I will eventually enter the stable iodine pool. The total amount of iodine that can be absorbed into the thyroid is under metabolic control and is limited to about 0.012 g. Iodine-129 cannot deliver a significant dose to the thyroid because this would require deposition of 34 g of ^{129}I, several thousand times the average normal value (NCRP, 1983). However, if one assumes that the 2500 Ci of ^{129}I produced by the nuclear power industry up to the year 2000 will disseminate throughout the environment, it can be calculated that the collective dose equivalent to the world's population will be about 10^8 person–thyroid rem. The ^{129}I can be removed during fuel reprocessing, but whether this will significantly affect the long-range dispersal of ^{129}I in the environment is uncertain.

Chapter 12

Fallout from Nuclear Explosions:
I. Short-Term Effects

The use of nuclear weapons in war can produce radioactive contamination on a scale that, when compounded with the effects of blast and fire, would create wartime problems for which at present there is no foreseeable solution. Even the peacetime testing of nuclear weapons in the atmosphere is capable of producing worldwide contamination, and it was the subject of intense worldwide concern between 1954 and the signing of a test-ban agreement in 1963. In this chapter, we will first discuss the subject of nuclear and thermonuclear explosions in general and then review the short-term radiological consequences of the use of such weapons in war. The long-term consequences of nuclear weapons use will be reviewed in Chapter 13.

Most of what we know about fallout from the explosion of nuclear weapons has been learned from studies of the effects of test explosions that have taken place in various parts of the world since the first atomic bomb was detonated on a New Mexico desert in July 1945. In the intervening 40 years, hundreds of explosions have been conducted by the United States and the Soviet Union and also by the United Kingdom, India, France, and the People's Republic of China. Table 12–1 summarizes the yields of nuclear explosions that have been announced by the various nations. The

TABLE 12–1

ESTIMATED YIELDS OF ATMOSPHERIC NUCLEAR WEAPONS TESTS[a]

		No. of tests	Estimated yield (MT)	
	Period		Fission	Total
United States	1945–1962	193	72	139
USSR	1949–1962	142	111	358
United Kingdom	1952–1953	21	11	17
France	1960–1974	45	11	12
China	1964–1980	22	13	21
Total		423	218	547

[a]From UNSCEAR (1982).

tests have been conducted on land and sea, hundreds of miles above the earth's surface, and thousands of feet underground.

The first U.S. tests of nuclear weapons after World War II took place in the Marshall Islands in 1946 and 1948. The Soviet Union conducted its first tests in 1948, following which the United States accelerated its rate of testing and constructed a second proving grounds near Las Vegas, Nevada.

Testing nuclear weapons in the open atmosphere began to arouse worldwide concern in the mid-1950s, and fallout became a highly emotional and controversial subject. It was the widespread interest in the subject of nuclear fallout that alerted the world to problems of global pollution generally and led in the 1960s to a greater awareness of the dangers of pollution by halogenated hydrocarbons, lead, the products of fossil fuel combustion, and other waste products of technology.

Responding to worldwide popular pressure, the United States, United Kingdom, and Russia declared a moratorium in the fall of 1958 on further weapons testing. By that time the three nuclear powers had conducted 38 separate series of tests having a total of at least 227 detonations. France, which did not participate in the moratorium declaration, later became the fourth nuclear power by conducting a small series of nuclear tests on the Sahara Desert in 1960.

In 1961, without advance warning, the Soviet Union broke the moratorium agreement and exploded about 50 devices. The United States responded, and the two major powers began a frenetic competition which led to additional worldwide concern. The accelerated pace of weapons testing is illustrated in Fig. 12–1, which shows the rapidly increasing quantities of ^{90}Sr produced by nuclear explosions conducted up to 1962.

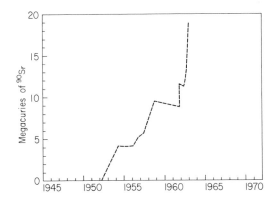

Fig. 12–1. Production of ^{90}Sr by nuclear weapons tests, 1952–1963.

A nuclear weapons test-ban agreement was signed by the United States, United Kingdom, and Soviet Union early in 1963 and to this writing has succeeded in eliminating further testing in the open atmosphere by the three signatory powers (Seaborg, 1981). However, France did not sign this agreement, nor did China or India, all of which have conducted tests in recent years.

The test-ban agreement did not rule out underground explosions but stipulated that venting to the atmosphere must not be detectable beyond the borders of the nation that conducts the test. Accordingly, signing of the test-ban agreement initiated a new era of nuclear weapons technology in which a variety of devices have been tested in underground cavities by methods which, with only a few minor exceptions, have prevented atmospheric pollution by radioactive debris. Hundreds of underground tests have been conducted by the United States and USSR during the past 20 years. Whether the underground accumulations of radioactive debris will in time prove significant as a form of environmental pollution remains to be seen. The quantities of debris that remain underground after such tests are huge, but objective evaluation of possible long-range off-site risks has not been possible because few of the basic data have been made available. By the end of 1984, the United States had conducted a total of 635 tests at the proving grounds near Las Vegas, Nevada. Of these, 105 were tests in the atmosphere and the remainder were conducted underground (U.S. DOE, 1985d).

Fallout of radioactive debris from explosion of a nuclear weapon first occurred after the 1945 test in New Mexico, in which a 19-kiloton device was fired from a 30-m steel tower. Some of the larger particles of radioactive debris fell on cows grazing about 20 miles downwind and produced

Fig. 12–2. X-ray film marred by exposure to contaminated interleaving paper processed August 6–10, 1945. (Courtesy of Eastman Kodak Company.)

skin burns (Lamont, 1965). The finer particles drifted across the Middle West, and enough fallout occurred in Indiana (Webb, 1949) to contaminate cornstalks that ultimately found their way into a papermaking process used by the photographic industry. Radioactive particles in the finished product eventually caused damage to X-ray film that had been packaged with contaminated interleaving paper (Fig. 12–2).

Fallout near the Nevada Test Site caused considerable local concern during the 1950s. More recently, there have been reports of increases in the incidence of leukemia (Lyon *et al.*, 1979) in nearby communities, but these reports have not been substantiated (Land *et al.*, 1984). Estimates of the collective dose to these populations suggest that any increases in cancer would not be detectable (Beck and Krey, 1983).

Physical Aspects of Nuclear Explosions

Nuclear energy can be released from a bomb by means of either the fission or fusion process. It has been customary to equate the explosive yields of nuclear and thermonuclear detonations to the equivalent amount of TNT. Thus, a 20-kiloton nuclear bomb is said to have an explosive yield equivalent to 20,000 tons of TNT, and a bomb having an explosive yield equivalent to 1 million tons of TNT would be said to have a yield of 1 megaton.[1]

The pure fission (nuclear) bomb obtains its energy from either ^{235}U or ^{239}Pu. Uranium-233 is also fissionable. Though it does not occur naturally, it can be produced in reactors by irradiation of ^{232}Th. Fissionable material may be made critical either by quickly joining two or more subcritical masses, which then become supercritical, or by implosive compression of a hollow subcritical mass. Both processes may be accomplished with the aid of chemical explosives. The complete fission of about 56 g of material will produce an explosion equivalent to 1 kiloton of TNT with a thermal yield of 10^{12} cal.

The nuclear bomb can be thought of as a fast reactor in the prompt critical condition in which neutron multiplication proceeds with a generation time of about 10^{-8} sec. Since 99.9% of the explosive yield is developed in the last 0.07 μsec, it is important that the shape of the supercritical mass be preserved for as long a time as possible by means of a surrounding tamper. In this way, the maximum number of fission generations can be achieved before the explosion effects change the shape of the device to the extent that criticality can no longer be sustained.

Approximately 50% of the energy from a nuclear explosion is released in the form of blast, 35% as thermal radiation, and the remaining 15% as ionizing radiation. In considering the total consequences of a nuclear explosion, the effects of blast and fire may be of even greater importance than the effects due to ionizing radiations, but only the latter aspect will be discussed in detail in this text. The reader is referred elsewhere (Glasstone and Dolan, 1977) for a review of the effects due to blast and fire.

---

[1]It may help the reader to appreciate the size of thermonuclear explosions if it is noted that 10 megatons of TNT would be a mass 3 m wide by 2 m high and 1000 km in length.

Of the ionizing radiations, one-third is prompt radiation produced within a few seconds after detonation, and two-thirds, or 10% of the total energy released by the explosion, is in the form of delayed ionizing radiation produced by the decay of fission products and induced radionuclides. The prompt ionizing radiations, consisting of γ rays and neutrons which are released at the time of detonation, produce their effects to approximately the same distance as the blast and thermal effects.

The fusion (thermonuclear) bomb utilizes the fusion reactions of light elements such as deuterium, tritium, or lithium. Several different reactions may occur, but all require that the nuclei have energies which can only be obtained with the aid of temperatures of several million degrees. Since this heat can be achieved in a fission bomb, such a device may be used as a trigger for a thermonuclear explosion. The thermonuclear reactions which can take place include the following:

$$^2H + {}^2H \rightarrow {}^3He + n + 3.2 \text{ MeV}$$

$$^2H + {}^2H \rightarrow {}^3He + {}^1H + 4.0 \text{ MeV}$$

$$^3H + {}^2H \rightarrow {}^4He + n + 17.6 \text{ MeV}$$

$$^3H + {}^3H \rightarrow {}^4He + 2n + 11.3 \text{ MeV}$$

$$^6Li + n \rightarrow He^+ + {}^3H + 4.8 \text{ MeV}$$

The last reaction involving ^6Li is of particular importance. The ^6Li can be used as $^6Li^2H$ (lithium deuteride, which has the advantage that the ^3H produced can react with the deuterium, ^2H) in the $^3H + {}^2H$ reaction given above (Glasstone and Dolan, 1977).

These reactions are sources of fast neutrons that cause fission of ^{238}U and can thus be utilized to increase the explosive yield by surrounding the fusion weapon with natural uranium. The fission of ^{238}U contributes a major fraction of the energy released by some thermonuclear weapons.

The terms "clean" and "dirty" have sometimes been used to describe the relative amounts of radioactivity produced by bombs. Those in which the energy is obtained primarily from fusion yield comparatively less radioactivity than weapons whose energy is derived primarily from the fission reactions. However, even a pure fusion device will produce some radioactivity by means of neutron activation.

Explosion of a nuclear or thermonuclear device produces a cloud of incandescent gas and vapor called the fireball, which is manyfold brighter than the noonday sun. Although the brightness begins to diminish after only 1 msec, the fireball continues to grow, reaching a final diameter of $D = 180W^{0.4}$ (Glasstone and Dolan, 1977), where D is in feet and W is the energy in kilotons.

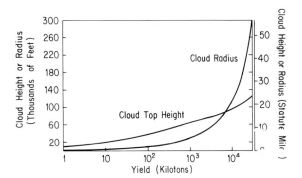

Fig. 12–3. Approximate values of stabilized cloud height and radius as a function of explosion yield for land surface or low air bursts. (From Glasstone and Dolan, 1977.)

In about 1 sec, when the fireball from a 20-kiloton explosion has reached its maximum size, it will be about 1400 ft in diameter. In 1 min, the fireball will have cooled sufficiently that it no longer glows, and by this time it will have risen to a height of about 4.5 miles. The size and height of stabilization are given as a function of yield in Fig. 12–3.

Convective forces initiated by the fireball result in enormous amounts of air and debris being sucked upward. Figure 12–4, which is a photograph of a relatively small nuclear explosion over the Nevada desert, shows the way desert sands are being convected into the fireball, which has assumed a toroidal shape. Particles which enter the fireball sufficiently soon after its formation are vaporized and mixed within the fireball. Later, as the fireball cools appreciably, convected particles will no longer be volatilized but may serve as nuclei on which condensation of the radioactive constituents of the fireball can occur. Some of these particles may be as large as grains of sand and possess sufficient mass that fallout will occur in a matter of minutes. However, if the fireball is sufficiently high off the ground that large particles are not sucked into it, the vapors will condense into a fume in which the particles are very small and will fall more slowly.

The radioactive debris from nuclear explosions divides into three fractions, depending on the height of burst and explosive yield. The first fraction consists of the larger particles, which fall out close to the site of detonation within hours and are intensely radioactive. The second fraction is dispersed into the troposphere, but may not result in fallout during the first day because the particles are sufficiently small to behave somewhat like aerosols and be subject to the laws of dispersion and rainout that govern small particles. The third fraction penetrates the stratosphere and will deposit worldwide over a period of many months. The tropospheric

Fig. 12–4. Explosion of a nuclear weapon, showing the toroidal structure of the fireball shortly after its formation. The fireball did not touch the ground, but it is sufficiently low that dust dislodged from the surface by the blast waves is being sucked into the fireball. (From U.S. Atomic Energy Commission.)

fallout tends to be distributed in bands at the latitude of detonation, whereas the stratospheric debris distributes itself globally, as will be described in the next chapter.

Debris from bombs smaller than about 100 kilotons detonated in temperate latitudes tends to remain in the troposphere, whereas stratospheric injection is almost complete for detonations greater than 500 kilotons.

The physical and chemical characteristics of the particles have been observed to be highly variable in several respects (Crocker *et al.*, 1966). Depending on the temperature–time history of the particle, the radioac-

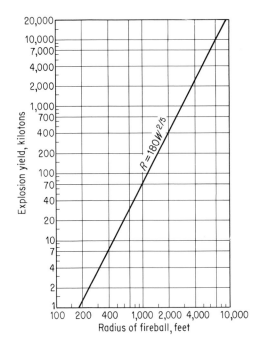

Fig. 12–5. Approximate height of burst below which a bomb of a given yield will produce a fireball that touches the ground. (From Glasstone, 1962.)

tivity can be coated on the surface or distributed throughout. The radioactivity of a particle can thus be proportional to either d^2 or d^3, depending on how the radionuclides are distributed. Intermediate distributions are possible owing to mixtures of the two kinds of particles or to gas inclusions in the larger particles.

The extent to which a given explosion will produce radioactive fallout in its immediate vicinity depends on the size of the explosion and its height above ground. The most fallout will be produced when the fireball touches the ground. If only 5% of a 1-megaton bomb is spent in volatilizing soil with which the fireball is in contact, about 20,000 tons of debris will be added to the fireball (Glasstone and Dolan, 1977). Figure 12–5 shows the relationship between explosive yield in kilotons and the height above ground below which the fireball may be expected to touch the ground and produce heavy local fallout.

The fireball will rise to a height determined by the explosive yield, the height at which the detonation occurred, and the meteorological conditions at the time of detonation. It is seen in Fig. 12–3 that in temperate latitudes fireballs from explosions under 100 kilotons stabilize below the

TABLE 12–2

APPROXIMATE YIELDS OF THE PRINCIPAL
NUCLIDES PER MEGATON OF FISSION

Nuclide	Half-life	MCi
^{89}Sr	53 days	20.0[a]
^{90}Sr	28 years	0.1[a]
^{95}Zr	65 days	25.0[a]
^{103}Ru	40 days	18.5[a]
^{106}Ru	1 year	0.29[a]
^{131}I	8 days	125.0[b]
^{137}Cs	30 years	0.16[a]
^{131}Ce	1 year	39.0[a]
^{144}Ce	33 days	3.7[a]

[a]From Klement (1965).
[b]From Knapp (1963).

tropopause (about 15 km in the mid-latitudes), and detonations in the megaton range penetrate well into the stratosphere.

Whether underground bursts produce fallout depends on whether the explosions vent through the surface. Many explosions have occurred in Nevada in which all the radioactive debris has been contained below the surface. On the other hand, underground bursts that penetrate above ground may have a very great potential to produce surface contamination.

The radioactive debris from a nuclear detonation originates in a number of ways. The principle source is the production of fission products in the relative amounts given by the fission-product mass yield curves in Fig. 9–8. The initial fission-produce mixture contains more than 200 isotopes of 35 elements. Most of the isotopes are radioactive, and most of them have very short half-lives, so that the diminution in radioactivity is very rapid immediately after fission. The yields of the principal fission products of concern are listed in Table 12–2.

The radioactivity A at any given time t after a nuclear explosion may be approximated if the radioactivity at unit time A_0 is known:

$$A = A_0 t^{-1.2} \qquad (12\text{-}1)$$

This equation provides estimates of the radioactivity for periods of less than 6 months. When the radioactivity decays according to this law, the levels diminish approximately 10-fold for every 7-fold increase in time since the explosion. Thus, if the ambient radiation level is 3.0 R/hr at 4 hr after the burst and if all fallout has ceased, the radiation levels can be

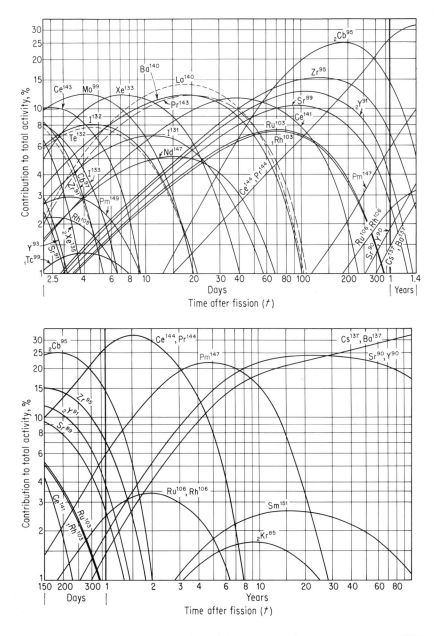

Fig. 12–6. Yields of the principal radionuclides from the slow-neutron fission of ^{235}U. (From Hunter and Ballou, 1951.)

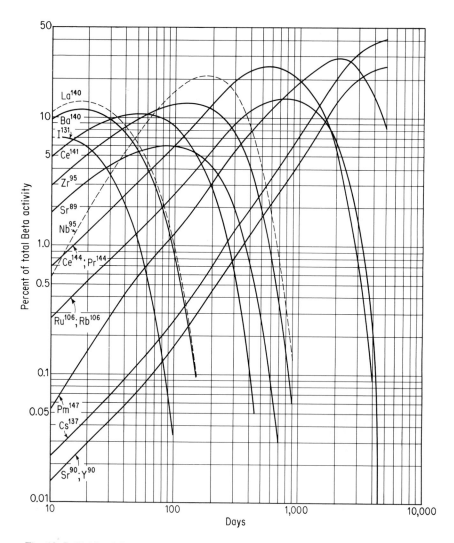

Fig. 12–7. Yields of the principal radionuclides in the debris from megaton weapons. The differences from Fig. 12–6 occur because in such weapons fission occurs from both fast and thermal neutrons and in ^{238}U and ^{239}U as well as ^{235}U. (From Hallden et al., 1961.)

expected to diminish to about 0.3 R/hr at the end of 28 hr. This decay equation is only approximate, since fission yields vary from burst to burst, depending on the weapon design.

The fission product composition produced by the slow neutron fission of ^{235}U at various times after production is given in Fig. 12–6. Hallden et al.

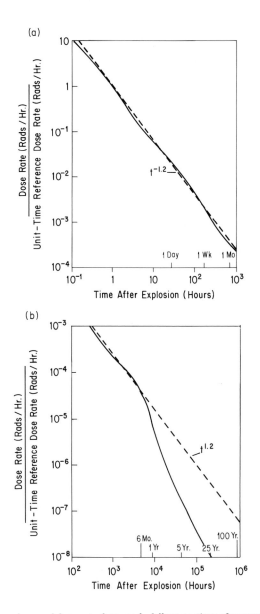

Fig. 12–8. Dependence of dose rate from early fallout on time after explosion from (a) 0.1 to 1000 hr and (b) up to 10^6 hr. (From Glasstone and Dolan, 1977.)

TABLE 12–3

PRINCIPAL RADIONUCLIDES INDUCED IN AIR[a]

Isotope	Half-life (years)	Ci/megaton
^{3}H	12.3	<1
^{14}C	5600	3.4×10^4
^{39}A	$\sim$260	59

[a] Adopted from Klement (1959).

(1961) have reported on the isotopic composition of the debris of megaton weapons in which fast fission of ^{238}U is presumed to have occurred along with fission of ^{235}U and ^{239}Pu. Their data, which are given in Fig. 12–7, cover the period from 10 to 5000 days and are based to some extent on radiochemical analysis of the debris from megaton tests. Although some differences exist between the distributions in Figs. 12–6 and 12–7 when the various radionuclides in each case are summed, the total β activities at any given time and the rates of decay are almost identical.

The manner in which the radioactivity of fission-product mixtures may depart from the rate described by Eq. (12–1) is shown in Fig. 12–8a,b. It is seen that the actual decay is more rapid than that described in Eq. (12–1).

In addition to fission products, a number of induced radionuclides are produced by nuclear bombs, including substantial amounts of ^{239}Pu. Other radionuclides are produced by neutron interactions with nonradioactive elements of the bomb, the atmosphere, and in some instances seawater or soil. Libby (1958) estimated that the interactions of neutrons with atmospheric nitrogen produces 3.2×10^{23} ^{14}C atoms per kiloton of yield. We will see in Chapter 13 that these reactions have been sufficient to produce a marked increase in the natural background of radiocarbon and tritium.

The complex neutron spectrum is capable of producing a variety of other induced activities in the atmosphere, and some of these are given in Table 12–3.

When nuclear weapons are detonated close to the ground, a number of radionuclides may be produced by neutron reactions in soil. These are summarized in Table 12–4, which shows that ^{45}Ca and ^{55}Fe are the only long-lived nuclides produced in significant amounts. Other activation products may also be important, such as those produced in the weapon itself (^{239}Np) and by structural materials in which ^{60}Co can be produced.

A number of attempts have been made to estimate the extent of induced radioactivity at Hiroshima and Nagasaki. From the estimated neutron

TABLE 12-4

<small>PRINCIPAL RADIONUCLIDES INDUCED IN SOIL[a]</small>

Isotope	Half-life	Ci/megaton
^{24}Na	15 hr	2.8×10^{11}
^{32}P	14 days	1.92×10^{8}
^{42}K	12 hr	3×10^{10}
^{45}Ca	152 days	4.7×10^{7}
^{56}Mo	2.6 hr	3.4×10^{11}
^{55}Fe	2.9 yr	1.7×10^{7}
^{59}Fe	46 days	2.2×10^{6}

[a] Adopted from Klement (1959).

spectrum of the two bombs and radiochemical analyses of soils and building materials from the two cities, it has been estimated that the dose commitment to individuals who entered the Hiroshima hypocenter 1 day after the bombing and remained for 8 hr would have been 3 rad (Hashizume *et al.*, 1969). The dose due to induced radioactivity at 500 and 1000 m from the hypocenters would have been 18 and 0.07% of the hypocenter doses in Hiroshima and Nagasaki, respectively.

It is possible that under certain conditions fractionation of fission products will occur, so that debris falling out in different places will be enriched or depleted in certain radionuclides. One reason for this fractionation is that among the fission products are noble gases such as xenon and krypton, which, though short-lived, exist sufficiently long that the radionuclides to which they decay may be formed relatively late in the life of the fireball, in some cases after condensation of debris has already begun to take place. For example, there are several isotopes of krypton and xenon that decay to rubidium and cesium, respectively, and these, in turn, decay to strontium and barium. Thus ^{90}Sr, which is the granddaughter of ^{90}Kr, may exist in less than theoretical amounts in fallout close to the site of detonation and may be correspondingly enriched in the debris that falls out at a later time. The decay scheme of ^{90}Kr is as follows:

$$^{90}\text{Kr} \xrightarrow{(33\ \text{sec})} {}^{90}\text{Rb} \xrightarrow{(2.7\ \text{min})} {}^{90}\text{Sr} \xrightarrow{(28\ \text{yr})} {}^{90}\text{Yr} \xrightarrow{(65\ \text{hr})} {}^{90}\text{Zr}\ (\text{stable})$$

It is seen that the ^{90}Sr will not be formed completely for several minutes after detonation.

Disequilibria may also occur because the different nuclides condense from the fireball at different temperatures. The more volatile elements, which condense last, are deposited on the surfaces of small particles, which settle more slowly.

From the data obtained during weapons tests, it is generally accepted that about 90% of the radioactivity produced in a detonation is to be found in the head of the mushroom-shaped cloud formed by the fireball as it cools. The remaining 10% is contained in the cloud stem.

The particle size of the debris is highly dependent on the type of explosion. Near-surface detonations produce large glassy masses that are highly radioactive, as well as a lognormally distributed spectrum of smaller particles. Bursts that take place high in the atmosphere produce smokelike particles that remain suspended for considerable periods of time. Analyses of many samples from air-burst clouds indicated a modal particle diameter (Freiling and Kay, 1965). It has been estimated on theoretical grounds that air-burst particles cannot grow to more than 0.3 μm by condensation and that larger particles are probably the results of coalescence of smaller ones.

The radioactivity of the debris produced in a nuclear explosion diminishes by a factor of about 20 from the first hour to the end of the first day. In addition, the radioactivity of the cloud becomes more diffuse during this period as a result of meterological dispersion. The potential hazards of radioactive fallout, therefore, vary greatly, depending on whether the debris falls out within a few hours after the detonation or over a longer period of time.

RADIOACTIVE FALLOUT

Our knowledge of fallout has been gathered from weapons tests and is limited by the fact that conditions were necessarily very different from those that would be faced in time of war. Tests in the United States were conducted in Nevada, where, for reasons of public safety, it was necessary in general to restrict the size of atmospheric tests to less than 50 kilotons. Bombs larger than 100 kilotons were tested in the Marshall Islands, where much information about fallout was obtained, but the circumstances were somewhat less than ideal because of the enormous expanses of water over which it was necessary to make observations. The Marshall Islands are thin coral reefs, and large land surface bursts could not be simulated. In fact, many of the devices were mounted on barges floating in lagoons. Other detonations were on reefs just barely above the surface of the water, with cratering extending well below the water line.

As a result of these limitations, many questions remain unanswered. Very little is known about the physical and chemical properties of the fallout that would be encountered under the wide range of conditions that might prevail during a nuclear war. Bombs of various sizes would be detonated at various heights, the terrain would vary from place to place, and it might be raining or snowing. The only thing about which one could be quite certain is that the conditions would be very unlike those in either Nevada or the Marshall Islands.

Nevertheless, much has been learned. The sequence of events from the moment of detonation until the time of fallout is known, and a first approximation can be made of the probable extent of the area to be affected by fallout if the conditions of detonation are known. The remainder of this chapter will be concerned with fallout effects in the immediate vicinity of the explosion. More remote effects or "worldwide fallout" will be discussed in Chapter 13.

The levels of radioactivity that will exist at any given location will depend on the amount of radioactivity produced in the burst, the cloud height, the distribution of radioactivity within the cloud, and the manner in which this radioactivity is distributed as a function of particle size. Of course, the fallout pattern will also be influenced markedly by the wind velocity and direction along the full height of the cloud, as well as by the presence of precipitation.

Fallout patterns can be predicted from knowledge of the characteristics of the explosion and meterological conditions, and this information can be useful for forecasting the general areas in which fallout is most likely to occur. However, any reliance on fallout predictions beyond the need to decide that fallout may occur in a given sector and that the public should seek the protection of evacuation or shelters would be extremely dangerous in view of the lack of precision of the fallout prediction methods even under the best conditions. The only experience thus far has been from weapons tests for which there was far more exact knowledge of the explosive yield, the conditions of detonation, and the prevailing meteorology than would be available under war conditions, when it might be necessary to utilize meteorological data several hours old that might have been collected at some location other than the target area. Even under the comparatively ideal conditions of the weapons tests, the fallout predictions were only an approximation of the actual fallout patterns. It seems doubtful that under wartime conditions it would be possible to know any more than that fallout is likely to occur in a given general area. Information as to exactly when, where, and how much fallout will occur would have to await actual observations. Figure 12–9 (Ferlic, 1983) compares the predicted and actual fallout from the test of a 43-kiloton device fired from a 500-ft tower in Nevada in 1955. The prediction was made 8 hr before the test. The maximum cloud height was 42,000 ft. While this is an extreme case in which the wind structure proved to be grossly different from what was predicted, Fig. 12–9 does serve to illustrate the difficulties involved in

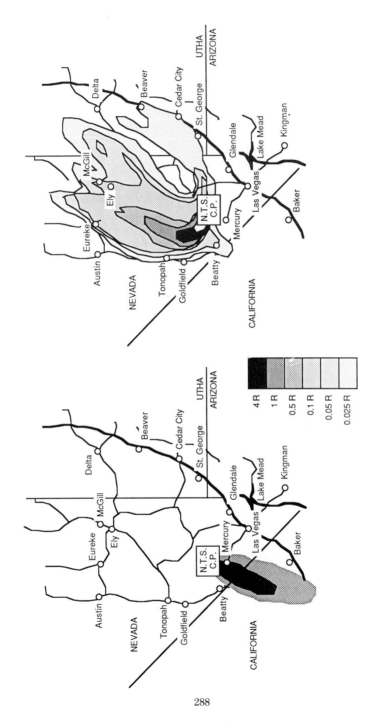

Fig. 12–9. Predicted fallout and actual infinite isodose contours for TURK shot, Nevada, 1955. (From Ferlic, 1983.)

■	4 R
(dark gray)	1 R
(medium gray)	0.5 R
(light gray)	0.1 R
(lighter gray)	0.05 R
(lightest)	0.025 R

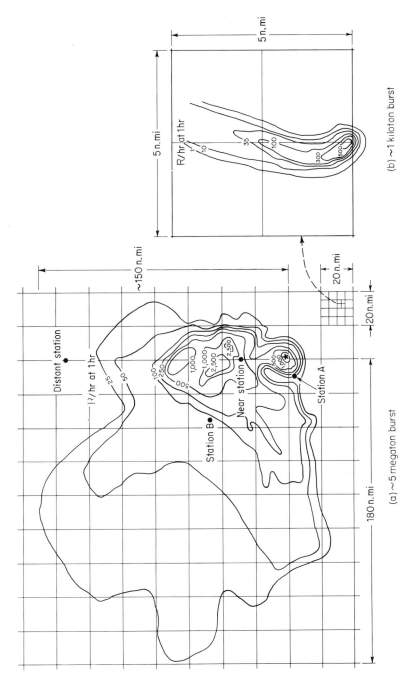

Fig. 12–10. Observed fallout following a 5-megaton burst on a Pacific island (n.mi, nautical miles). (From Triffet, 1959.)

289

making fallout predictions. The problem would be even more complicated for the overlapping fallout patterns of megaton explosions. The complexity of the fallout pattern from a 5-megaton test on a coral reef in the Pacific is shown in Fig. 12–10. Note the irregularity of the isodose contours and the two "hot spots" 40 and 60 nautical miles north of the hypocenter (Triffet, 1959).

In general, one may expect that relatively small detonations in the kiloton range will result in fallout patterns that more closely resemble the idealized cucumber-shaped patterns. However, fallout predictions following megaton bursts may be expected to be subject to great error because the greater cloud height subjects the debris to more wind shear.

There is as yet no way of forecasting the effects of natural terrain features or man-made structures on fallout. For example, it is possible that great differences may exist from one side of a building to another, particularly if the buildings are multistoried, reinforced-concrete structures.

METHODS OF ESTIMATING FALLOUT EXPOSURE

For the purpose of plotting data and standardizing the types of dose calculations that must be made in order to estimate dose rates and dosages over various intervals of time, it is customary to forecast the fallout levels at a given point as though all the fallout occurred 1 hr after the detonation. The fallout patterns in Fig. 12–9 are plotted in this way despite the fact that the contours cover such a large area that fallout at distant locations could not begin until 6 or 7 hr after detonation.

The plots of radiation levels based on the activity at 1 hr grossly overestimate the magnitudes of levels actually encountered, but they do serve the purpose of facilitating computation of dose rates and dose commitments during any given interval of time.

The dose calculations are based on the approximation that a fallout of 1 MCi/mi^2 will result in a dose of about 4 R/hr at 3 ft above the ground.

Given the dose rate at any time after a detonation and assuming that radioactive decay occurs at a rate proportional to $t^{-1.2}$, the various required calculations can be performed quickly with the aid of nomograms, curves, and slide rules. Some useful examples of these calculations are the following, which have been adopted from a U.S. government manual, "The Effects of Nuclear Weapons" (Glasstone and Dolan, 1977).

Example 1

Given: The radiation dose rate due to fallout at a certain location is 8 R/hr at 6 hr after a nuclear explosion.

Find: (a) The dose rate at 24 hr after the burst. (b) The time after the explosion at which the dose rate is 1 R/hr.

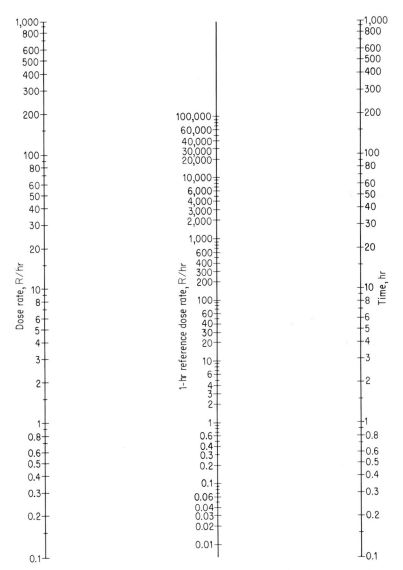

Fig. 12–11. Nomogram for the calculation of dose rate from fission products in fallout. (From Glasstone, 1962.)

Solution: By means of a straight edge, join the point representing 8 R/hr on the left scale of Fig. 12–11 to the time 6 hr on the right scale. The straight line intersects the middle scale at 70 R/hr; this is the 1 hr reference value of the dose rate.

(a) Using the straight edge, connect this reference point (70 R/hr) with

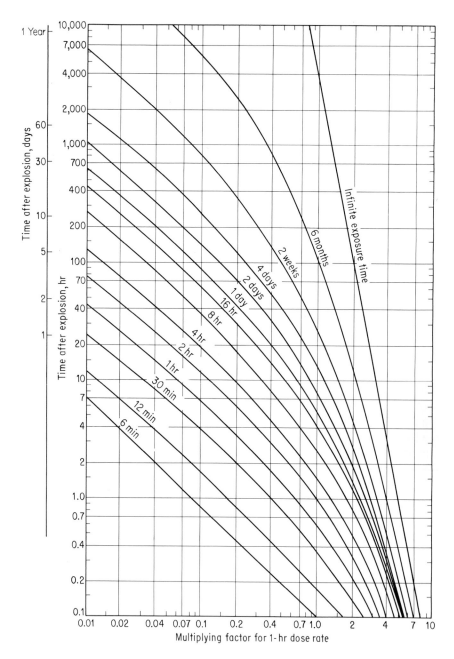

Fig. 12–12. Method of computing accumulated dose based on 1-hr reference dose rate. See Example 2 in text. (From Glasstone, 1962.)

that representing 24 hr after the explosion on the right scale and extend the line to read the corresponding dose rate on the left scale, i.e., 1.5 R/hr.

(b) Extend the straight line joining the dose rate of 1 R/hr on the left scale to the reference value of 70 R/hr on the middle scale out to the right scale. This is intersected at 34 hr after the explosion.

Example 2

Given: The dose rate at 4 hr after a nuclear explosion is 6 R/hr.

Find: (a) The total dose received during a period of 2 hr commencing at 6 hr after the explosion. (b) The time after the explosion when an operation requiring a stay of 5 hr can be started if the total dose is to be 4 R.

Solution: The first step is to determine the 1-hr reference dose rate. From Fig. 12–11, a straight line connecting 6 R/hr on the left scale with 4 hr on the right scale intersects the middle scale at 32 R/hr per hour; this is the reference dose rate at 1 hr.

(a) Enter Fig. 12–12 at 6 hr after the explosion (vertical scale) and move across to the curve representing a time of stay of 2 hr. The corresponding reading on the horizontal scale, which gives the multiplying factor to convert the 1-hr dose rate to the required total dose, is seen to be 0.19. Hence, the total dose received is

$$0.19 \times 32 = 6.1 \text{ R}$$

(b) Since the total dose is given as 4 R and the 1-hr dose rate is 32 R/hr, the multiplying factor is $4/32 = 0.125$. Entering Fig. 12–12 at this point on the horizontal scale and moving upward until the (interpolated) curve for 5-hr stay is reached, the corresponding reading on the vertical scale, giving the time after the explosion, is seen to be 19 hr.

Example 3

Given: On entering a contaminated area 12 hr after a nuclear explosion, the dose rate is 5 R/hr.

Find: (a) The total radiation dose received for a stay of 2 hr. (b) The time of stay for a total dose of 10 R.

Solution: Start at the point on Fig. 12–13 representing 12 hr after the explosion on the vertical scale and move across to the curve representing a time of stay of 2 hr.

(a) The multiplying factor for the dose rate at the time of entry, as read from the horizontal scale, is seen to be 1.9. Hence, the total dose received is

$$1.9 \times 5 = 9.5 \text{ R}$$

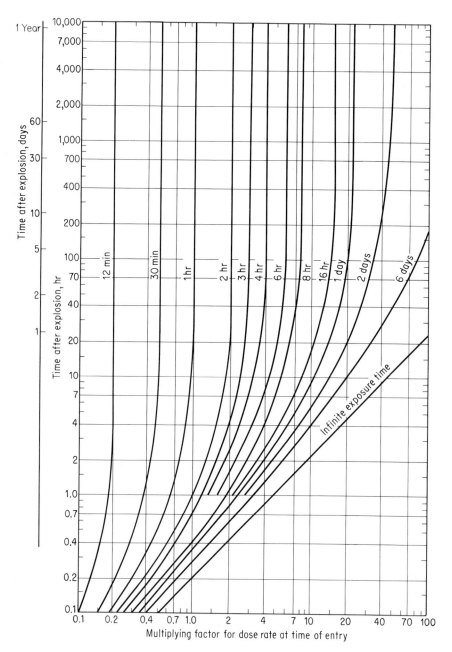

Fig. 12–13. Method of computing accumulated dose based on dose rate at time of entry into contaminated area. See Example 3 in text. (From Glasstone, 1962.)

(b) The total dose is 10 R and the dose rate at the time of entry is 5 R/hr; hence, the multiplying factor is $10/5 = 2.0$. Enter Fig. 12–13 at the point corresponding to 2.0 on the horizontal scale and move upward to meet a horizontal line which starts from the point representing 12 hr after the explosion on the vertical scale. The two lines are seen to intersect at a point indicating a time of stay of about $2\frac{1}{3}$ hr.

The Short-Term Radiological Effects of Nuclear War

There have been many reports of the radiological effects of nuclear war, each based on a selected scenario that describes the number of weapons, their size, heights of detonations, and meterological conditions. In one 1959 Congressional hearing, memorable because it was the first to receive the attention of the public, many experts presented their assessments of the consequences of an attack on the United States (U.S. Gov. Print. Off., 1959). A hypothetical attack was considered in which a total of 223 targets in this country were struck by nuclear and thermonuclear bombs, each of which had an explosive force of 1 to 10 million tons of TNT. The total explosive yield of all bombs dropped on the United States was equivalent to 1453 megatons of TNT, and it was assumed that additional bombs delivered outside the continental United States had a total explosive yield equivalent to 2500 megatons of TNT. The fallout patterns in the continental United States at 7 and 48 hr after an attack are shown in Figs. 12–14 and 12–15.

This hypothetical attack, which may be presumed to be a realistic appraisal of the ability of a potential enemy to deliver nuclear weapons within our borders in 1959, would have resulted in death to an estimated 42 million Americans and injury to an additional 17 million. Almost 12 million dwellings would have been so badly damaged that they could not have been salvaged, and an additional 8 million dwellings would have had to be evacuated for major repairs. Thus, more than 30% of the U.S. population would have been killed or injured, and more than 40% of the homes would have been destroyed or badly damaged. The stockpiles of nuclear weapons in the arsenals of the two superpowers have greatly increased since the 1959 hearings. Ten years later (NAS–NRC, 1969) the assumed attack was equivalent to 12,000 megatons, an order of magnitude greater than the earlier assumption. The fallout patterns in Figs. 12–14 and 12–15 must, therefore, be taken as a lower estimate of the radiological problems that would be faced in the event of a massive nuclear attack. More recent scenarios (Office of Technology Assessment, 1979; Barnaby, 1982) have involved even greater explosive yields. It is estimated that there are now 50,000 warheads in the world, most of them in the stockpiles of the United States and the Soviet Union.

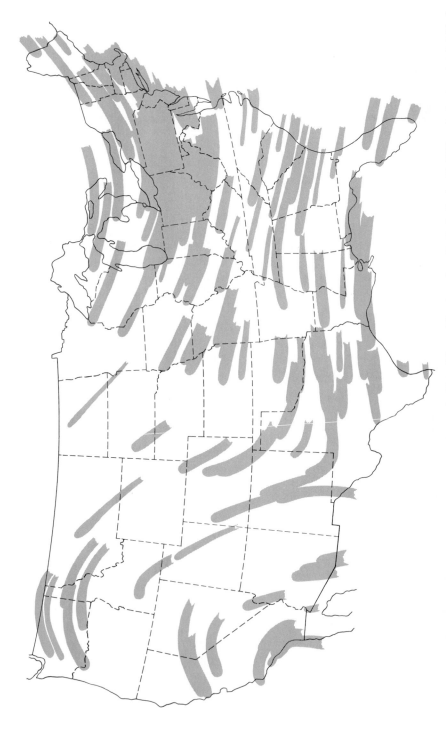

Fig. 12–14. Pattern of near-lethal levels of radioactive fallout in the United States 7 hr after a hypothetical attack with 223 bombs having a total yield of 1453 megatons. (From Shafer, 1959.)

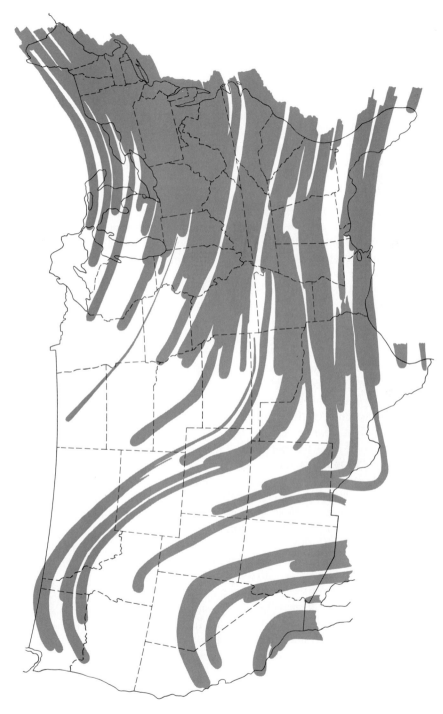

Fig. 12–15. Fallout patterns 48 hr after hypothetical attack of Fig. 12–14. (From Shafer, 1959.)

The above casualty estimates are based on an analysis of the radiation levels, the blast effects, and the effects of fires produced by the bombings. The extent to which additional casualties would be produced among survivors by infectious disease, starvation, deterioration of law and order, and exposure to the elements was not assessed. The vast numbers of injured and diseased people, the enormous destruction of production facilities and housing, interruptions of power and communications, and disruption of the methods of distributing essential goods and services are only a few of the factors that would interact in so complicated a manner as to place the true consequences of a nuclear attack beyond comprehension. In the above hypothetical attack on the United States, about 96 million people lived outside the areas of likely blast damage. These people would be exposed to radioactive fallout at various times, from a few minutes to a few hours after the detonation, depending on their location with respect to the bursts. The principal immediate problem that would face this portion of the population would be the potentially lethal levels of radioactive fallout. Although they would be beyond the range of the effects of blast and thermal radiation, the problems of dealing with the radiation effects would nevertheless be complicated greatly by the enormous social stresses that would be produced throughout the nation as a whole.

It has recently been suggested (Turco et al., 1983; Ehrlich and Sagan, 1984) that by far the most devastating effect of nuclear war, previously overlooked, would be the absorption of solar radiation in the upper atmosphere by large quantities of soot and dust that would be disseminated in the atmosphere. This could result in severe long-term climatic effects (from a 5000-MT nuclear exchange) that would result in subfreezing temperatures and darkness for periods of many months, with catastrophic effects on vegetation and widespread death by starvation.

It is most likely that in areas where the fallout is acutely hazardous the particles would be so large as to be visible. Japanese fishermen who were trapped by fallout on their boat, the Lucky Dragon, about 80 miles downwind of the large thermonuclear detonation on March 1, 1954, reported that the fallout of white dust resembled snow, and the deck was said to have been covered to such an extent that footprints were clearly visible (see Chapter 14). Japanese scientists who investigated this accident (Japan Society for the Promotion of Science, 1956; Kumatori et al., 1980) reported that the fallout at the time of deposition amounted to 38 to 85 g of dust per square meter of deck surface. The fallout was observed to be in the form of dust having a particle size of 0.1–3 μm, but agglomerated into granules of about 300 μm. At closer distances the particles might range up to several millimeters in size (Triffet, 1959).

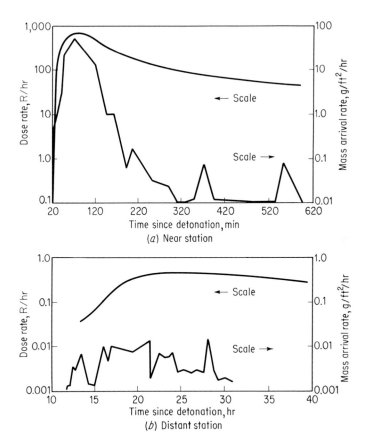

Fig. 12–16. Arrival times of fallout at two stations downwind of a large thermonuclear explosion. This is only one example, and it cannot be assumed that the curves would be similar for another explosion. The shapes of the curves will depend on many factors, including size and type of burst and meterology. (From Triffet, 1959.)

These particles become sources of external γ radiation and, in addition, can produce β burns on the skin or can be inhaled or ingested.

γ *Radiation*

The rate of rise of the γ-radiation levels is most likely to be variable, depending on distance from the detonation. Experiments during U.S. nuclear weapons testing programs in the Pacific provided data such as those shown in Fig. 12–16 (Triffet, 1959), in which the γ-radiation levels as a function of time after burst are plotted. The data were collected at two typical stations, one just beyond the zone of blast and thermal damage and the other sufficiently far downwind that it took several hours for the fallout cloud to arrive.

The short-term biological effects to be expected from fallout would depend in a complex way on the rate at which the total dose was delivered and on whether there were any concomitant effects such as those from blast and thermal radiation. The effects of acute exposure were discussed in Chapter 2 and summarized in Table 2–1.

β *Burns of the Exposed Skin*

A subsequent chapter will describe the detailed findings of investigators who studied the results of a 1954 episode in which a group of inhabitants of Rongelap Atoll in the Marshall Islands and 22 Japanese fishermen aboard a nearby vessel were inadvertently exposed to heavy fallout. In both series of cases, skin burns were produced by the β activity of dust deposited on the skin. This effect was limited to exposed portions of the body, particularly portions on which the dust particles tended to collect. The wearing of clothing, the precaution of remaining indoors during periods of fallout, and reasonable standards of personal hygiene should make it possible to avoid injuries of this type.

Inhalation and Ingestion of Radioactivity

The question of what levels of radioactive contamination of foods or water would be acceptable for consumption in the postattack period is one which cannot be answered in advance. The permissible levels of contamination in peacetime are properly conservative, so that food and water are safe for lifetime consumption or, if not absolutely safe, would result in injury on a scale that would be undetectable against the frequency with which the effects occur normally in the population. In contrast, the levels of contamination that would be permissible in a postattack period would vary, depending on many factors. It would be difficult to withhold food from a starving population, even though it might be contaminated. The harsh realities of the postattack period might make it relatively unimportant that delayed radiation effects may occur in a substantial fraction of the population. The risk of not surviving the postattack period owing to factors totally unrelated to radioactive contamination might be so great that a higher risk of developing cancer in future years might not seem important. It would be necessary for local authorities to determine, on a day-to-day basis, what levels of contamination should be permitted, depending on the number of survivors and the amount of food known to be available. All of this presumes a degree of social order that might not exist.

In the immediate postattack period, there would be two principal dangers from contaminated food. Fresh foods that contain radioiodine could result in massive doses to the thyroid, and contamination of leafy vegetables or water by intensely radioactive fallout particles could result in high exposure of the lining of the intestinal tract.

Of the several radioactive isotopes of iodine that are produced in nuclear explosions, the most significant is ^{131}I, with a half-life of 8.1 days. The most important way in which radioiodine passes to humans is by fresh dairy products, because of the relative speed with which the short-lived isotope can move from contaminated forage to humans via this route.

In wartime, thyroid doses among the general population could exceed thousands of rads, assuming that no countermeasures are taken, despite the fact that weathering and radioactive decay combine to reduce radioiodine deposited on foliage with a half-time of about 5 days. The requirement for restrictions and countermeasures might thus be limited to the first few weeks of the postattack period, so far as ^{131}I is concerned.

Individuals who survived the period of acute radiation danger would emerge into an environment that would be contaminated with radioactivity for the rest of their lives, but many of the fission products that have intermediate or long half-lives are relatively inactive biologically and would not appear as important contaminants of foods. These isotopes, which include ^{95}Zr, ^{106}Ru, and ^{144}Ce, would result in elevation of the external γ-radiation background, but would contaminate food only by being deposited on the surface of plants and would not be absorbed significantly by ingestion. However, these radionuclides could result in high doses to the gastrointestinal tract.

Considerably more significant would be the effects of ^{89}Sr and ^{90}Sr. Strontium-89 has a half-life of only 59 days, but in fresh fallout the ratio of ^{89}Sr to ^{90}Sr may be as much as 200 to 1, diminishing to about 10 to 1 at the end of 1 year and to an insignificant fraction after 2 years. These strontium isotopes can enter human food supplies either by foliar deposition or by absorption from soil. The former mechanism is dominant when the rate of fallout is relatively high but becomes less important when the rate of fallout diminishes (see Chapter 6).

The dose from ^{89}Sr is self-limiting because of its short half-life, but it has been estimated (Dunning, 1962) that during the first year of exposure this nuclide may deliver as much as 3.4 times the dose delivered by ^{90}Sr. At the end of the second year this isotope will have decayed to such an extent that the ^{89}Sr component of the skeletal dose will become a relatively unimportant fraction of the total dose commitment.

The extent to which fallout would cause significant contamination of water supplies would depend on such factors as the amount of fallout directly on the reservoir, the rate of runoff from the watershed, the reservoir volume, the time it takes the water to pass from the reservoir to the consumer, and the extent to which decontamination takes place. Factors that assist decontamination would include the settling of suspended solids to the bottom of the reservoir, adsorption of radionuclides on suspended solids, and water-treatment processes.

If one makes the assumption that uniform mixing of the fallout occurs and that no decontamination procedures exist, and if it is further assumed that there is no delay between the reservoir and the consumer, it is possible to calculate the dose a consumer would receive from any given fallout level. If the fallout level on an infinite smooth plane is 3000 R/hr at 1 hr and this amount of fallout is received by a reservoir 10 ft deep, a dose of 305 rad would be received by the thyroid of an average person drinking the water for 30 days (Hawkins, 1961). The dose commitment from ingestion during the first day after the fallout occurs would be 142 rad. If the consumer abstained from drinking the reservoir water for 9 days, the dose would be less than 15 rad.

The extent to which decontamination would take place in water-treatment plants of various conventional types is shown in Fig. 12–17 (Lacy and Stangler, 1962). The conventional process of coagulation, chlorination, and filtration removes 70–80% of the radioactivity.

On balance, the problem of water contamination is one which must certainly be considered, but in the postattack period this factor would be secondary, since a population could survive on contaminated water, assuming other sources of exposure were under control.

It is generally accepted that the long-term risk of somatic injury from residual radiation would result mainly from exposure to ^{90}Sr. In the 1959 Congressional hearings mentioned earlier, the effects of a hypothetical 1453-megaton attack on the United States were considered. It was further assumed that the total explosive yield of bombs delivered outside the United States was equivalent to 2500 megatons of TNT. Thus, the total explosive yield was equivalent to a little less than 4000 megatons of TNT, and it was assumed to be equally divided between fission and fusion. The total production of ^{90}Sr during such an exchange can be estimated from knowledge that the yield of this nuclide is about 0.1 MCi/megaton of fission. Since the fission yield was about 2000 megatons, ^{90}Sr production would be 200 MCi.

A 4000-megaton exchange would be an enormously destructive war, which would result in hundreds of millions of casualties and could destroy civilization as we know it. The dangers from ^{90}Sr contamination of food following such an attack would be one of the minor problems when considered against the total consequences of a war of such magnitude. One can gain perspective from experience with the ^{90}Sr produced in past weap-

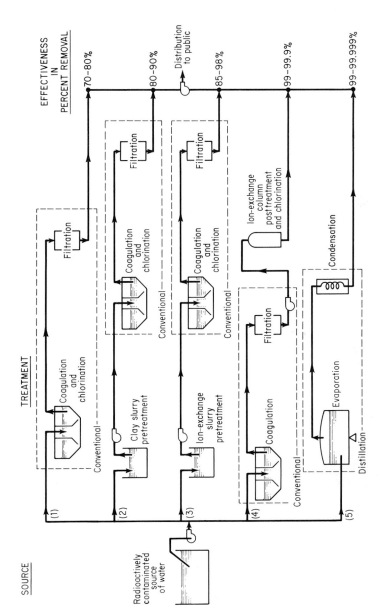

Fig. 12–17. Decontaminating effectiveness of various water-treatment processes. (From Lacy and Stangler, 1962; reproduced from *Health Physics*, vol. 8, by permission of the Health Physics Society.)

ons tests, in which the fission yields have totaled about 200 megatons. The tests have thus injected about 20 MCi of ^{90}Sr into the atmosphere. We will see in Chapter 13 that the global fallout of ^{90}Sr has totaled about 12 MCi, or about 55% of the total. The remainder can be assumed to have been deposited in the immediate vicinity of the explosion.

We will also see in Chapter 13 that the "average" 50-year skeletal dose commitment from this deposit has been estimated to be about 100 mrem in the northern hemisphere. Thus, the skeletal dose commitment from ^{90}Sr might reach 1000 mrem following a 2000-megaton attack. This estimate applies to areas beyond a few hundred miles of the targets and is an average figure. The dose from ^{90}Sr, being a 50-year dose commitment, would be superimposed on a dose of about 5000 mrem due to the natural background.

The ^{90}Sr levels in the most contaminated areas immediately downwind of the targets would reach about 300,000 mCi per square mile, about 100 times the deposition in the regions just discussed (Machta, 1959). Assuming that these areas were farmed after the attack, the skeletal burden of children deriving their calcium from such food would be about 12,000 pCi/g Ca, equivalent to about 36 rem during the first year and about 520 rem over a lifetime. An additional bone dose would be received during the first year because of the presence of ^{89}Sr, and this is estimated to be about 3.4 times the dose from ^{90}Sr, or about 120 rem. The total dose to the bone during the first year would thus be about 150 rem, and over a lifetime it would be about 640 rem. To these bone doses, derived from contaminated food, it is necessary to add the dose from external sources, which might be larger than the dose from ^{90}Sr. In these heavily contaminated areas, severe restrictions on agricultural practices would be required. These would consist primarily of prohibiting the use of the land for dairy pastures or crops consumed directly by humans. It would, however, be less hazardous to grow corn or other food for hogs, steers, or other animals used for meat.

LIVESTOCK AND WILDLIFE DAMAGE

Since farm animals are generally about as radiosensitive as humans, radiation alone would decimate the livestock of the warring nations (Bensen and Sparrow, 1971; NAS–NRC, 1963). The lethality of radiation exposure of farm animals is given in Table 12–5, but would no doubt be higher under wartime conditions as a result of the additive effects of blast, fire, starvation, disease, internal exposure to radionuclides ingested by foraging animals, and skin burns caused by fallout deposited directly on the animals. For human and animal alike, the tables of lethality may under-

TABLE 12–5

Lethal Response of Mammals and Poultry to Brief Exposures to Nuclear Radiations[a]

Species	LD$_{50/30}$[b]	Rate (R/hr)
Burro	784	50
Burro	651	18–23
Burro	585	19–20
Swine	618	50
Sheep	524	20
Cattle	540	25
Swine	555	180
Swine	388	90
Burro	402	—
Poultry		
Males	600	50
Females	1000	50
Chicks	900	Very short

[a] National Academy of Sciences—National Research Council (1963).
[b] The dose that produces 50% mortality in 30 days.

estimate the effects of radiation because they do not take into consideration the effects of other concomitant types of stress.

Although decontamination of heavily contaminated farmlands would be desirable, it is doubtful that under the conditions that exist in the first few years after attack manpower would be profitably expended if used in this way. It would seem far wiser to manage these contaminated farmlands in some other way, for example, by denying the land to agriculture, growing foods low in calcium, or growing fodder for stock animals that are raised for meat.

Some Problems of Recovery from Nuclear Attack

Although nuclear attacks may produce complete destruction by blast and fire over an area many miles in diameter, even a saturation attack would leave much of the country relatively intact except for the effects of fallout. For example, woodframe buildings would not be severely damaged beyond 40,000 ft from the detonation of a 1-megaton bomb, and fires would not be produced beyond about 20 miles. Although the combination of blast, thermal, and radiation effects would make any discussion of countermeasures in the immediate target area of doubtful value, there is much that could be done to ameliorate the problems that would be faced by the large numbers of people outside the zone of blast and thermal damages but well within the region of potentially lethal fallout. During Operation Alert, 1959, a nationwide civil defense exercise in which a saturated attack on the United States was simulated, there were 5 million estimated fatalities in the state of New York. Of this number, approximately 1 million would have been killed by blast and heat, and 4 million fatalities would have been caused only by fallout and not the primary blast or thermal effects. It is understandable that public officials should give attention to the possibility of fallout shelters.

Home shelters were widely recommended up to 1961 (NAS–NRC, 1964; New York State, 1959), after which there developed a greater recognition of the advantages of community shelters (U.S. DOD, 1961). The advocates of shelters noted that millions of people would find themselves in localities in which the radiation levels are lethal but which are beyond the zone of physical destruction. These people would require radiation shielding for their survival. Thus, the 4 million radiation fatalities in New York State during the above civil defense exercise could have been saved from radiation injury had they had access to shelters which could attenuate the radiation levels by a factor of 100. Assuming that these inhabitants took no precautions other than to remain indoors in frame houses, thereby attenuating the radiation by a factor of 2, the vast majority of inhabitants would have received more than disabling doses during the first 24 hr after the attack. By taking advantage of the basements of ordinary houses, an attenuation factor of 10 would have been gained, and the number of casualties would have been reduced by 50%. By using improved basement shelters offering an attenuation factor of 100, almost all the 4 million radiation casualties would have been prevented.

After a full-scale nuclear onslaught, survival for millions of families would require that each be protected for as long as several weeks after the fallout begins. Shelters would be needed that are equipped with food, water, essential drugs, sanitary facilities, and radio communication. Some people would be required to remain in the shelters longer than others. At first, it might be safe to leave the shelter for a few minutes a day. In time, longer periods out of doors would be permitted, and as these excursions are extended in time, the survivors could begin to assume duties that would assist the community as a whole to begin functioning once again.

In some areas, the radiological situation would be such that people could be instructed to leave their shelters to be transported to less contaminated areas. It is possible that large areas may be uninhabitable for many months. However, the most crucial survival problems would be those encountered during the first hours or days following attack. It is in this, the acute phase, that small groups of people must be self-sufficient and equipped with the basic tools for survival.

The acute phase following attack would present government with problems of unprecedented complexity. It would be necessary for officials to assume the most complicated tasks in all history, but these tasks would be faced at a time of unprecedented disruption of manpower, communications, and transportation and overwhelming destruction of the material resources of government.

Although the shelter inhabitants might be self-sufficient so far as food, clothing, water, and drugs are concerned, there would be many things that government must do to ensure their survival. The civil defense organization must have the capacity to assemble the information needed to reach decisions that will affect everyone. If able-bodied people are kept in their shelters too long, they cannot be used to assist the stricken areas. If the people are told to leave their shelters too soon, they will become casualties. Thus, the civil defense officials must have accurate knowledge of radiological conditions throughout the area for which they are responsible. Use of radiation detectors by the family would not be practical except in unusual cases. Particularly in urban areas, the radiation status of a neighborhood should be evaluated by skilled monitors rather than many individuals using many instruments of doubtful performance under conditions in which it would be very difficult to integrate the changing dose rates.

Rapid radiation monitoring in the postattack period would be essential to competent decision making. There would be decisions such as whether a population in a given sector can safely leave their shelters or whether that section of the community can serve as a receiving area for people to be evacuated from regions of blast or fire. Accurate information would be required on which would be based difficult decisions such as whether the children in a given school may be released to rejoin their families, whether power lines can be serviced, streets can be patrolled, or warehouses filled with food can be made available as sources of relief.

Unfortunately, these decisions could not be made competently on the basis of contemporary methods of collecting information. Present techniques for gathering radiation data would be ineffective because they depend largely on manual observations, which would result in radiation monitors receiving lethal doses of radiation in only a few minutes. People on foot or in conventional vehicles could not make measurements in heavily contaminated communities, as is now intended by most civil defense organizations.

A network of automatic, continuously recording radiation monitors would be required that would transmit data to a central headquarters, where the data could be fed into computers which would provide continuous and accurate radiological information. Only in this way could a civil defense commander obtain the information needed in sufficient time to make the decisions that would be necessary for the management of large populations beyond the blast area. Until this is accomplished, the shelters cannot be fully effective.

Following an explosion, so long as the radioactive cloud has not reached a given community, the people could prepare themselves for the days of shelter life that may be ahead of them. These would be difficult minutes in which one could only plan if the community authorities have the means with which to maintain discipline among the populations and the knowledge and ability to keep the people continually informed. If the radiological situation permits during the first hour or two after attack, people could be permitted to remain above ground to assemble their families and to gather the various paraphernalia that may assist their survival.

The arguments for and against shelters as part of a civil defense program are complicated by political, moral, and sociological questions as well as by the incomprehensible consequences of nuclear war. These factors are outside the scope of this text, which has dealt only with the rationale based on the need to protect people who would be outside the zones of destruction and for whom protection against radiations from fallout would be an essential part of a survival program.

The foregoing has constituted a straightforward summary of what is known about the immediate radiological effects of nuclear weapons, particularly when the explosions occur at or near ground level. It is in all respects a superficial treatment that deals only with certain very obvious aspects of the radiological effects of nuclear weapons. Little or nothing has been said about the social, logistic, and medical implications of massive fallout. The problems that would face people at every level of the social structure would be so complicated as to defy meaningful analysis. The problems become more complex by additional orders of magnitude when we consider the concomitant effects of blast damage and fire superimposed on the radiological problems. The physical dimensions of the matter are illustrated by Fig. 12–18, which depicts the extent of the damage that would be produced if a 10-megaton bomb were exploded in the middle of Manhattan in New York City. Stonier (1964) has considered the consequences of such a catastrophe in vivid detail.

The radiation problems cannot be considered without understanding the social effects of blast damage and extensive conflagrations, but the effects of nuclear war that cannot be calculated are probably as great as those for which calculations have been attempted (Office of Technology

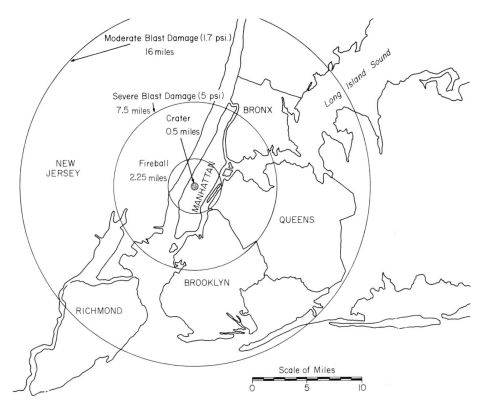

Fig. 12–18. Consequences of a 20-megaton ground burst in mid-Manhattan. (From Stonier, 1964.)

Assessment, 1979). A wide spectrum of individual factors such as low water pressure, radiation damage to livestock, burned fields and forest subject to rapid erosion, acute food shortages, social disorganization, mass casualties, severe ecological disruption, disease, and pollution would create a hundred new problems for every one foreseen. The disastrous immediate consequences of such a war would be followed by a prolonged period of social retrogression from which recovery would be slow and uncertain.

Chapter 13

Fallout from Nuclear Explosions: II. Worldwide Effects

The radioactive debris from a nuclear or thermonuclear explosion is apportioned among three fractions: large particles that deposit from the atmosphere within hours, smaller particles that remain in the troposphere, from which they are removed on a time scale of days, and the fraction injected into the stratosphere, from which they are removed on a time scale of months. The first fraction, which includes the highly radioactive short-lived nuclides, is responsible for the patterns of lethal fallout that were discussed in the last chapter.

The importance of the fact that the radioactive debris is partitioned between the troposphere and the stratosphere was not appreciated until the advent of thermonuclear explosions equivalent to megatons of TNT. Prior to 1952, all the nuclear explosions were in the kiloton range, and after each series of weapons tests the atmospheric radioactivity diminished at a rate corresponding to the half-life of dust in the lower atmosphere, which was shown to be about 20 days (Stewart *et al.*, 1957). Essentially all the debris from kiloton bombs is deposited within about 2 months following injection into the atmosphere. The dust from such explosions is carried by the winds characteristic of the latitude in which injection takes place, and deposition ultimately takes the form of bands, as

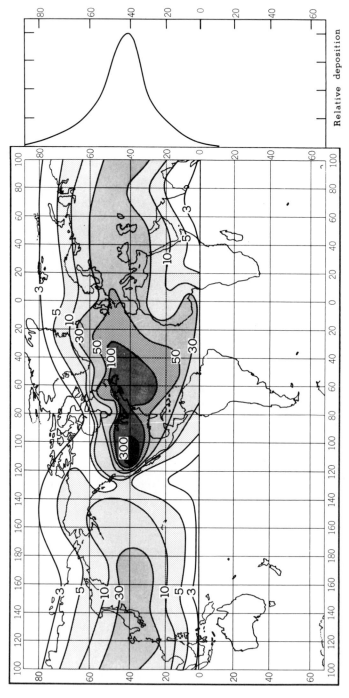

Fig. 13–1. Worldwide radioactive fallout from nuclear weapons tests in Nevada in 1953. The explosions were in the kiloton range of yields, and debris was confined to the troposphere. The intensity of fallout is shown in relative units. (From Machta *et al.*, 1956.)

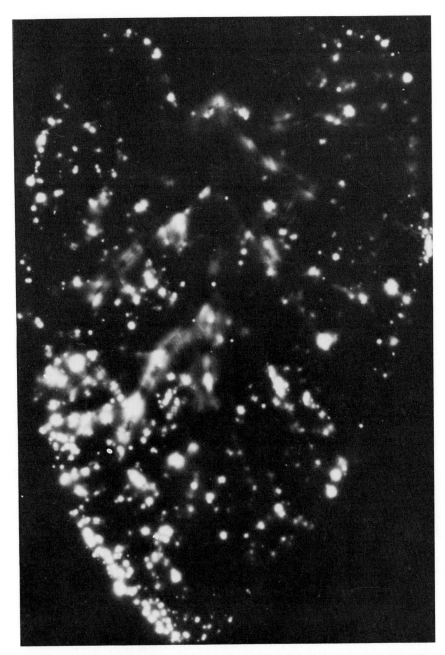

Fig. 13–2. Autoradiograph of a leaf following a fallout of radioactive dust in Troy, New York. The dust originated from an explosion in Nevada about 36 hr previously. (Courtesy of Herbert M. Clark.)

Fig. 13–3. Autoradiograph of adhesive film exposed to the atmosphere for 24 hr at a fallout monitoring station operated by the Health and Safety Laboratory of the U.S. Atomic Energy Commission in 1953. The sample was collected several hundred miles from the test site.

shown in Fig. 13–1. The fallout from any one explosion is often spotty, because rainfall may coincide in some areas with passage of the cloud of radioactive dust. After one Nevada explosion in April 1953, the highest fallout recorded anywhere in the United States was at Troy, New York, more than 2000 miles from the Nevada Test Site, where rain caused precipitation of the radioactive dust to the extent that it was estimated (Clark, 1954) that the cumulative γ dose received by the inhabitants was about 100 mrad. This proved to be a higher dose than was received anywhere in the United States, except in the immediate vicinity of the test site, for the

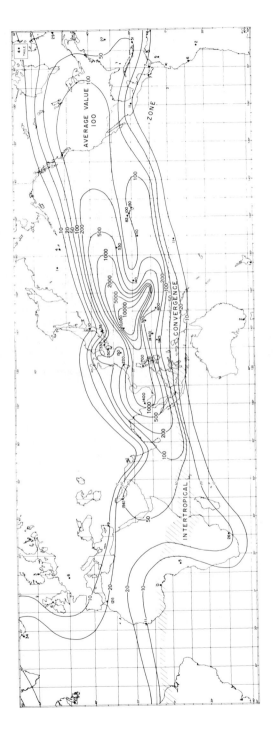

Fig. 13–4. Radioactive fallout from a multimegaton thermonuclear explosion in the Marshall Islands in November 1952. The values shown are millicuries per 100 square miles between 2 and 35 days after the explosion. (From L. Machta *et al.*, copyright 1956, AAAS.)

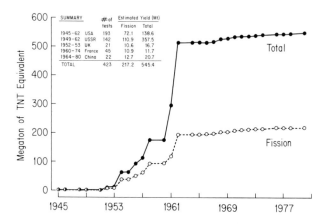

Fig. 13–5. Estimated yields of atmospheric tests of nuclear weapons, 1945–1980. (Plotted from data provided in UNSCEAR, 1982.)

entire 1953 series of tests. Figures 13–2 and 13–3 illustrate the extent to which exposed surfaces can be contaminated by radioactive particles at considerable distances from nuclear explosions.

As shown in Fig. 13–4, the tropospheric fallout from a thermonucléar explosion in the Pacific Ocean was, within 35 days, distributed widely in both hemispheres. The fraction of the debris that was injected into the stratosphere was deposited from pole to pole within 1 or 2 years.

Our knowledge of global fallout phenomena is largely the result of the collaborative efforts of many nations. In particular, the United Nations Scientific Committee on the Effects of Atomic Radiation (UNSCEAR), created by the U.N. General Assembly in 1955, has reported on the data accumulated by member countries. The periodic reports of UNSCEAR present excellent summaries of the subject of worldwide fallout from nuclear weapons tests.

The cumulative yields of tests conducted in the atmosphere since 1945 are shown in Fig. 13–5. Of the many radionuclides produced in nuclear and thermonuclear explosions, the nine listed in Table 12–2 are the most important. Deposition from the stratosphere takes place so slowly that many radionuclides injected into that compartment decay before they settle to the lower atmosphere. However, since the average life of dust in the troposphere is about 1 month, many of the short-lived nuclides (such as ^{131}I) deposit on the surface of growing crops and thus enter the human food supply. However, the short-lived nuclides become relatively unimportant when deposited on soil because of the length of the growing time of most crops.

How the Radioactive Debris Is Partitioned among the Three Components of Fallout

An understanding of the dose commitments that result from nuclear explosions requires knowledge of the manner in which the radioactive debris is transported by the atmosphere, the mechanisms by which it is deposited on the surface of the earth, and other pathways of human exposure. One of the first steps requires that estimates be made of how radioactive debris is apportioned among three fractions: (1) the fallout in the immediate vicinity of the explosion, (2) the debris injected into the troposphere, and (3) the debris injected into the stratosphere. Surprisingly, there is comparatively little reliable information on this subject. Devices in the megaton range that were exploded near the surface of the Marshall Islands are believed to have deposited as much as 80% of the fallout within about 100 km, but this is subject to great uncertainty, and for any given explosion depends on the size of the explosion, height above ground, type of terrain, and meteorological factors. The tropospheric component is estimated to account for no more than about 5% of the radioactive yield of megaton surface explosions (Machta and List, 1959) and less in the case of explosions well above the surface. The principal source of worldwide contamination by long-lived radionuclides from bombs in the megaton range of yield is the component of the debris that was injected into the stratosphere. Tropospheric fallout contains only a small fraction of the long-lived radionuclides from bombs in the megaton range but can be responsible for heavy exposure from the fallout of short-lived tropospheric debris, notably ^{131}I.

The first estimates of the amounts of radioactive dust injected into the stratosphere were made by subtracting the estimated amount of close-in fallout from estimates of the total amount of debris produced. The fraction of the debris that deposited close to the blast was approximated from field measurements that were often relatively crude. The amount of debris produced was estimated from knowledge of the type of device and its explosive yield. Local fallout was on average found to be 80% for land surface explosions, 20% for explosions on the surface of the water, and 10% for explosions in the air. The stratospheric inventory was not believed to be influenced significantly by the many detonations in the kiloton range since, for the most part, these did not penetrate the stratosphere.

On examination, the material balance studies of debris produced in the Pacific detonations from 1952 to 1958 prove to be subject to many uncertainties. In order to ascertain the amounts deposited "close" to the site of detonation, it was necessary to collect fallout samples over an area of many thousands of square miles of the Pacific Ocean under operating conditions that defied the ingenuity of a number of well-equipped teams of investigators. Even the best data left considerable uncertainty.

Given an estimate of the amount of a given nuclide produced in a megaton-size explosion and an estimate of the amount deposited in the immediate vicinity, the balance, within the limits of uncertainty of the estimates, can be said to have been injected into the stratosphere. This follows from the observation that the tropospheric component from very large near-surface explosions is only about 5% of the total. If an estimate is available of the total amount of debris that has been deposited globally at any given time following an explosion, the stratospheric inventory can thus be approximated. This is the method that was used prior to 1959 by American investigators (Libby, 1956; Eisenbud, 1957).

Recognizing that the method of material balancing resulted in very uncertain estimates of the stratospheric inventory, a number of attempts were made to collect samples of stratospheric dust. The Atomic Energy Commission in 1953 succeeded in collecting about 12 dust samples at an altitude of 80,000 to 100,000 ft, using electrostatic precipitators carried aloft by balloons. Although fission products were found in these samples and served for the first time to demonstrate the presence of radioactive dust in the stratosphere, the data were at best only semiquantitative because little was known about the performance of the electrostatic precipitator at the altitudes involved (Loysen et al., 1956; Loysen, 1965).

Beginning in 1956, the AEC undertook Project ASHCAN, in which balloon-borne filters sampled the stratosphere at Minneapolis, Minnesota; San Angelo, Texas; Panama; and Sao Paulo, Brazil (Holland, 1959b). Samples at each of these stations were collected monthly at four altitudes, from 50,000 to 90,000 ft, and the filters analyzed for gross β activity and six radionuclides; ^{140}Ba, ^{95}Zr, ^{144}Ce, ^{137}Cs, ^{89}Sr, and ^{90}Sr. The shorter-lived isotopes made it possible to ascertain the age and origin of the debris.

In August 1957 the U.S. Air Force began to undertake a series of long-range high-altitude flights with Lockheed U-2 aircraft equipped with filtering equipment (Feely, 1960). The planes monitored the atmosphere along the 70th meridian west, from 66°N latitude to 60°S latitude, and by mid-1959 more than 1400 samples had been collected and analyzed (Stebbins, 1961).

With the data from ASHCAN and the high-altitude sampling program (Project HASP), it became possible for the first time to make reliable estimates of the stratospheric inventory of bomb debris. The most striking finding was that the inventory of ^{90}Sr was very much less than had been previously estimated. At the time of the October 1958 short-lived moratorium on weapons testing, immediately following heavy test series by both the United States and the Soviet Union, the stratospheric inventory

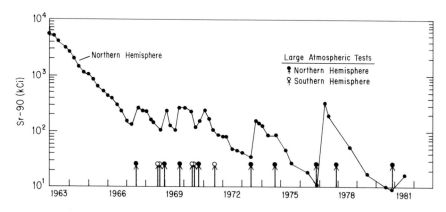

Fig. 13–6. Stratospheric inventory of ^{90}Sr from 1963 to 1982. (From Leifer *et al.*, 1984.)

was estimated by Eisenbud (1959) to be about 4.3 MCi and by Libby (1959) to be between 4.5 and 5.5 MCi. The HASP investigators (Shelton, 1959; Feely, 1960; Stebbins and Minx, 1962) concluded from an analysis of data collected from January to August 1959 that the stratospheric inventory during that period was only 0.8 MCi, less than 20% of the previous estimates. By examining the rate of fallout from the tests by the Soviet Union in the fall of 1958 and from the measured ^{90}Sr stratospheric inventory in the early fall of 1958, prior to the tests by the Soviet Union, Feely concluded that more than half of the debris from the tests by the Soviet Union was deposited from the stratosphere within 6 months and that at least half of the debris injected by the United States and Great Britain during 1958 was deposited within 12 months. These values would be equivalent to a mean residence time of about 8 months for Soviet debris injected at high latitudes and about 18 months for debris injected into the stratosphere near the equator. These residence times were very much less than the estimates of 5–7 years that had been made by others.

Martell (1959) came to a similar conclusion by studying the relative contribution of debris of tests made by the Soviet Union and the United States during late 1958 and 1959. He found it possible to apportion the radioactivity in rainwater between the U.S. tests in the spring of 1958 and the Soviet tests in the fall of 1958. This was done by taking advantage of the fact that the debris from tests in the Marshall Islands between May and July 1958 were uniquely labeled with ^{185}W. About 250 MCi of this 75-day half-life isotope were formed as of August 1, 1958 (Libby, 1959). Martell estimated that 40% of this total was retained in the stratosphere, where it coexisted initially with other debris from the U.S. tests. By studying the ratio of [185]W to other nuclides such as ^{89}Sr, ^{90}Sr, and [140]Ba, Martell was able to trace the debris from the U.S. tests and differentiate it from the debris of the Soviet explosions, which was characterized by similar fission product spectra but did not contain [185]W. Kuroda *et al.* (1962), by analyzing the diminution of ^{90}Sr in rainfall during the test moratorium, concluded that the overall stratospheric residence time during the period 1958–1960 was 0.7 ± 0.1 year. The estimated stratospheric inventories of ^{90}Sr from 1963 to 1982 are shown in Fig. 13–6 (Leifer *et al.*, 1984).

The initial overestimation of the stratospheric inventory was due to underestimates of the amount of fallout in the vicinity of the explosions. As noted earlier, the fraction of close-in fallout may vary considerably, depending on the type of burst, and the estimates of the total amount of close-in fallout from the various shots were based on very scanty data.

Behavior of Individual Radionuclides

STRONTIUM-90

Studies conducted shortly after World War II identified ^{90}Sr as the most hazardous radionuclide in fallout and concluded that there was a limit to the amount of ^{90}Sr that could be disseminated, above which bone cancer would be caused to develop in the world's population. A study conducted in 1953 concluded that the limit would be greater than 2.5×10^4 MT (Rand Corp., 1953), based on the then predominant concept that bone cancer was a nonstochastic effect and that a skeletal burden of 1.0 μCi was safe. This conclusion resulted in an expansion of studies of both the biological effects of ^{90}Sr and the behavior of strontium in the environment.[1] Those investigations have made it possible to predict the manner in which this nuclide is transported through the atmosphere and enable one to predict the pattern of global contamination that will result from the injection of a given amount of ^{90}Sr into the atmosphere. It is also possible to estimate how deposition of ^{90}Sr results in contamination of foods and, finally, given a knowledge of the amount of contamination in food, it is possible to forecast the skeletal burden of ^{90}Sr in individuals consuming the food.

Global Inventory of Strontium-90

Essentially all of the ^{90}Sr injected into the atmosphere during the period of weapons testing prior to the test-ban agreement was deposited on the

---

[1]This program was known as Project SUNSHINE, and for this reason the unit "picocuries of strontium-90 per gram of calcium" was originally designated, somewhat incongruously, the Sunshine unit. The unit was gradually replaced by *strontium unit*, retaining the same abbreviation SU.

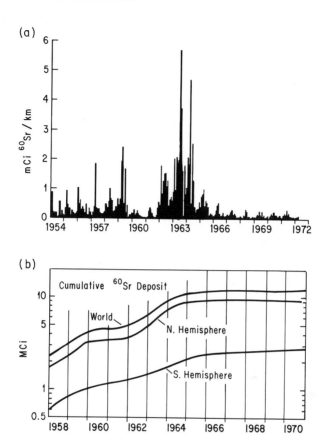

Fig. 13–7. (a) Monthly deposition of ^{90}Sr in New York City, 1954–1971. (U.S. Department of Energy, Environmental Measurements Laboratory.) (b) Cumulative ^{90}Sr deposition on the earth's surface, 1958–1970. Note that the deposition is plotted logarithmically. (From Volchok and Kleinman, 1971.)

earth's surface by 1970. This is seen in Fig. 13–7a, which shows the monthly deposition of ^{90}Sr in New York City, and Fig. 13–7b, which shows the worldwide deposition of ^{90}Sr through 1971 (Volchok and Kleinman, 1971). Worldwide deposition reached a peak of about 12.5 MCi by late 1967, at which time southern hemisphere deposition was less than one-third that in the northern hemisphere. The amount of ^{90}Sr deposited has been diminishing at a rate of 2.5%/year because of radioactive decay, which has been only partially offset by occasional tests by France and China (Volchok, 1970; UNSCEAR, 1982).

The estimated stratospheric inventories (Leifer *et al.* 1984) are shown in Fig. 13–6. Except for minor perturbation due to atmospheric tests by

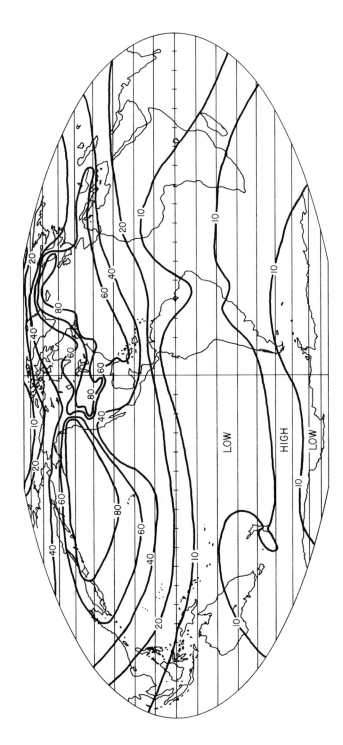

Fig. 13–8. Isolines of cumulative ^{90}Sr deposits based on analyses of soils collected 1965–1967 (in millicuries per square kilometer). (From UNSCEAR, 1969.)

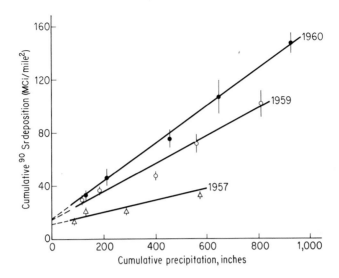

Fig. 13–9. Dependence of fallout on precipitation on the Olympic Peninsula. (From E. Hardy and L. T. Alexander; copyright 1962, AAAS.)

France and China after 1967, the exponential diminution that followed cessation of tests by the United States and the Soviet Union in 1963 is clearly evident, with a half-life of about 1 year.

The deposition of ^{90}Sr was far from uniform during the period of maximum deposition, as can be seen from Fig. 13–8. The band of relatively heavy fallout (60–80 mCi/km^2) in the northern midlatitudes is conspicuous and is believed to be due to meteorological factors that result in increased stratospheric–tropospheric transfer in the spring months. This phenomenon was discussed in Chapter 5.

The deposition is affected at any given latitude by the amount of rainfall. Analysis of soil samples (Alexander, 1959) from the latitudinal band 20°S to 30°S showed a range of ^{90}Sr values from 0.4 mCi/square mile at Antofagasta, Chile, to 11.6 mCi/square mile in Brisbane, Australia. The mean of samples collected at seven localities was 6.1 mCi/square mile. The low value at Antofagasta is explained by the fact that no precipitation was recorded for several years prior to sample collection. The dependence of fallout on the amount of precipitation is illustrated elegantly in Fig. 13–9, in which cumulative ^{90}Sr fallout is plotted against rainfall at five sites on the Olympic Peninsula in Washington State (Hardy and Alexander, 1962).

There have been suggestions that deposition of ^{90}Sr (and other dust-borne substances) may be greater on ocean surfaces than on land, because ocean spray acts as a scavenging mechanism (UNSCEAR, 1969). This possibility was proposed initially because analyses of ocean water indicated that the oceanic inventory of ^{90}Sr might be higher than could be explained on the basis of fallout measurements made on land (Belyaev *et al.*, 1965). However, field and laboratory studies by Freudenthal (1970a,b) and by Kupferman *et al.* (1979) seem to indicate that the phenomenon of spray scavenging is not an important mechanism for oceanic deposition.

The deposition of ^{90}Sr has been well documented on a global scale. The fallout rate is influenced by latitude, season, and the total and fission yields of each burst and is amenable to construction of models that provide deposition estimates that are in reasonable agreement with observed accumulations (Peterson, 1970).

Radiostrontium in Food

It was seen in Chapter 6 that the ^{90}Sr content of plants can be due either to direct uptake from soil or to foliar deposition. The relative proportions of the two components depend on the rate of fallout in relation to the cumulative soil deposit, as well as the time of fallout relative to the growing season and the amount of plant surface area exposed to fallout.

The dietary sources of ^{90}Sr depend in part on the food consumption habits of the population, including the kinds of foods eaten and the way in which the food is processed or prepared. Because ^{90}Sr becomes part of the pool of Ca in the biosphere, the amount of this element contained in food is critical. Daily per capita consumption of calcium in the United States is about 1 g, but in other countries this may vary, as discussed in Chapter 6 and summarized in Table 6–4.

Since 1960, the U.S. Department of Energy, Environmental Measurements Laboratory (formerly the Atomic Energy Commission, Health and Safety Laboratory) has reported on the whole-diet ^{90}Sr content in two cities, New York and San Francisco. (Chicago was originally included but was discontinued in 1967.) There have been significant differences between the two cities, as can be seen from Fig. 13–10 (Klusek, 1984a). The daily ^{90}Sr intake in New York has been consistently higher than in San Francisco. A comparison of the ^{90}Sr content of typical diets in New York and San Francisco during 1982 is given in Table 13–1. This difference has persisted since 1960 and is believed to be due mainly to lower annual rainfall in the regions that supply food to the San Francisco area (Harley, 1969).

It is seen from Table 13–1 that during 1982, dairy products were responsible for 32% of the daily ^{90}Sr intake in New York, compared to 21% in San Francisco. The ^{90}Sr content of cow's milk can be predicted accurately from knowledge of deposition rate (Fig. 13–11), and the total intake from all dietary sources can be predicted from data on ^{90}Sr in milk (Bennett, 1972; Klusek, 1984a).

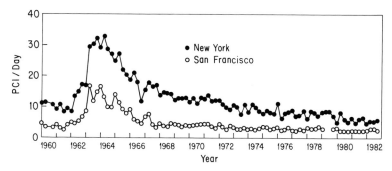

Fig. 13–10. Results of quarterly analyses of ^{90}Sr in total diet in New York and San Francisco. (From Klusek, 1984a.)

The concentrations of ^{90}Sr in the diet can be predicted from known rates of fallout, using a basic model that can be written as

$$M_m = p_1 F_n + p_2 F_{n-1} + p_3 \sum_{m=1}^{\infty} F_{n-m} \, e^{-m\mu} \qquad (13\text{-}1)$$

where M_m is the average ^{90}Sr : Ca ratio in milk in the year n (pCi/g Ca), F_n the deposition of ^{90}Sr in year n (mCi/km^2), p_1 the proportion attributable to current year deposition, p_2 the proportion attributable to the preceding year, p_3 the proportionate reduction due to reduced annual availability for m preceding years, and μ the reduced availability per year.

Annual reduced availability has been observed to be 8% per year in addition to 2.4% due to radioactive decay, corresponding to a mean residence time of 10.1 years. Thus, μ has a value of 0.11. Klusek has assigned

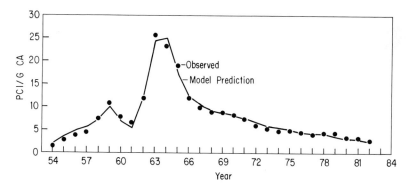

Fig. 13–11. Observed and predicted average annual ^{90}Sr concentration in milk, 1954–1982, in New York City based on annual ^{90}Sr deposition. (From Klusek, 1984a.)

TABLE 13–1

STRONTIUM-90 IN THE DIET DURING 1982[a]

Diet category	kg/yr[b]	g Ca/yr[b]	New York City — Percent yearly intake Ca	New York City — pCi ^{90}Sr/kg	New York City — pCi ^{90}Sr/yr	New York City — Percent yearly intake ^{90}Sr	San Francisco — pCi ^{90}Sr/kg	San Francisco — pCi ^{90}Sr/yr	San Francisco — Percent yearly intake ^{90}Sr
Dairy products	200	216.0	58	3.2	641	32	1.0	200	21
Fresh vegetables	48	18.7		8.8	422		2.4	116	
Canned vegetables	22	4.4		5.4	119		2.9	64	
Root vegetables	10	3.8		3.4	34		3.8	38	
Potatoes	38	3.8		2.3	88		2.1	79	
Dry beans	3	2.1		15.9	48	36	7.9	54	36
Fresh fruit	59	9.4	9	2.6	152		1.3	77	
Canned fruit	11	0.6		1.1	12		0.8	9	
Fruit juice	28	2.5		1.7	48		1.4	40	
Bakery products	44	53.7	3	3.0	131	11	1.9	84	13
Flour	34	6.5		4.5	153		3.5	119	
Whole grain products	11	10.3		6.2	69		2.9	32	
Macaroni	3	0.6		2.4	7		2.3	7	
Rice	3	1.1		0.6	2		0.8	2	
Meat	79	12.6	20	0.4	35	18	0.4	31	25
Poultry	20	6.0		0.3	6		0.3	5	
Eggs	15	8.7		0.6	10		0.6	8	
Fresh fish	8	7.6		0.2	1		0.1	1	
Shell fish	1	1.6	10	0.2	<1	3	0.7	1	5
Yearly intake		370 g			1978 pCi			967 pCi	
Intake					54 pCi/g Ca			2.6 pCi/g Ca	
					54 pCi/day			2.6 pCi/day	

[a] From Klusek (1984a).

[b] From U.S. Department of Agriculture (1967).

values of $p_1 = 0.68$, $p_2 = 0.26$, and $p_3 = 0.16$. It is remarkable that the constants in this model, which were first proposed in 1970, have been found to be very stable with the addition of data for all succeeding years through 1982. The agreement between the observed and predicted concentrations in milk is shown in Fig. 13–11.

Milk is an important source of ^{90}Sr and a good index of exposure of growing children in many countries where dairy products are the main source of calcium. Many countries maintained milk monitoring networks during and after the period of atmospheric testing.

The longest series of milk samples is from the program of the Department of Energy Environmental Measurements Laboratory, which has been analyzing fresh milk sold in New York City since early 1954. The average annual ^{90}Sr content of liquid whole milk through 1982 is given in Fig. 13–11, from which it is seen that the milk reached a concentration greater than 25 pCi ^{90}Sr/g Ca in 1963. The tap water in New York City has also been monitored since 1954, and reached a peak of 2.12 pCi/liter in 1963.

Harley (1969) has compared the New York diet with similar data reported for the Soviet Union by Petukhova and Knizhnikov (1969) (Table 13–2). During 1966 the ^{90}Sr contained in a representative Soviet diet totaled about 13,000 pCi, compared to about 7000 pCi ^{90}Sr for a New York diet. Since the daily calcium intake is higher in New York than in the Soviet Union (1 g/day compared to about 0.7 g/day), the difference is even greater when expressed as strontium units, about 19 SU for the New York diet and 54 SU for the Soviet Union.

Figure 13–10 shows that the ^{90}Sr content of U.S. diets diminished for several years with a half-time of 3½ to 4 years, following the peak values of 1963–1964, when direct deposition on foliage was a major contributor to dietary ^{90}Sr. Dietary ^{90}Sr in recent years has originated almost entirely from soil. The data of Fig. 13–10 suggest that the mean residence time of ^{90}Sr in soil is about 9 years after the initial year of deposition. This results from the 8% per year removal from soil and the 2.4% due to radioactive decay as predicted by Eq. (13–1).

The models developed to predict the concentrations of ^{90}Sr in milk and total diet from the rates of deposition were among the first examples of environmental modeling and have evolved during the past 30 years, with major contributions by Knapp (1961) and Bennett (1972) that have been summarized by Klusek (1984a).

The amount of ^{90}Sr contributed by grains is influenced by the milling practices employed. In the United States, Canada, and the United Kingdom, the ^{90}Sr : Ca ratio in flour has been approximately one-third to one-half of that in the whole grain and one-quarter of that in bran (UNSCEAR, 1962). Milling reduces the ratio ^{90}Sr : Ca in rice grain to one-fifth to one-

TABLE 13-2

DIETARY ^{90}Sr IN THE UNITED STATES AND SOVIET UNION, 1966[a]

	United States diet (New York City)				Soviet Union diet (country)[b]			
Diet category	Intake (kg/year)	^{90}Sr (pCi/kg)	^{90}Sr (pCi/year)	Percent of total intake	Intake (kg/year)	^{90}Sr (pCi/kg)	^{90}Sr (pCi/year)	Percent of total intake
Milk	200	13.4	2,970	42	110	14	1,540	12
Bread	89	17.4	1,580	22	220	33	7,250	55
Meat	114	2.8	320	4	60	10	600	5
Cereals	6	6.3	40	1	20	18	360	3
Fish	9	1.8	20	—	4	30	120	1
Potatoes and vegetables	219	8.3	1,800	25	220	12	2,640	20
Water	400	1.1	440	6	440	1.1	480	4
Yearly intake			7,170				12,990	

[a] Harley (1969).
[b] Based on data given by Petukhova and Knizhnikov (1969).

tenth of the value in whole rice. Harley (1969) attributed the relatively large contribution of bread to the daily intake of ^{90}Sr in the Soviet Union to the use of black bread that includes ^{90}Sr deposited on the surfaces of wheat and rye.

Strontium-90 in Human Bone

In parallel with the food sampling network, many countries have for many years reported on the ^{90}Sr content of human bone. It has been found (Klusek, 1984b) that a relatively simple model for the uptake of ^{90}Sr can account for the observed concentrations in adult bones:

$$B_n = cD_n + g \sum_{m=0}^{\infty} D_{n-m} e^{-m\lambda} \qquad (13\text{-}2)$$

where B_n is the ^{90}Sr concentration in vertebrae in year n from m years of accumulation (pCi), D_n the ^{90}Sr concentration in diet from midyear in the year n-1 to midyear in year n (pCi), c the fractional short-term retention, g the fractional long-term retention, and λ the annual fraction removed from the skeleton (including radioactive decay).

The predicted and observed values for adult vertebral ^{90}Sr in New York City for 1954–1982 are given in Fig. 13–12.

The concentration of ^{90}Sr in children's bone can be estimated by a modification of Eq. (13–2) in which the constants, c, g, and λ vary with age. Current estimates of these parameters are given in Table 13–3. The modified equation is (Klusek, 1984b):

$$B_{i,n} = (c_i + g_i) D_{i,n} + (B_{i-1,n-1} - c_{i-1} D_{i-1,n-1}) e^{-\lambda_i} \qquad (13\text{-}3)$$

The subscript i refers to the value for any given age. The nomenclature in Eq. (13–3) is otherwise that of Eq. (13–2).

Because of the structural and physiological complexity of bone, calculation of the dose delivered by ^{90}Sr is not a simple matter (Bjornerstedt and Engstrom, 1960; Spiers, 1966; UNSCEAR, 1969, 1982). Two tissues are of interest: the marrow and the bone cells lining the cavity within which the marrow is contained (the endosteum). The dose estimate is not simple if one assumes uniform skeletal deposition of ^{90}Sr, and it is even more difficult in the case of nonuniform deposition. The latter case is encountered in practice owing to the fact that the pattern of skeletal deposition is dependent on the variations of dietary ^{90}Sr with time and variations in the rate of skeletal calcium accretion and turnover with age.

Taking all available information into consideration, UNSCEAR (1982) estimated that the 50-year dose commitment from ingested ^{90}Sr by inhabi-

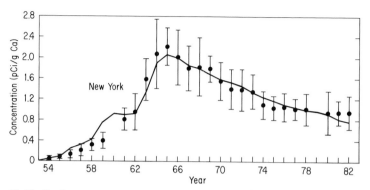

Fig. 13–12. Predicted and observed ^{90}Sr in adult vertebrae in New York City, 1954–1982. (From Klusek, 1984b.)

TABLE 13–3

PARAMETERS OF THE BONE MODEL[a]

Age (yr)	λ_i	c_i	g_i	Skeletal calcium Ca_i (g Ca)	Dietary calcium D Ca_i (g Ca yr^{-1})	Relative retention $(c_i+g_i)/Ca_i$ ($\times 10^{-3}$)	Turnover rate $(1-e^{-\lambda_i})-0.024$ (% yr^{-1})
0–1	6.79	0.000	0.076	100	320	0.76	97
1–2	0.99	0.000	0.075	147	328	0.51	60
2–3	0.81	0.000	0.69	179	336	0.39	53
3–4	0.72	0.000	0.057	201	344	0.28	49
4–5	0.59	0.000	0.048	219	352	0.22	42
5–6	0.61	0.000	0.050	239	360	0.21	43
6–7	0.61	0.000	0.052	264	368	0.20	43
7–8	0.62	0.000	0.062	297	376	0.21	44
8–9	0.64	0.017	0.056	341	384	0.21	45
9–10	0.52	0.43	0.44	396	387	0.22	38
10–11	0.41	0.39	0.72	463	387	0.24	31
11–12	0.44	0.46	0.77	539	387	0.23	33
12–13	0.39	0.32	0.109	624	387	0.23	30
13–14	0.40	0.019	0.127	715	387	0.20	31
14–15	0.41	0.34	0.119	806	387	0.19	31
15–16	0.40	0.28	0.140	894	387	0.19	31
16–17	0.36	0.62	0.97	973	386	0.16	28
17–18	0.32	0.78	0.88	1035	376	0.16	25
18–19	0.29	0.81	0.79	1073	365	0.15	23
19–20	0.25	0.76	0.62	1078	354	0.13	20
Adult	0.24	0.71	0.57	1078	342	0.12	19

[a]From Klusek (1984b).

tants of the North Temperate Zone has been 94 mrem to the bone marrow and 210 mrem to the endosteal cells. The dose commitments to residents of the South Temperate Zone were estimated to be about 25% of these values.

STRONTIUM-89

Strontium-89 behaves like ^{90}Sr in all respects, except that it is relatively more important as a foliar contaminant because of its short (50.5-day) half-life. It has been shown (Dunning, 1962) that ^{89}Sr may deliver as much as 3.4 times the dose from ^{90}Sr during the first year of a child's life. However, unless fresh ^{89}Sr is added during the second year, this isotope will have decayed almost completely, so that the ^{89}Sr component would become unimportant over the whole lifetime of the individual. The effective dose equivalent commitment from ^{89}Sr deposited in the northern hemisphere has been estimated to be about 0.2 mrem (UNSCEAR, 1982).

CESIUM-137

Cesium-137 has a 30-year half-life, compared to 28 years for ^{90}Sr, and it is produced somewhat more abundantly, at a rate of about 1.6 Ci ^{137}Cs per curie of ^{90}Sr. As noted in Chapter 6, ^{137}Cs is tightly bound by soil and thus is not readily incorporated metabolically into vegetation, but does contaminate human foods by foliar absorption (NCRP, 1977a).

Measurement of ^{137}Cs in both food and people was greatly simplified by the development of γ spectrometry in the 1950s. It has proved useful to report the presence of ^{137}Cs in biological material as picocuries of ^{137}Cs per gram of potassium, although it is known that the metabolisms of cesium and potassium are, in fact, somewhat different from food to food and organ to organ (Yamagata and Yamagata, 1960). For the purpose of dose calculation, ^{137}Cs is usually assumed to be distributed uniformly throughout the body.

The distribution of ^{137}Cs in the stratosphere and the pattern of terrestrial deposition are similar to those for ^{90}Sr. A number of investigators have shown that fractionation of ^{137}Cs with respect to ^{90}Sr does not occur (UNSCEAR, 1969). Gustafson (1969) used ^{90}Sr deposition measurements at various stations in the United States to estimate the ^{137}Cs deposition and concluded that over most of the country the accumulation by the end of 1965 ranged between 60 and 100 mCi/km^2; there was even less variation in the ^{137}Cs body burdens of the American population, owing to the fact that blending occurs because many important items of food originate from many different areas of the country.

The ^{137}Cs content of various foods from the Chicago area in 1968 is

TABLE 13–4

[137]Cs IN CHICAGO DIET IN OCTOBER 1968[a]

Diet category	Intake (kg/yr)	Potassium			[137]Cs		
		g/kg	g/yr	Percent total intake	pCi/kg	pCi/yr	Percent total intake
Dairy products	200	1.4	280	21	18	3,600	29
Fresh vegetables	48	2.3	110		2	100	
Canned vegetables	22	1.3	29		9	200	
Root vegetables	10	2.9	29	29	7	70	8
Potatoes	38	4.5	171		17	650	
Dried beans	3	13.9	42		5	10	
Fresh fruit	59	1.9	112		4	240	
Canned fruit	11	1.2	13	13	18	200	10
Canned fruit juices	28	1.9	53		26	730	
Bakery products	44	1.2	53		21	920	
Flour	34	1.0	34		33	1,120	
Whole grain products	11	3.5	38	10	30	330	20
Macaroni	3	1.8	5		19	60	
Rice	3	n.d.[b]	—		n.d.[b]	—	
Meat	79	3.3	261		26	2,060	
Poultry	20	2.7	54	25	15	300	20
Eggs	15	1.5	22		7	100	
Fresh fish	8	3.4[c]	27		194[c]	1,550	
Shellfish	1	n.d.[b]	—	2	n.d.[b]	—	13
Water	400	n.d.[b]	—		(0.05)[d]	20	
Yearly intake (rounded)			1330			12,300	

[a]From Harley (1969).
[a]Not determined.
[c]Based on 90% ocean, 10% freshwater fish.
[d]Number in parentheses represents estimated value.

given in Table 13–4, which shows that dairy products, grains, and meat products are the most important source of this nuclide. Because cesium is bound so tightly in soil, the ^{137}Cs content of land-grown food is generally dependent on the rate of fallout rather than cumulative deposition. This is not so for freshwater fish: in one lake system, the half-life of the ^{137}Cs concentration in fish was about 2½ years (Gustafson, 1969).

For many years, Gustafson (1969; Gustafson et al., 1970) made careful

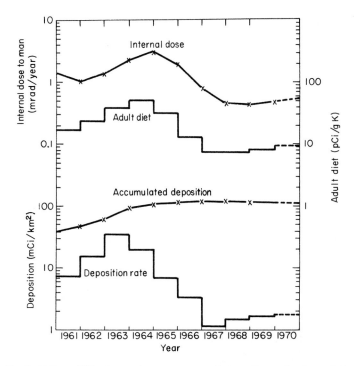

Fig. 13–13. Fallout ^{137}Cs deposition rate, accumulation, dietary levels, and dose to humans in the Chicago area, 1961 to mid-1970. (From Gustafson *et al.*, 1970.)

observations of ^{137}Cs in air, fallout, food, and humans in the Chicago area. His data from 1961 through 1970 are summarized in Fig. 13–13, which shows that ^{137}Cs deposition reached a peak of about 130 mCi/km^2 during 1965–1966, and this resulted in an annual dose of about 3 mrem due to internally deposited ^{137}Cs. The dose commitment from ingesting ^{137}Cs in the northern hemisphere is estimated to have been about 17 mrem on average (UNSCEAR, 1982). The ^{137}Cs in fallout delivers a somewhat higher dose by external radiation. By 1969, when the shorter-lived γ-emitting fallout radionuclides had decayed, the external radiation exposure owing to fallout on an open field in Illinois was almost entirely due to ^{137}Cs and was 13 mR/year. Gustafson *et al.* (1970) estimated that the indoor exposure, assuming 40% transmission, was about 5 mR/year, and if one applies a factor of 0.6 to estimate gonadal exposure, the absorbed gonadal dose would be about 3 mrem/year. UNSCEAR (1982) estimated that the 50-year external effective dose equivalent commitment from ^{137}Cs deposited in the North Temperate Zone was 40 mrem.

The biological half-life of Cs in humans is a function of sex and age.

Representative half-lives are 2 days for the first 10% of the absorbed quantity and 110 days for the remaining 90% (ICRP, 1979).

As discussed in Chapter 6, the ^{137}Cs absorption by humans can be increased markedly by unusual environmental factors. One example was in an area of potassium-deficient soils on the island of Jamaica, where the ^{137}Cs content of milk was 10–100 times that of milk from farms with normal soil (Broseus, 1970). The most remarkable anomaly is in the sub-Arctic, where the body burdens of ^{137}Cs in individuals eating large quantities of moose or caribou meat have been more than 10 times higher than the local population average (UNSCEAR, 1969). In studies of Alaskan Eskimos, Eckert et al. (1970) found less ^{137}Cs in inhabitants of villages subsisting on reindeer meat than in villages where the main source of meat was moose or caribou (Chapter 6). The high doses to consumers of reindeer and caribou are attributed to the high consumption by these animals of lichens, which tend to concentrate ^{90}Sr and ^{137}Cs.

CARBON-14

It was seen in Chapter 7 that cosmic-ray reactions in the upper atmosphere result in transmutation of atmospheric nitrogen to ^{14}C and that this nuclide has been in secular equilibrium in the biosphere at a concentration of 7.5 ± 2.7 pCi ^{14}C/g C. This equilibrium concentration is believed to have been unchanged for at least 15,000 years prior to 1954, when large thermonuclear explosions resulted in the production of additional ^{14}C that perturbed the natural equilibrium. The ^{14}C content of the atmosphere is believed to exist as CO_2.

Because the average life of ^{14}C is 5730 years, the dose from ^{14}C introduced into the environment will be delivered for many generations. Since the genetic effects may be cumulative even in such small doses, there has been some concern about the effects of increasing the size of the ^{14}C pool in the biosphere (Totter et al., 1958; Leipunsky, 1957; Zelle, 1960). A special reason for concern is that ^{14}C can be incorporated in the molecules of which the genes are formed. Thus, in addition to the mutations due to the energy deposited by the decay of ^{14}C, mutations may result from the transmutation of carbon atoms incorporated in the genetic material.

The best estimates of the amount of ^{14}C produced in weapons tests are those made by actual measurement of the ^{14}C activity in stratospheric air by using high-flying aircraft and balloons. This quantity is estimated to have been 9.6×10^6 Ci (NCRP, 1985b).

Carbon distributes itself quickly among the major environmental compartments—the stratosphere, troposphere, biosphere, and surface ocean waters. Transfer among these compartments takes place with time constants on the order of a few years, but transfer to the deep oceans proceeds more slowly. The biospheric ^{14}C response to a stratospheric injection will reach equilibrium after relatively few years and decrease slowly thereafter at a rate determined by transfer of the ^{14}C to deep ocean water and possibly humus (NCRP, 1985b).

By the end of 1967, the tropospheric ^{14}C content had increased about 60% above natural levels in the northern hemisphere and a little less in the southern hemisphere (Nydal, 1968). Because of the short time constants involved in transfer from the atmosphere to biosphere, the ^{14}C content of human tissues and foods increased rapidly following the heavy testing schedules of 1961 and 1962. The dose equivalent from ^{14}C in fallout is estimated to have reached a peak of 0.96 mrem/year in 1965 and had diminished to 0.37 mrem/year by 1984 (NCRP, 1985b). Because of the long half-life, bomb-produced ^{14}C will persist in the environment for many thousands of years.

PLUTONIUM-239/238

Plutonium injected into the atmosphere by nuclear explosions originates from volatilization of both unfissioned plutonium and plutonium produced by neutron irradiation of ^{238}U. Several isotopes are produced, including ^{238}Pu (87.7 years), ^{239}Pu (24130 years), ^{240}Pu (6570 years) and ^{241}Pu (14 years). The last nuclide decays by beta emission to produce ^{241}Am (430 years) (Hardy, 1974b; Fisenne, 1982). Plutonium-239 and ^{240}Pu are the most abundant of these nuclides, and because they cannot be distinguished by alpha spectrometry, the two nuclides are usually reported together as "plutonium." It has been estimated that about 320 kCi of Pu has been distributed globally, mainly from explosions of megaton-range weapons that took place before 1963. In 1974 it was reported that another 4 kCi remained in the stratosphere. Transfer from the stratosphere and deposition from the troposphere to the earth's surface proceed at the same rates as for ^{90}Sr, with the result that the ratio of ^{239}Pu to ^{90}Sr has been remarkably constant since the cessation of large-scale tests in 1963. About 9 kCi of ^{238}Pu has been injected in the atmosphere by nuclear tests, somewhat less than the 17 kCi from the abortive reentry of a satellite in 1964 (see Chapter 14).

Surface deposition has been estimated by relating the known ^{90}Sr deposition to the ratio ^{239}Pu : ^{90}Sr in stratospheric air, which has been observed to be about 0.017, corrected for decay to the time of production. Since deposition of ^{90}Sr in the North Temperate latitudes was about 80 mCi/km^2, one can in this way estimate the deposition of Pu to be about 1.4 mCi/km^2.

Because Pu tends to bind tightly in soil, there is little plant uptake, and the main route of entry to the human body is by inhalation of suspended dust (Fisenne *et al.*, 1980).

Using the constant ratio ^{90}Sr : Pu, and from historical ^{90}Sr measurements, Bennett (1978) estimated that the concentration of Pu in the air of New York City reached a peak of 1.7 fCi/m^3 in 1963.

RADIOIODINE

Although several species of radioiodine are produced in fission, only one, ^{131}I (half-life 8.1 days), is of major significance as far as worldwide fallout is concerned (Holland, 1963). This nuclide is produced copiously, about 64 Ci/megaton of fission. It was seen earlier that the fission yields of all nuclear weapons tests conducted in the atmosphere through 1980 totaled about 217 megatons. Thus, in excess of 14 billion Ci of ^{131}I was released to the atmosphere, most of which was produced by explosions in the megaton range. The bulk of the ^{131}I was injected into the stratosphere, where it decayed substantially before transferring to the troposphere and depositing on the earth's surface. However, the ^{131}I produced by explosions having yields less than 100 kilotons remained in the troposphere, together with a small fraction of the debris from megaton-yield weapons. Substantial amounts of radioiodine were distributed throughout the world in this manner.

Van Middlesworth (1954) demonstrated the presence of ^{131}I in cattle thyroids throughout the United States during the period of nuclear weapons testing in Nevada in 1953. This was the first indication that this radionuclide might be present in the human food chain during periods of nuclear weapons testing. Comar *et al.* (1957) subsequently undertook *in vitro* ^{131}I measurements of human and cattle thyroids during a 23-month period from January 1955 to December 1956.

The paucity of ^{131}I data during the 1950s, when extensive open-air testing of nuclear weapons was taking place in the continental United States, was due to the fact that radioiodine was relatively difficult to measure with the instrumentation then available. Measurement was to become a relatively simple procedure by means of γ spectrometry, but this technique was not generally available prior to the first test moratorium of 1958, and no systematic measurements of radioiodine in human tissues were made during the period 1950–1958, when copious amounts of radioiodine were being released to the atmosphere from tests in Nevada. Radioiodine contamination of milk during that period may have reached higher levels in many parts of the country than at any time since. Lewis (1959), based on study of the published values of ^{131}I in cow's milk, concluded that over the

5-year period prior to 1958 the average accumulated dose to the thyroids among children in the United States was 0.2–0.4 rad.

Following weapons tests in Nevada in the early 1950s, it was known that relatively heavy fallout occurred at distances of a few hundred miles from the test explosions (Eisenbud and Harley, 1953, 1956). More than a decade later, when it was realized that radioiodine was a significant source of exposure at some distance from the tests, efforts were made to reconstruct the thyroid doses in several communities where relatively heavy fallout was known to have occurred (Knapp, 1963; Pendleton *et al.*, 1963, 1964). Pendleton and associates used 24-hr air samples made in a number of communities to estimate the ^{131}I content of cow's milk and the thyroid dose to infants who consumed the milk. It was concluded that the exposures ranged from about 1 rad to as high as 84 rad in the community of St. George, Utah, following a test on May 19, 1953. These estimates appear credible when all available information is taken into consideration, and they are supported by more recent analyses of existing data by Tamplin and Fisher (1966), who concluded that the thyroid doses to children in Washington County, Utah exceeded 100 rad.[2]

By the time the Soviet Union resumed weapons testing in 1961, γ spectrometry had progressed to the point where *in vivo* measurements of ^{131}I in human thyroids were feasible (Laurer and Eisenbud, 1963). Measurements could also be made of human thyroid tissue that became available at autopsy, as well as of the ^{131}I content of cow's milk and other foods. Measurements made in New York City during weapons tests in the Marshall Islands and Siberia showed that during two 3-month periods in 1961 and 1962 the thyroid doses in children averaged 50 and 140 mrem, respectively (Eisenbud *et al.*, 1962, 1963a).

Calculation of the population thyroid dose from the ^{131}I (Eisenbud *et al.*, 1963b) was greatly facilitated by the remarkable uniformity of the ^{131}I concentration in the milk distribution system of New York City, where the above studies were made (Eisenbud *et al.*, 1963c). The variations from dairy to dairy were so random that at the end of an extended period of sampling, such as 1 month, the mean concentrations of the milk from the various dairies that supplied New York City with milk were not significantly different. It was concluded that, at least during the period when the study was conducted, the practice of analyzing single, daily, randomly selected samples of the milk in New York City provided a satisfactory estimate for the milk supply as a whole. However, the fallout then occur-

[2]Studies of 4000 Utah and Arizona children in 1965–1968 (Weiss *et al.*, 1971) showed no difference in the incidence of nodules among those exposed to the high levels of fallout and those who lived outside that area. A major study of the thyroid disease is now being conducted at the University of Utah, but the findings have not yet been published.

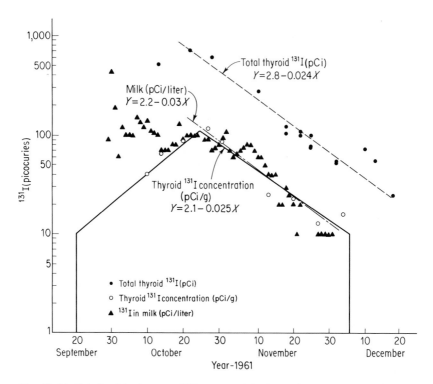

Fig. 13–14. Relationships between ^{131}I content of fresh cows' milk and human thyroids in New York City during the Soviet tests of 1961. (From M. Eisenbud *et al.;* copyright 1962, AAAS.)

ring was from weapons tests conducted nearly halfway around the world, and the levels of contamination would be expected to be more uniform at such a distance than would be the case at closer distances.

The excellent correlation between thyroid radioiodine and concentration of radioiodine in milk is shown in Fig. 13–14, which shows the total thyroid ^{131}I content in a number of individuals during a 3-month period in 1961 when Soviet testing was in progress. On the same chart are the thyroid concentrations (picocuries per gram of thyroid) and the milk concentration of radioiodine. The thyroid dose to maximally exposed persons was estimated from the integral under the solid line, which is the envelope of the plotted concentration of thyroidal ^{131}I; i.e., each of the thyroid measurements that forms the envelope is the highest value reported for a given day. The area under the envelope can be shown to be 3400 pCi day/g of thyroid tissue, from which the thyroid dose can be determined as follows:

$$1 \; \frac{\text{pCi day}}{\text{g}} = \left(2.2 \; \frac{\text{disintegrations}}{\text{min}} \right) \left(0.189 \; \frac{\text{MeV}}{\text{disintegration}} \right)$$

$$\times \left(1.6 \times 10^{-6} \; \frac{\text{erg}}{\text{MeV}} \right) \left(1440 \; \frac{\text{min}}{\text{day}} \right) \left(100 \; \frac{\text{erg}}{\text{g}} \right)^{-1}$$

$$= 9.7 \; \frac{\mu\text{rem/day}}{\text{pCi}}$$

and the integrated thyroid dose is, therefore,

$$\left(3400 \; \frac{\text{pCi day}}{\text{g}} \right) \left(9.7 \; \frac{\mu\text{rem/day}}{\text{pCi/g}} \right) \left(10^{-3} \; \frac{\text{day mrem}}{\mu\text{rem}} \right) = 40.7 \; \text{mrem}$$

These data are for a large metropolitan area located thousands of miles from the test site. Pendleton *et al.* (1963) estimated the doses to children in Utah, a few hundred miles from the scene of nuclear explosions in Nevada, to average 1 rem, with peak doses as high as 14 rem. However, there have been few studies such as these.

MISCELLANEOUS OTHER NUCLIDES

Tritium (Moghissi and Carter, 1973; NCRP, 1975b) is produced copiously in thermonuclear explosions, and its incorporation in precipitation caused the tritium content of surface waters in the North Temperate Zone to reach several nanocuries per liter in the mid-1960s (Wyerman *et al.*, 1970). Moghissi and Lieberman (1970) measured the tritium content of body water in children by means of tritium analyses of urine samples during the 2-year period 1967–1968, during which time the ^{3}H content of urine decreased from 1.5 to 0.2 nCi/liter.

The dose commitment from ^{3}H produced by nuclear weapons tests is estimated to have been about 5 mrem in the North Temperate Zone (UN-SCEAR, 1982).

Manganese-54 is produced as an activation product in thermonuclear explosions, as are ^{55}Fe, ^{59}Fe, ^{65}Zn, and a number of other nuclides. These are relatively short-lived nuclides that are readily detectable in the environment during and immediately following nuclear weapons testing, but they have not contributed significantly to the dose received by humans. A possible exception is ^{55}Fe, which was first detected in fallout by Palmer and Beasley (1965) and was shown by Wrenn and Cohen (1967) to have delivered doses of about 1 mrem to the erythrocytes of inhabitants of the New York City area. The dose to ferritin aggregates in hemoglobin was calculated to be 235 mrem by including the energy of the short-range Auger electrons from ^{55}Fe (Wrenn, 1968).

LUNG DOSE DUE TO INHALATION OF DUST

Radioactive dust particles that are too large to be respirable deposit quickly, but at distances beyond 200 miles from near-surface explosions the mass median diameter of fallout particles has been found to be 2 μm, which is close to the optimum particle size for lung retention (Eisenbud and Harley, 1953). Shleien et al. (1965) found that 88% of the total activity of airborne dust was contained in particles less than 1.75 μm in diameter. The radioactivity of the dust depended on the age of the debris and distance from the test site. In one city a few hundred miles from the Nevada test site, a 24-hr average concentration of beta activity of 24 nCi/m^3 was recorded following a test in 1952 (Eisenbud and Harley, 1953). The extent to which radioactive dust was then present in the atmosphere is illustrated in Figs. 13–2 and 13–3.

There have been very few estimates of the dose to the human lung during the period of active testing. Wrenn et al. (1964) undertook a series of human lung measurements which showed that the principal dose was received from the ^{95}Zr–^{95}Nb pair. It was estimated that the lung dose was about 3 μrad/day from lung burdens that ranged from 210 to 450 pCi at the time of death. These lung burdens were associated with average air concentrations over a 6-month period of 2–4 pCi/m^3, several orders of magnitude lower than the peak concentrations reported above.

Wegst et al. (1964) autoradiographed the ashes from one human lung obtained in 1963 and found 264 radioactive particles having a total activity of 436 pCi. As noted in Chapter 2, there is great uncertainty about the method of calculating absorbed dose from β particles deposited in the lung, and the significance of the high doses in the vicinity of a β-radiation point source is not understood. Wegst et al. calculated that the tissue within 10 μm from the particle would receive 2×10^3 rem if the particle had a 120-day half-life in the lung. This volume would contain 16 cells. However, the dust particles are apt to be in motion in the lung, in which case the energy would be deposited in a larger volume of tissue with a corresponding reduction in the absorbed dose.

EXTERNAL RADIATION

Many investigators have calculated the effect of delayed fallout on ambient gamma radiation, using data on the quantities of the various radionuclides known to be present in fallout. These calculations give estimates of the γ dose, assuming deposition of a given mixture of radionuclides distributed uniformly on an infinite smooth plane. Among the difficulties in calculations of this kind are the following:

1.   The dose integral is sensitive to the time of fallout, particularly if the fallout occurs within 2 or 3 days after the explosion. The time of arrival of the fallout is not always known.

2.   The surface of the earth is not an infinitely smooth plane but is highly irregular, particularly in inhabited areas. The natural and man-made irregularities tend to absorb γ radiation and may modify the distribution of fallout to the extent that it is not correct to assume that the deposit is uniform.

3.   The effect of weathering is difficult to evaluate. Rain will in most cases lessen the dose received by washing the fallout from places of habitation, but in some cases the rain may tend to concentrate deposits and thereby increase the dose.

4.   Buildings provide an uncertain degree of shielding, which depends on structural factors as well as the fraction of the day spent indoors.

Nevertheless, Beck (1966) has shown that estimates of the γ-radiation exposure rates based on rainfall deposition data are reasonably consistent with measurements made *in situ* by γ-spectrometric techniques.

Gustafson *et al.* (1970) measured the external radiation dose from fallout near Argonne National Laboratory since 1960. They reported that the maximum open-field exposure was 56 mR/year in 1963, which diminished to 13 mR/year by 1969. They considered that a more realistic estimate of the exposure would take into consideration the time people spent indoors, in which case they estimated the exposure rates to have been 22 and 5 mR/year, respectively. However, these exposures, as noted earlier, are to dust from explosions many thousands of miles away. Localities closer to the test sites undoubtedly received very much higher doses, but very few measurements were made at the time. The *average* dose commitment to inhabitants of the North Temperate Zone from external radiation due to tests conducted up to 1980 has been estimated to be 107 mrem (UNSCEAR, 1982).

The measurements of Beck included estimates of the contribution from the various nuclides present in fallout (Beck, 1980). In a total exposure of 76 mR during the period 1960–1964, the ^{95}Zr–^{95}Nb pair contributed about 37 mR, and ^{137}Cs accounted for 13 mR. The long life of ^{137}Cs results in a larger dose commitment from this nuclide.

Chapter 14

Accidents That Resulted in Contamination of the Environment

   A number of accidents that have occurred during the 45-year history of the nuclear industry have resulted in radioactive contamination of the environs. Those accidents, which varied considerably in severity, will be reviewed in this chapter. Not included in this chapter are a number of accidents that resulted in injuries and death to radiation workers, but did not result in environmental contamination. These accidents have been reviewed by Lushbaugh *et al.* (1980). Considering the size of the U.S. atomic energy industry, the occupational safety record has been excellent. There have been seven deaths due to accidents, the last of which occurred in 1964 in a privately operated industrial plant (Karas and Stanbury, 1965). Three of the seven fatalities occurred at a reactor accident in Idaho in 1961 and will be included in this chapter. All but one of the accidents in the United States occurred in connection with research or development activities in support of the country's military program for the military applications of nuclear energy. Most of the accidents that involved exposure to external sources of gamma radiation in this country and abroad have been from the mishandling of industrial radiographic sources (Lushbaugh *et al.*, 1980). There have been about one dozen injuries from such accidents in the United States and several more in foreign countries (Hubner and Fry, 1980).

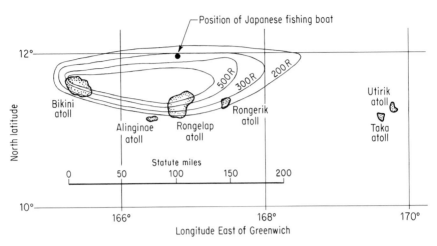

Fig. 14–1. Fallout pattern from the thermonuclear explosion of March 1, 1954, as reconstructed from various sources.

Fallout from the Thermonuclear Detonation of March 1, 1954

Prior to March 1, 1954, there had been some speculation among the very few people who were then informed about the subject of fallout as to whether surface and near-surface bursts in the megaton range could produce lethal amounts of fallout over large areas. The many doubts that existed even as late as March 1, 1954 were quickly dispelled when extensive fallout occurred following detonation of a 15-megaton device mounted on a barge over the reef of Bikini Atoll in the mid-Pacific Ocean. The barge was located in shallow water and a large amount of coral was incorporated in the fireball. There was little wind shear, and it is estimated that from 50 to 80% of the radioactivity from the explosion fell out in the pattern shown in Fig. 14–1.

The first indication that radioactive fallout was occurring was an increase in a recording gamma detector located on Rongerik Island about 160 miles east of Bikini. This island was inhabited by 28 American servicemen, who were operating a weather station. The fallout began about 7 hr after the detonation, and 30 min later the recording γ detector went off scale at 100 mR/hr. Aerial reconnaissance subsequently confirmed that fallout had occurred on Rongerik and that even heavier fallout had occurred on the Rongelap Atoll, about 105 miles east of Bikini, as well as the atoll of Ailinginae, located about 75 miles to the east-southeast. Evacuation procedures were put into effect, beginning about 30 hr after the detonation, with the removal of the 28 Americans by air. Rongelap, Utirik, and

TABLE 14–1

SUMMARY OF DOSES RECEIVED FOLLOWING TEST EXPLOSION OF MARCH 1, 1954

Group	Number exposed	Time fallout started (hr after detonation)	Exposure duration	Whole-body dose (estimated rem)	Iodine dose to thyroid (rem)
Rongelap[a]	67	4–6	Evacuated in about 50 hr, returned in 38 months	175	<10 yr old–810–1800 10-18–334–810 <18–335
Rongerik[a]	28	7	Permanently evacuated in about 30 hr	78	50
Alinginae[a]	19	4–6	Evacuated in about 50 hr, returned to Rongelap in 38 months	69	<10–275–450 10-18–190 <18–135
Utirik[a]	163	22	Evacuated in 55–78 hr, returned to Utirik	14	<10–60–95 10-18–30–60 <18–30
Fukuru Maru[b]	23	4	Remained on boat for 13 days	170–590	296–1026

[a]Conard et al. (1960, 1980).
[b]Kumatori et al. (1980).

Ailinginae were evacuated subsequently. Figure 14–1 shows the relative locations of these atolls.

It was later found that Rongelap Island was so heavily contaminated that the natives would be unable to return for an extended period, and they remained on another atoll of the Marshall Islands for 38 months until July 1957. The 18 natives evacuated from Ailinginae were actually natives of Rongelap who had been visiting the normally uninhabited island. These 18 natives eventually returned to Rongelap with their kin.

The fallout on Utirik was insufficient to require prolonged evacuation, and the natives of this atoll were returned to their homes shortly after the incident (Conard et al., 1980).

It was not known until some days later that a Japanese fishing vessel, the Fukuru Maru, was located in the path of the fallout about 80 miles to the east of Bikini. The boat had somehow not been observed during the aerial searches that preceded each detonation. When the explosion occurred, the men hauled in their lines and proceeded away from the detonation, but visible fallout began to occur about 4 hr later. These men lived on the vessel for 13 days until they returned to their home port of Yaizu (Kumatori et al., 1980; Jpn. Soc. Promotion Sci., 1956).

Although the fallout was visible on both Rongelap and the Japanese fishing boat, neither the Marshallese nor the Japanese took any precautions to minimize the amount of their exposure. This is understandable in view of their ignorance of the subject. Thus, the 64 natives of Rongelap and the 23 fishermen of the Japanese fishing vessel lived for 50 hr and 15 days, respectively, in intimate contact with the most radioactive inhabited environment that is known ever to have existed. Large particles of fallout fell into their hair and came in direct contact with their skin. The Japanese fishermen described the dust deposit as so heavy as to seem like a coating of snow (Jpn. Soc. Promotion Sci., 1956).

The whole-body and thyroid doses received by the several groups for whom estimates are available are given in Table 14–1. The dust filled the air and fell on the bodies of the victims as well as into their food and water. The Marshallese sat on the contaminated ground, and the Japanese sat on the contaminated deck. No special precautions were taken in the consumption of food, and no special hygienic procedures were followed, except that a number of Marshallese swam in the surf during the period of exposure.

MEDICAL AND ENVIRONMENTAL INVESTIGATIONS

Excellent reports of the American and Japanese studies of this accident are available (Conard et al., 1980; Jpn. Soc. Promotion Sci., 1956; Kumatori et al., 1980). The Marshallese were evacuated to the island of Kwajalein, where an emergency hospital was established and teams of specialists from the United States were made available to provide medical care for the natives. Survey teams were sent to the islands for radiological measurements.

The Japanese did not enter their home port of Yaizu until March 14, by which time they were already suffering from the effects of exposure, but it was another 2 days until it became known that their sickness was due to their having been subjected to radioactive fallout. The 23 men were hospitalized in Yaizu and later transferred to Tokyo.

The early effects of their exposure may be described under (1) effects on the skin, (2) hematological effects, and (3) effects due to internal emitters.

1. *Skin effects.* Itching and burning sensations were experienced from 1 to 2 days after the fallout by all groups except the Utiriks. After various time intervals, ranging from 3 days in the case of the Japanese fishermen to 21 days at Rongerik and Ailinginae, skin lesions and epilation began to develop. The lesions became ulcerous in 70% of the Japanese fishermen and in about 25% of the Rongelaps. The investigators found that the severity of the injuries at Rongelap and on the Fukuru Maru was definitely related to obvious factors such as whether outer garments were worn or whether the individuals went swimming. For example, Rongelap children who spent much time wading in the lagoon had fewer foot injuries, and the worst burns among the Japanese fishermen were on two men who did not wear hats.

2. *Hematological effects.* The detailed hematological findings will not be reviewed here except to say that the blood changes were related to the severity of the whole-body γ dose. The diminution in white blood count was most marked in the Japanese, among whom the values diminished by as much as 50% in about 28 days, at which time slow recovery began. The values in the Marshallese varied from 55% of normal at 44 days in the case of the Rongelaps to about 84% of normal in the case of the Utiriks, who received the smallest dose.

3. *Dose from internal emitters.* Until they were removed from the contaminated environment, the Marshallese and Japanese existed under conditions which maximized the opportunity for inhalation and ingestion of the radioactive particles. Life on an atoll and on a fishing vessel in the tropics is largely an outdoor existence. The Marshallese drank from exposed cisterns and ate food which was exposed to the open air. The Japanese ate raw fish which were certainly in contact with the contaminated deck. Almost everything on the ship was found to be contaminated when it arrived in Yaizu 13 days later, and the contamination must have been even greater during the first few days after exposure.

TABLE 14-2

DISTRIBUTION OF RADIOACTIVITY IN DECEASED FISHERMAN[a]

Fraction	Probable nuclide	Liver (pCi/g)	Fresh tissue			
			Kidney	Lung	Muscle	Bone
Ru + Te	^{106}Ru + ^{106}Rh, ^{129}Te	<0.1	0.9	<0.1	0.2	2
Zr + Nb	^{95}Zr + ^{95}Nb	1	1	0.4	0.3	2
Rare-earth elements	^{144}Ce + ^{144}Pr	2	1	0.5	0.5	20
Sr	^{90}Sr + ^{90}Y	0.6	0.4	~0.1	<0.1	1

[a] Tsuzuki (1955).

Urine analyses and body scans of both the Marshallese and Japanese were positive for absorbed fission products. In the Marshall Islands, the first samples collected for radiochemical analysis were obtained 15 days after the detonation. The principal absorbed isotope was ^{131}I, and two independent estimates of the amount initially deposited were 6.4 and 11.2 µCi (Conard *et al.*, 1980). The dose from the short-lived nuclides [132]I, [133]I, and [135]I was estimated to be two to three times the dose from [131]I. These data served as the basis for the thyroid dose estimates shown in Table 14–1.

Samples of urine were also collected from the Japanese fishermen and were sent to the Atomic Energy Commission Health and Safety Laboratory (now the Department of Energy Environmental Measurements Laboratory) for radiochemical analysis (Kobayashi and Nagai, 1956). Except for ^{131}I, only minimal urinary excretion of fission products was found, despite the fact that the men lived for 13 days on the contaminated fishing vessel. Urinary excretion of fission products was less than that reported for the Marshall Islanders. The β activity of oxalate precipitates from samples collected on April 21, 1954, ranged from 10 to 110 pCi/liter, compared with values as high as 37 pCi/liter obtained on five unexposed Americans and analyzed at the same time.[1]

One of the fishermen died of a liver disease 6 months after the accident. The results of radiochemical analysis of his tissues as reported by Tsuzuki (1955) are given in Table 14–2. The amount of radionuclides in the tissues of this man was so low as to rule out the possibility of a significant dose due to absorption of intermediate and long-lived radionuclides. However, as noted earlier, the dose from radioiodine was certainly significant in all these cases.

A second fisherman died in 1975, 21 years after the accident. Radiochemical analyses of lung, liver, spleen, kidney, pancreas, and bone were negative for two long-lived nuclides of concern, ^{90}Sr and [239]Pu (Kumatori *et al.*, 1980).

Follow-Up Studies

The exposed Marshallese were evacuated by plane and ship about 2 days after the fallout and were taken to Kwajalein, where they remained for about 3 months, during which time extensive medical studies were possible. The Rongelap natives were then taken to a temporary village on Majuro Atoll, where they lived for 3½ years, by which time the radiation levels of Rongelap had decreased to a point where they could be returned.

---

[1]These data were obtained prior to the availability of modern methods of γ spectrometry, which permit more sophisticated analysis.

The Utirik and Ailinginae inhabitants were allowed to return to their atolls following a 3-month stay on Kwajalein. Annual medical examinations of the Marshallese, undertaken under the direction of Robert A. Conard of Brookhaven National Laboratory, have revealed a number of the late effects of irradiation. The findings of Conard and associates are reviewed in a summary report published 26 years after the fallout occurred (Conard *et al.*, 1980).

The principal late effect of the fallout on the Marshallese has been the development of thyroid abnormalities, including nodules, carcinoma, and hypothyroidism. These effects were seen mainly in the residents of Rongelap who were less than 10 years of age at the time of exposure (Conard *et al.*, 1980). Of three persons who were exposed *in utero*, two have developed benign thyroid tumors. Several children became hypothyroidal, and growth was stunted in two boys who received doses of about 1800 rad at 1 year of age.

During the first 4 years following exposure, the frequency of miscarriages and stillbirths was higher among exposed women, but no difference has been observed since that time.

Some low-level residual radioactivity remained on Rongelap at the time the natives returned in 1957. It was found that the coconut crab, a much-favored food among the Marshallese, tended to concentrate ^{90}Sr, and consumption of local crabs was banned. After 15 years the concentration of ^{90}Sr in the flesh of these crabs was about 700 pCi/g Ca. Whole-body counts of the natives indicated elevated levels of ^{137}Cs and ^{60}Co. By 1980 the body burdens of ^{137}Cs averaged about 190 nCi, having decreased from about 680 nCi in 1961 (Lessard *et al.*, 1984). A number of individuals then living on Rongelap were not present at the time of fallout. The ^{137}Cs burdens of the two groups were indistinguishable, implying that the ^{137}Cs was absorbed after the natives returned to Rongelap. The ^{137}Cs body burdens increased from the time of their return in 1957 to about 1965 and then began to decline. It is estimated that the 50-year dose commitment from ^{137}Cs is about 2 ± 3 rad. Other nuclides (^{60}Co, ^{65}Zn, ^{90}Sr, and ^{55}Fe) accounted for much less than the dose from ^{137}Cs.

It is estimated from urine analysis (Lessard *et al.*, 1984) that the mean ^{90}Sr burden of the Rongelap people increased to about 8 nCi by about 1961. Analysis of bone samples from a deceased Rongelap woman in 1962 corroborated this estimate (Conard *et al.*, 1970). The body burdens of both ^{137}Cs and ^{90}Sr began to drop after 1963, but it is not known if this has been due to lowered radioactive contamination of food from the island of Rongelap or to dilution of the diet by an increasing amount of imported food.

The body burdens of ^{239}Pu among the Rongelaps are low, but indeterminate as of this writing. Early measurements of urine samples showed traces of ^{239}Pu, but this may have been due to sample contamination. Samples of feces and urine were recently collected and are being analyzed by the improved methods now available (Lessard *et al.*, 1984).

Follow-up studies of the surviving Japanese fishermen were negative 28 years postexposure, except that the frequency of chromosomal changes in blood cells remained abnormally high (Ishihara and Kumatori, 1983).

KNOWLEDGE GAINED

The events subsequent to the fallout of March 1, 1954 led to research that has produced important information about the behavior of radioactivity in the environment:

1.   The detonation of March 1, 1954 demonstrated conclusively that fallout is a major consequence of the use of megaton nuclear weapons at or near the surface.

2.   Severe exposure to residual radiation was associated only with the fallout of visible dust. Although it cannot be concluded that this would always be the case regardless of the type or place of detonation, this observation must be taken into consideration in the indoctrination of civilians and military who may someday be involved in nuclear war.

3.   The β-ray burns owing to deposition of fallout on the skin may be avoided by simple procedures such as wearing clothes, staying indoors, and washing.

4.   Inhalation and ingestion of radioactive substances appear to be very much less of a problem in fallout than had been previously thought. Both the Marshallese and Japanese lived intimately with their contaminated environment during and immediately after the period of exposure, but the amounts of radionuclides deposited in their tissues did not contribute appreciably to the overall effects observed.

The Accident to Windscale Reactor Number One in October 1957

The Windscale (now called Sellafield) works of the U.K. Atomic Energy Authority is located on a low-lying coastal strip in the northwest of England. At this site were two air-cooled graphite-moderated natural-uranium reactors employed primarily for plutonium production. The core of one of these reactors was partially consumed by combustion in October 1957, resulting in the release of fission products to the surrounding countryside. The Windscale accident resulted in the first release of radioactive material from a reactor accident.

TABLE 14–3

PRINCIPAL FISSION PRODUCTS RELEASED
DURING THE WINDSCALE FIRE[a]

Isotope	Estimated releases (Ci)	
	Dunster et al.[b]	Clarke[c]
^{131}I	20,000	16.200
^{137}Cs	600	1,240
^{89}Sr	80	137
^{90}Sr	9	6

[a]In addition, an estimated 240 Ci ^{210}Po
were released as discussed in the text.
[b]Dunster et al. (1958).
[c]Clarke (1974).

The interaction of neutrons with the crystalline form of graphite displaces carbon atoms from their normal positions in the molecular lattice. Vacancies in the normal crystalline structure occur, and atoms of carbon appear interstitially. This has a number of effects on the physical properties of the graphite, including dimensional growth, which may occur linearly to the extent of about 3%. The ultimate effect of continued irradiation is breaking of the crystalline structure, with the carbon appearing as carbon black after 10^{21} neutrons/cm^2 (Harper, 1961; Wittels, 1966).

If the crystalline structure has not been destroyed, the original molecular form can be restored by an annealing process, which permits the interstitial atoms to diffuse to vacant positions in the crystalline lattice. The disarray of carbon atoms represents stored energy, which is released during the annealing process. As much as 500 cal/g of graphite may be released, enough to raise the temperature of the graphite to a dangerous level.

The 1957 accident was caused by the release of this stored energy at an excessive rate during the regular annealing procedure, in which the temperature of the core was being raised with nuclear heat. The release of stored energy was excessive in portions of the core, but the local increases in temperature that were occurring went undetected because of insufficient core instrumentation (United Kingdom Atomic Energy Office, 1957, 1958). Failure of a fuel cartridge evidently resulted from this factor. The metallic uranium and graphite began to react with air, and from the time combustion began on the morning of October 12 a substantial portion of the core was destroyed. The isotopes released to the environment were estimated originally by Dunster *et al.* (1958) and the estimates were subsequently reviewed by Clarke (1974; Crick and Lindsley, 1984). The two sets of data are in reasonable agreement, as can be seen in Table 14–3.

The original reports of the Windscale accident indicated that at the time of the accident the reactor was being used for production of plutonium. However, in a reassessment of the accident published in 1983 it was reported that the reactor was also being used as an irradiation facility for the production of ^{210}Po by neutron bombardment of bismuth. The ^{210}Po is used, with beryllium, to produce neutron sources in nuclear weapons. Table 14–3 does not include the estimate by Crick and Lindsley (1983) that about 240 Ci of ^{210}Po were released during the accident.

The first evidence that a mishap had occurred was the observation of elevated β activity of atmospheric dust collected by an air sampler located in the open about 0.5 mile from the reactor stack. A concentration of about 1.4 nCi/m^3 of air was observed, this being 10 times the level normally present from the atmospheric radon and thoron daughter products. Air samples collected elsewhere in the vicinity of the reactor confirmed that a release of radioactivity to the atmosphere was occurring.

Visual inspection through a plug hole in the face of the reactor revealed that the uranium cartridges were glowing at red heat in about 150 fuel channels. Because of distortion that had already occurred, these cartridges could not be discharged, but the fuel was removed from channels adjacent to the affected area, thereby creating a fire break which served to limit the extent of the mishap. For several hours, various schemes were devised for extinguishing the slowly burning core, but none was effective. On the following day, what must have been a most difficult decision was reached, and the graphite core was cooled by flooding the core with water. The reactor was cold by the afternoon of October 13.

ENVIRONMENTAL SURVEY PROCEDURES

When the mishap was discovered, procedures were implemented to determine the extent of exposure from the following sources: (1) external radiation, (2) inhalation of radioactive dust or vapor, and (3) contaminated food and water.

Vehicles equipped with radiation detection equipment were dispatched downwind from the stack. It was observed that the highest radiation level, 4 mR/hr, occurred directly under the plume at a point about 1 mile downwind. It was subsequently determined that the maximum dose of external radiation which would be received by a person remaining out-of-doors for 2 weeks following the accident would be in the range 30–50 mR.

During the period of release, about 12,000 air samples were collected on the site and about 1000 were collected in the environs. As would be expected, they revealed wide variations, with concentrations ranging as high as 0.45 pCi/ml. The average concentration during the period of the incident was approximately 4.5×10^{-3} pCi/ml, which is about 50% greater than the ICRP standard for permissible continuous exposure to ^{131}I (ICRP, 1960). As described by Dunster *et al.* (1958), the atmospheric contamination from the incident "rose on occasion to worrying but not dangerous levels on the site, while the dilution resulting from wind variations considerably reduced the hazard in the district."

It was found that, beginning on the afternoon of the first day, milk from cows in the vicinity of Windscale was contaminated with ^{131}I. Up to that time no emergency level had been established for short-term permissible exposure to radioiodine in food, but on consideration of the problem the Medical Research Council promptly recommended that the maximum permissible concentration (MPC) be 0.1 μCi/liter and that all milk containing more radioiodine than this be discarded. The manner in which the council arrived at this figure is of some interest in illustrating the thinking of such groups 30 years ago and the ingenuity with which workable guidelines can be arrived at on short notice when the occasion demands.

The council started with the knowledge that cancer of the thyroid in children had been known to occur following X-ray doses greater than 200 rad. Although no cases were known to have occurred following exposure to smaller doses, the data were insufficient to permit the conclusion that 200 rad was actually the threshold for tumor production. It was decided to limit the dose to children to a maximum of 20 rad. The amount of radioiodine in milk that would produce this dose in children became the permissible level for the entire population, children and adults alike.

Constants furnished in the ICRP tables were used to relate the concentration of radioiodine in milk to thyroid dose. One microcurie of ^{131}I/g of thyroid was calculated to result in a dose commitment of 130 rad. The mass of the child's thyroid was taken to be 5 g, and the thyroid was assumed to retain 45% of the ingested iodine. The limiting concentration in milk was computed to be 0.15 μCi/liter, which was rounded off to 0.1 μCi/liter.

Milk measurements were originally hampered by the fact that radiochemical methods were used at Windscale in the absence of spectrometric equipment, but beginning on the fifth day, spectrometric analyses were undertaken with equipment located in various laboratories in the United Kingdom. Up to 300 samples/day were analyzed by γ spectrometry. The distribution of ^{131}I in milk on October 13 is shown in Fig. 14–2. Altogether, the milk exceeded 0.1 μCi/liter in an area of approximately 200 square miles stretching in a southeasterly direction from Windscale. The irreg-

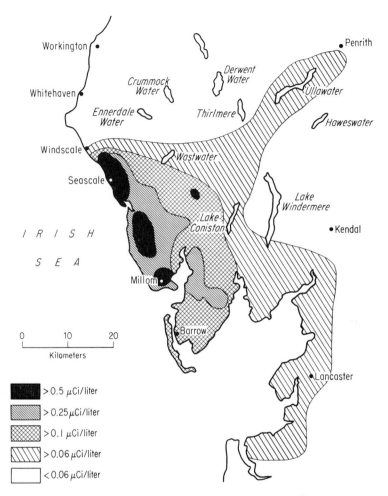

Fig. 14–2. Geographic area surrounding Windscale, showing the ^{131}I concentration in the milk from various districts 5 days after the accident. (From Dunster *et al.*, 1958.)

ularities can be explained by the changing meteorological patterns that existed during the period of emission (Crabtree, 1959). The highest concentration of radioiodine was 1.4 μCi/liter in milk obtained from a farm about 10 miles from Windscale. The amount of radioiodine in milk was found to correlate with the γ-radiation levels in the area. The concentration of ^{131}I in milk exceeded 0.1 μCi/liter in pastures where the γ radiation exceeded 0.035 mR/hr.

The criteria for restricting the sale of milk remained unchanged throughout the episode. By November 4, the permissible level was exceeded only in a region extending about 12 miles southward from Windscale. This area remained under restrictions until November 23.

Drinking-water samples were collected from reservoirs and streams and no concentrations of radioiodine or other radionuclides were found to exceed the MPC permitted by ICRP. Children and adults living downwind as far as 24 miles from Windscale were scanned for iodine uptake with a scintillation counter. Among 19 children studied, the highest dose received was estimated to be 16 rad. The highest adult dose was estimated to be 9.5 rad.

Milk and other foods in the vicinity were analyzed for ^{89}Sr. It was found that the concentrations of these isotopes in foods did not exceed the levels known to exist in the area prior to the accident.

LESSONS LEARNED FROM THE WINDSCALE EXPERIENCE

This accident would not have occurred if the fuel had been fabricated from uranium oxide rather than metallic uranium, which is pyrophoric. The incident emphasizes the need for great caution in the annealing of graphite-moderated reactors, but aspects related to design and operation of reactors are beyond the scope of this book. This discussion will be limited to the evaluation and management of the contaminated environment.

The following may be enumerated as being the significant conclusions derived from this accident.

1. In the slow oxidation of the Windscale fuel, radioiodine was released preferentially from the core. Because the stack effluents were filtered, other isotopes were not released in important amounts.

2. The dose to individuals from inhalation of the iodine or from direct exposure to the plume or deposited radioiodine was negligible compared to the dose received from dairy products.

3. The extent of iodine contamination could be estimated readily by scanning the γ-radiation levels in the area.

4. When the ^{131}I concentration in milk exceeds 0.1 μCi/liter, the dose to the thyroids of children may exceed 20 R. As a first approximation, it may be assumed that the concentration of radioiodine in milk will exceed 0.1 μCi/liter if it is produced by cows grazing in an area where the γ-radiation levels exceed 0.05 mR/hr.

In addition to the above, one cannot read the accounts of the British experience without seeing the need for extensive advance planning. An incident of this kind taxes the health physics capacity of the local organization to the extent that only by pooling resources on a regional or national basis can the necessary technical assistance be brought to bear. It is of interest in this connection that at the height of the Windscale survey about 15 vehicles, each having a team of two men, were used for radiation surveying and sampling. In addition, about 20 persons were employed in handling and recording the samples, and about 150 radiochemists throughout the United Kingdom assisted in the analysis of samples.

The Oak Ridge Plutonium Release of November 1959

The radiochemical processing pilot plant at Oak Ridge National Laboratory was built in 1943 and was being used for the chemical processing of highly irradiated fuels. In November 1959 a chemical explosion occurred in one of the shielded cells during a period when the process equipment was being decontaminated. No one was injured, and the monetary loss because of damaged equipment was relatively minor; however, the explosion resulted in plutonium contamination of the pilot-plant building, nearby streets, and building surfaces. The ensuing cleanup operations were costly, and the contaminated areas could not be usefully employed for many weeks while the cleanup was in progress.

The chemical explosion occurred during decontamination of an evaporator and is thought to have resulted from the formation of compounds such as picric acid when concentrated hot nitric acid was mixed with a proprietary decontaminating agent containing phenol. A small quantity of this solution had been left in the equipment because a normal water wash was omitted, and the explosion occurred when nitric acid was later introduced into the evaporator and was brought to the boiling temperature (King and McCarley, 1961).

The explosion breached a door leading from the cell directly to the outside of the building, and plutonium released through this doorway contaminated nearby streets and building surfaces. The adjacent air-cooled graphite-reactor building became contaminated when plutonium was drawn into the ventilation system. In addition, plutonium was forced through penetrations in the concrete cell walls into the remainder of the chemical-processing building. In all, an area 1000 ft in diameter was contaminated. Quantities of ^{95}Zr and ^{95}Nd were also released, but were of secondary importance.

As reported by King and McCarley, the radiological safety procedures consisted of the following: (a) immediate containment of the radioactivity to prevent it from spreading to other laboratory areas; (b) decontamination of areas that were slightly contaminated and could easily be made

TABLE 14–4

Summary of Decontamination Treatments for Various Surfaces[a]

Surface	Character of contamination	Primary decontamination treatment	Clean-up rate (ft^2/man-hr)	Other decontamination treatment
All (walls, floor, ceiling)	Transferable	Scrubbed with detergent and water and brush or sponge	27	Dusty areas vacuumed
Painted metal (walls and ceiling)	Fixed	Paint removed with paint remover and scrapers, and surface scrubbed with soap and water	4	Outer layer of paint removed with sandpaper
Concrete (floor)	Fixed	Ground with terrazzo floor grinding machine	5	"Hot" spots chipped out and vertical surfaces washed with dilute hydrochloric acid
Bare metal (SS[b] piping and tanks)	Fixed	Rinsed with dilute nitric acid and scrubbed with steel wool	—	Surfaces abraded with emery paper
Bare metal (other than SS[b])	Fixed	Abraded with emery paper or ground to remove pits	—	—
Lead shielding	Fixed	Rinsed with dilute nitric acid	—	—
Oily metal (pumps)	Fixed	Washed with Gunk, a commercial solvent	—	—

[a] King and McCarley (1961).
[b] SS: stainless steel.

available for service; (c) decontamination of streets and exterior surfaces of buildings; and (d) decontamination of the graphite-reactor building and the radiochemical-processing pilot-plant building.

Contamination within the cell in which the explosion took place was from 10^4 to 10^8 pCi/100 cm^2. A penthouse above the cell was contaminated

TABLE 14–5

Target Levels for Alpha Decontamination

	Direct reading	Transferable
Maximum	300 dpm/100 cm^2	30 dpm/100 cm^2
Average[a]	30 dpm/100 cm^2	3 dpm/100 cm^2

[a]At least 10 samples were used to derive an average sample, and there was at least one sample from each square meter of the projected surface area.

from 2 to 20 nCi/100 cm^2. Contamination of streets and buildings in the immediate vicinity exceeded 50 nCi/100 cm^2 in spots, but for the most part was below 50 pCi/100 cm^2.

No employees were overexposed during the period of cleanup. Protective clothing consisted of two sets of coveralls, two pairs of shoe covers, two pairs of rubber gloves, an assault mask, and a hood. Wrists and ankles were sealed with masking tape. Employees leaving contaminated areas passed through monitoring stations, where their protective clothing was removed. They then showered and received a final examination for α contamination before being released.

Containment of the radioactivity was achieved initially by fixing the surface contamination in a number of ways. Washdown was not attempted because of the possibility that the plutonium might wash into inaccessible places and because the accumulation of wash water would have created a problem.

Fixation was accomplished by resurfacing roads and by painting roofs, walls, and equipment. Paint was even sprayed on grass lawns and sidewalks that were found to be contaminated.

The decontamination procedures consisted of brushing and sponging, scraping, grinding, and various other techniques, which are enumerated in Table 14–4. The target levels for alpha decontamination are shown in Table 14–5.

When decontamination to this extent was not practical, but where the levels were less than 10 times the target levels, the surfaces were covered with brightly colored enamel and then covered with either paint or concrete. The brightly colored paint would serve as a warning in the future should the protective coating be removed.

The decontamination proceeded smoothly, but the incident resulted in costly interruptions to normal operations. The Oak Ridge National Laboratory graphite reactor was not in operation from November 20 to December 22. The processing cells directly involved in the incident could not be cleaned up for about 8 months until building modifications could be made to provide for improved containment of radioactivity.

It is estimated that only 15 g (about 1 Ci) of plutonium was blown out of the evaporator subcell. This small amount of material resulted in contamination that could only be removed after an expenditure of hundreds of thousands of dollars and a costly interruption to an important research reactor and other activities in the immediate vicinity. All this occurred, after 16 years of safe operation, as a result of the rather subtle error of introducing phenol into a decontaminating agent which was later to be mixed with nitric acid. The incident serves well to illustrate the extraordinary care which must be exercised in the design and implementation of operations such as these.

Management of the accident would have been more difficult had the accident occurred anywhere but in a National Laboratory which had had unique experience with many minor incidents of this type. A well-trained technical staff was available, backed up by a well-indoctrinated labor force and services of a type which can only be found at large centers such as Oak Ridge.

The Army Stationary Low-Power Reactor (SL-1)

An explosion of the Army Low-Power Reactor (SL-1) during January 1961 resulted in the death of three military personnel at the National Reactor Testing Station in Idaho. The reactor was a direct-cycle boiling-water unit designed to operate at a level of 3 MWt and was fueled with enriched uranium plates clad in aluminum. After a little more than 2 years of operation, the SL-1 reactor was shut down on December 23, 1960, having accumulated an operating history of approximately 950 MW-days. A 12-day maintenance program was contemplated, and the reactor was scheduled to resume full power on January 4, 1961 (Buchanan, 1963; Horan and Gammill, 1963). On the night of the accident the crew consisted of three subprofessional military personnel, of whom two were licensed reactor operators and one was a trainee.

The first indication of trouble at the reactor was given by remote radiation and thermal alarms, which caused the fire department and health physicists to respond. On arriving at the SL-1, they encountered radiation fields of 200 mR/hr extending for a few hundred feet from the reactor building.

On entering the buildings, the emergency crews found neither a fire nor

Fig. 14–3. The 13-ton core and pressure vessel of SL-1 being removed from the reactor building 11 months after the accident. The top and one side of the building have been removed. The reactor is to be lowered into the plastic-covered concrete cask on the flatbed trailer. (From U.S. Atomic Energy Commission.)

any of the three operators. A brief reconnaissance of the floor of the reactor building indicated radiation levels as high as 500 R/hr. After a search of about 1½ hr, the dead bodies of two of the personnel were found. The third man, who was still alive, was removed, but he died shortly thereafter. The two dead men were not removed for several days, during which time several hundred men were engaged in recovery operations. Twenty-two personnel received radiation exposures in the range of 3 to 27 R (U.S. AEC, 1961). This is an excellent record, considering the complexity of the rescue and recovery operations and the fact that the radiation levels in some areas approached 1000 R/hr. According to Horan, the bodies had been saturated with contaminated water and penetrated by particles of fuel. The radiation intensities at the bodies were 100 to 500 R/hr at 6 in.

The accident evidently resulted when withdrawal of a single control rod caused the reactor to go into the prompt critical condition. Why the rod was withdrawn may never be known.

Despite the fact that the excursion was a violent one and the reactor was not surrounded by a vapor container, the radiation levels outside the reactor building were minimal, about 5 mR/hr at 1000 ft. Essentially all the radioactive material with the exception of ^{131}I was contained within a 3-acre plot. Thus, despite the fact that the reactor core contained approximately 1 MCi of medium- to long-lived isotopes, it is thought that less than 10 Ci of ^{131}I was released to the environs.

Several months after the accident, it was decided to dismantle the reactor, an operation that presented many difficulties because of the high levels of radiation. Figure 14–3 shows the core being hoisted from the containment shell preparatory to removal to a hot cell for study.

The Houston Incident of March 1957

A relatively minor accident which occurred in Houston, Texas, in March 1957 and which received extensive publicity (*Look Magazine*, 1960) is worth discussing. From this incident one can learn very little of a positive nature, but one can profit from an examination of the details of the accident and subsequent events, because mismanagement of the circumstances provided many illustrations of what not to do.

The company was licensed by the AEC to encapsulate sources for gamma cameras. In March 1957, two men were opening a sealed can containing 10 pellets of ^{192}Ir, each pellet being a ⅛ × ⅛ in. right cylinder containing about 35 Ci.

It was necessary to remove a plug at the end of the can to have access to the pellets, and this was accomplished on a jeweler's lathe in a sealed Plexiglas box, which in turn was located in a hot cell. The operation was performed with master–slave manipulators through 33 in. of concrete.

When the small can containing the pellets was removed from its larger container and opened, two of the pellets were found to be in a loose dusty form. Some of the dust escaped from the Plexiglas box and hot cell, as evidenced by the fact that an air monitor in the laboratory indicated the presence of radioactive dust. One of the two employees involved in the operation was dressed in street clothes and left the laboratory. The second employee, who was dressed in work clothing and wearing a respirator, remained in the room for an unknown period of time.

The fact that the laboratory had become contaminated came to the attention of the plant management for the first time about 1 month after the incident occurred. It was then observed that contamination existed in the vicinity of the hot cell and that it had been spread to employees' street clothes, shoes, and even into their homes and automobiles. In the intervening weeks, various steps had been taken to encapsulate and store the sources, and items of equipment which became contaminated as a result of the incident had been stored away after unsuccessful attempts at decontamination.

The AEC was notified, somewhat belatedly, about 5 weeks after the accident, and shortly thereafter the company retained the services of a private company to survey and decontaminate the affected areas. Of 19 private homes that were checked, 8 showed evidence of contamination; 7 out of 53 automobiles were found to have traces of contamination.

Nineteen employees and members of their families were examined by a physician. One neighbor of an employee was also examined. Except for minor radiation burns on each of the two employees who were present at the time of the incident, the medical findings were negative.

Very few quantitative data are available concerning the levels of contamination encountered. However, it is reported that one of the employees received an exposure of 1.7 rem during 1 week when decontamination of the laboratory was in progress. His 13-week exposure was 3.9 rem, which exceeds the permissible level by 25%.

This incident was widely publicized in sensational newspaper accounts, and it became the subject of nationwide televised programs and nationally distributed magazine articles. One of the main themes of the publicity was that the affected individuals and their families became socially ostracized because of their "radioactivity" and that serious disabilities were developing not only in the two employees but also in their families.

The incident occurred very soon after the licensing procedure of the AEC was put into effect, and the licensee was apparently ignorant of the need to notify the AEC that an accident had occurred. Moreover, the regulatory apparatus of the AEC had not as yet been confronted with an incident of this type, and its handling of the incident indicated that the agency had much to learn about its role in incidents of this kind.

Many experts who might have been sent to Houston by the AEC to advise and assist its licensee to minimize the consequences of the incident were not made available, apparently because the regulatory function of the AEC in this particular instance was thought to forbid its participation in an advisory capacity. As a result, much potentially valuable information was never obtained. For example, urinalyses and whole-body γ-ray measurements would certainly have been useful for the two employees who were known to have been most heavily exposed. Other employees might also have been so examined. Depending on these findings, it might or might not have been desirable to conduct similar examinations of the families and neighbors.

Decontamination procedures in the homes of the employees, including such drastic procedures as cutting out portions of rugs to remove observable contamination, were undertaken without any basis for deciding that such measures were necessary. In summary, a valuable study which might have obtained useful quantitative information about the consequences of an incident of this kind was not undertaken, and in the absence of reliable information, rumor and sensationalism swelled in uncontrolled fashion about the unfortunate principals for months afterward.

In 1961, after 4 years of exaggerated stories about the effects of the accident, some of the principals were examined at AEC expense by physicians at the Mayo Clinic (Atomic Industrial Forum, 1961). It was found that none was suffering from the radiation effects that were repeatedly alleged. However, as is so often true in incidents of this kind, the factual announcement of the findings at the Mayo Clinic did not receive nearly as much public notice as the more sensational claims of injury.

Fermi Reactor Fuel Meltdown

The Enrico Fermi reactor was a sodium-cooled fast breeder located on the western shore of Lake Erie near Monroe, Michigan. The first core was loaded in 1963, and the full power of 66 MWe was authorized in December 1965 (Scott, 1971).

The reactor core consisted of 105 subassemblies containing uranium oxide fuel enriched to 25.6% and assembled in the form of pins clad with zirconium. The cooling system was basically that of the LMFBR, described in Chapter 9.

On October 5, 1966, at a power level of 27 MW, radiation monitors in the building's ventilation exhaust ducts sounded an alarm which resulted in automatic isolation of the building, following which the reactor was shut down manually. At the time of the radiation alarm, no personnel were in the reactor building.

The source of the problem was not immediately apparent, but measurements soon revealed that about 20,000 Ci of fission products had been released to the primary sodium loop and the reactor cover gas. Radioactive noble gases, but no iodine, were found to be present. No particulate or gaseous contamination was observed outside the containment structure.

Many months of careful work, during which the sodium loop was drained to permit inspection of the reactor vessel, disclosed that coolant flow had been blocked by pieces of zirconium that had become detached from a conical flow guide. As a result, partial meltdown of two fuel assemblies occurred. Repairs of the reactor were undertaken, and full-power operation was achieved in October 1970, 4 years after the incident.

The radioactivity released from the core was almost completely contained and no exposure to the employees or the public resulted. The highest radiation level as a result of this incident was 9 mR/hr at one point on the outer surface of the reactor building.

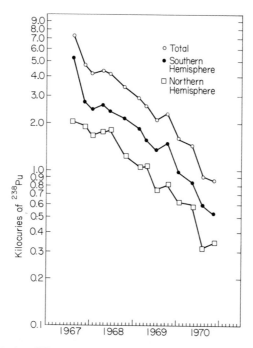

Fig. 14–4. Plutonium-238 stratospheric inventories, 1967–1970. (From Kleinman, 1971.)

Abortive Reentry of the SNAP 9-A

A navigational satellite launched on April 21, 1964, and carrying a radioisotope power generator failed to reach orbital velocity and reentered the atmosphere at about 150,000 ft over the Indian Ocean. The isotopic power unit was known as SNAP 9-A and contained about 17,000 Ci of ^{238}Pu (Krey, 1967).

Plutonium-238 was present in the upper atmosphere at the time as a residue from earlier nuclear weapons tests, but the system of high-altitude balloon sampling, mentioned in Chapter 13, demonstrated the sudden appearance of a new source of ^{238}Pu, which was first detected at an altitude of 108,000 ft 4 months following the abort. The stratospheric distribution of the debris during the next several years followed the predictions that had been made on the basis of transport models of the stratosphere developed through studies of the behavior of debris from weapons testing (Kleinman, 1971). However, the concentration of ^{238}Pu from SNAP 9-A was somewhat lower in ground-level air than had been predicted (Shleien *et al.*, 1970). About 16 kCi of ^{238}Pu, representing 95% of the amount originally injected, was estimated to have been deposited by the end of 1970. The estimated stratospheric inventories are given in Fig. 14–4, which shows that the debris apparently took about 2 years to diffuse to the sampling altitudes, following which depletion of the stratospheric inventory proceeded exponentially with a half-time of about 14 months (Krey et al., 1970). Particle-size analysis of the debris indicates that the particles ranged from 5 to 58 μm.

The [238]Pu content of ground-level air has been monitored by de Bartoli and Gaglione (1969) and by Shleien et al., 1970), and it has been concluded that the 50-year dose commitment to the pulmonary lymph node from [238]Pu exposure between 1961 and 1968 was 36 mrem (Shleien et al., 1970).

Plutonium Fire at Rocky Flats

A major fire occurred at the AEC's plutonium-processing plant operated by Dow Chemical Co. at Rocky Flats, Colorado, on May 11, 1969 (U.S. AEC, 1969). The extent of off-site plutonium contamination received national press attention several months later as a result of measurements made by a group of independent scientists (Colorado Committee for Environmental Information, 1970). A subsequent study of the plutonium content of soil in the vicinity of the plant was undertaken by the AEC (Krey and Hardy, 1970) and confirmed that plutonium was indeed present but that the distribution in the vicinity of the plant was not consistent with meteorological conditions at the time of the fire. The off-site inventory of [239]Pu, integrated within the 3 mCi/km^2 contour, was found to be 2.6 Ci and extended for a distance of more than 7 miles in a southeastern direction. However, the winds were generally blowing toward the southwest at the time of the 1960 fire. It was concluded that the observed plutonium deposition, although clearly from the Rocky Flats plant, was compatible with a gradual accumulation of plutonium owing to many small emissions from the plant. The pattern of plutonium deposition conformed to the annual distribution of wind direction at the site. The data also suggested that a very considerable area, up to about 40 miles east and north of the plant, was contaminated to the extent of an additional 1 mCi/km^2, accounting for an additional 3.2 Ci of Rocky Flats plutonium. The "background" distribution of [239]Pu deposited from weapons testing was estimated to be about 1.6 mCi/km^2. Subsequent measurements (Volchok et al., 1972) indicated that [239]Pu was present as airborne dust east of the plant site in concentrations ranging up to approximately 10^{-9} pCi/ml.

Considerable public concern developed as a result of these findings, and there was a loss of confidence in the credibility of the information released by the plant management and the government (Shapley, 1971). The source of the first public knowledge that there was off-site plutonium contamination was a citizens' group with access to highly trained radiochemists and highly sophisticated radiochemical equipment. The basic failure was in the lack of adequate reporting of the off-site levels of plutonium contamination by the government and its contractor. It is no longer possible to conceal information of this kind from the public, intentionally or otherwise.

Reentry of the Satellite Cosmos 954

A Soviet satellite, Cosmos 954, powered by a nuclear reactor reentered the atmosphere over the Canadian Northwest Territories on January 24, 1978, and spread radioactive debris over a 1000-km path stretching northeast from Great Slave Lake to Baker Lake (Tracy *et al.*, 1984). The satellite reactor was believed to have contained about 20 kg of highly enriched uranium. From radiochemical measurements of recovered fragments it was estimated that the burnup was 2×10^{18} fissions per gram U. At the time of reentry, the core was estimated to have contained about 84 Ci ^{90}Sr, 4900 Ci ^{131}I, and 86 Ci ^{137}Cs. An intensive search directed by the Canadian government led to the finding of about 65 kg of satellite structural components and equipment, for some of which dose rates of 500 rem/hr were recorded on contact. More than 4000 radioactive particles that ranged in size from 0.1 to 1.0 mm were also recovered, but it was estimated that only 0.1% of the dispersed particles were found. Seventy-five percent of the original material is estimated to have remained in the upper atmosphere.

The recovered particles were found to be highly insoluble in water and in $0.1M$ HCl. No detectable contamination was found in samples of air, water, and food supplies. The Soviet government has published no technical information about this incident.

Three Mile Island Unit 2

A relatively minor mechanical malfunction that occurred early in the morning of March 28, 1979, at the Metropolitan Edison Company (Met Ed) Three Mile Island Unit 2 (TMI-2) nuclear power reactor initiated the first nuclear accident at a power plant. Radiation exposure to the plant workers and the general public was insignificant, but the financial consequences of the accident were enormous for Met Ed's parent company, the General Public Utilities Corporation (GPU). The accident resulted in extensive psychological trauma among nearby residents of TMI, and the sociopolitical implications of the accident affected the nuclear industry to such an extent that 7 years after the accident, not a single new nuclear power plant had been ordered in the United States and orders were canceled for many nuclear plants that were being constructed.

DESCRIPTION OF THE ACCIDENT

Three Mile Island Unit 2 was one of a pair of pressurized-water reactors of Babcock and Wilcox design. The reactors each had a generating capacity of 850 MW, and TMI-2 began commercial operation in December 1978. The plant is located outside the town of Middletown, Pennsylvania, not far from Harrisburg, the state capital. The accident that occurred at TMI-2 has been the subject of many investigations by governmental and private organizations (Kemeny, 1979; Rogovin and Frampton, 1980; New York Academy of Sciences, 1981).

At 4 a.m. on Wednesday, March 28, 1979, the feedwater pumps that supplied the reactor's steam generators shut down, leading to automatic shutdown of the reactor and steam generating system. Heat from residual radioactivity in the core caused the temperature and pressure of the reactor coolant to rise and a relief valve atop the pressurizer to open (Jaffe, 1981).

The open valve should have closed in seconds, but did not do so. The valve was stuck in the open position and, since the reactor coolant water was draining, a loss-of-coolant accident was in progress.

The plant was equipped with three emergency feedwater pumps, and these started to operate automatically 14 sec after the first shutdown. The control room operator was aware that these pumps were operating, but he did not notice two panel lights that would have told him the emergency feedwater valves were closed and water was unable to reach the reactor.

The reactor safeguard systems responded properly to the emergency. Two minutes into the accident, high-pressure injection pumps began operating at a flow of 1000 gallons per minute (gpm), but 30 sec later a reactor operator shut one pump down and reduced the flow of the second to less than 100 gpm. As a consequence, steam bubbles began to form in the reactor coolant system, but the operators misdiagnosed the problem and, instead of restarting the emergency high-pressure injection system, they opened the reactor coolant letdown system, thereby draining additional water. By this incredible sequence of misjudgments, the water level continued to drop, the temperature rose, the fuel cladding failed, and the fuel itself was partially melted. In the process, radioactivity, which passed to the coolant water, was leaving the reactor coolant system via the relief valve and was accumulating in the reactor building basement.

The core temperature rose to the point at which the zirconium alloy cladding began to react with steam to produce hydrogen, some of which escaped into the reactor building. Early in the afternoon of the first day, sufficient hydrogen had accumulated in the reactor building to result in a low-level explosion, but no damage was done.

A backup valve that could be operated manually was located between the pressurizer and its open relief valve, but it was not until more than 2 hr after the faulty valve stuck in the open position that the backup was closed. This stopped further loss of coolant, but the core continued to be partially exposed and further damage was occurring. By 10:30 a.m., 6½ hr after the accident was initiated, the core was again fully covered with water.

THE IMMEDIATE POSTACCIDENT PERIOD

It is only possible in this chapter to provide the barest summary of the drama that unfolded during the first few days after the accident occurred. By 7 a.m., 3 hours after the first indication of trouble, a site emergency had been declared because of the elevated radiation levels detected by radiation monitors in the reactor auxiliary building. About ½ hr later, a general emergency was declared, and by 8:30 a.m. the utility emergency response teams were finding levels of 1–3 mR/hr immediately off-site. Two survey teams were dispatched from the Pennsylvania Bureau of Radiation Protection at about 10 a.m., at which time assistance was requested from the Brookhaven National Laboratory emergency response team, which was part of a network of emergency teams maintained by the Department of Energy. The Department of Energy sent a radiation monitoring helicopter, and Nuclear Regulatory Commission (NRC) inspectors began to arrive at the site.

The official agencies involved in the emergency management were: (1) the Pennsylvania Emergency Management Agency, located in the office of the Governor, (2) the Nuclear Regulatory Commission, (3) the Department of Energy, and (4) the Bureau of Radiological Health of the Food and Drug Administration.

The best information about off-site gamma-radiation levels during the first hours after the accident were obtained from 34 thermoluminescence dosimeters located around the plant to a distance of 15 miles. Additional dosimeters were emplaced by the federal agencies, but not until the third day of the accident (Gerusky, 1981).

During the first few days postaccident, a considerable number of food samples were collected for analysis and whole-body gamma scans were completed on 700 people. All of these tests were negative.

In a situation such as existed, emergency management depended on the interpretation of measurements of many kinds. A serious misunderstanding of events occurred on Friday, March 30, in connection with the transfer of radioactive waste gases from the makeup water storage tank to a waste gas decay tank located in the auxiliary building, outside the containment structure. The operator was aware that this transfer would release radioactive gases to the outside atmosphere, and requested that a helicopter be assigned to make radiation measurements in the vicinity of the ventilation stack. The helicopter reported levels of 1200 mR/hr at a point 300 ft above the stack.

Coincidentally, an NRC specialist had been calculating the ground level exposures that would be experienced if there should be a failure of the waste gas decay tank relief valve. A briefing was in progress at the time of the helicopter measurements, and the NRC specialist, literally within seconds of the 1200 mR/hr report from the helicopter, stated that the valve failure would result in a ground level exposure of 1200 mR/hr. The officials erroneously assumed that the report from the field was of ground level measurements and therefore assumed that the waste gas decay tank relief valve had failed. The gases, in fact, were not coming from the intact valve near ground level, but from minor leaks in the transfer system that were venting to the stack. Based on their misinterpretation of the 1200 mR/hr reading, NRC officials recommended an evacuation of people as far as 10 miles downwind from the site. The head of the Pennsylvania Bureau of Radiation Protection knew about the helicopter reading and concluded that the evacuation was unnecessary, but communications were jammed and he could not reach the Governor's office (Kemeny, 1979). In a telephone conversation later that morning, the Governor talked with the chairman of the Nuclear Regulatory Commission, who assured the Governor that no evacuation was needed. However, it was decided that the Governor would urge everyone who lived 5 miles downwind of the plant to remain indoors for the next half-hour. Shortly after noon, on the basis of another conversation with the NRC chairman, the Governor recommended that pregnant women and preschool children leave the region within a 5-mile radius of Three Mile Island. They also decided to close all schools within that area.

Another serious misunderstanding developed Friday afternoon, when it was learned that a 1000-ft^3 gas bubble that contained hydrogen had formed within the reactor vessel. Concern developed that an explosion might occur within the reactor and, as reported by the presidential commission, "that it was a groundless fear, an unfortunate error, never penetrated the public consciousness afterward, because the NRC made no effort to inform the public that it had erred." The hydrogen was produced as a result of high-temperature reactions between the hot steam and zirconium fuel cladding. However, the chemistry of the reactor system was such that excess oxygen could not accumulate. This was apparently known by the NRC experts, but during the weekend beginning March 31 no statement to this effect was made. Finally, on Monday, April 2, it was announced that the bubble had absorbed, and that perhaps the NRC calculations of oxygen generation rate were too conservative: however, at the time no public statement was made to this effect.

Acting on the assumption that a major release of radioiodine was possible, the U.S. Department of Health, Education and Welfare (HEW) took steps to obtain supplies of potassium iodide that could be used by the surrounding population. Early Saturday morning, the Mallinckrodt Chemical Company of St. Louis agreed to provide 250,000 1-oz bottles of the compound. The first shipment reached Harrisburg by early Sunday morning, and the balance was supplied by Wednesday, April 4. Fortunately, there was no need for the potassium iodide (Wald, 1980).

An incident that took place on the second day of the accident illustrates the difficulties of anticipating public and official reaction to actions that must be taken by those responsible on-site. About 400,000 gal of slightly contaminated wastewater had accumulated by the second day, and since the radioactivity of the water was well within the limits for discharge into the Susquehanna River that were permitted by the license, it was decided to discharge the wastes in the usual way. When informed of the proposed action, the local NRC officials gave their approval and the utility also notified the State Bureau of Radiation Protection that the water was being dumped.

The press and many public officials and members of the general public reacted adversely when they learned of the release to the Susquehanna. Although it was reasonable at the time for the plant personnel and NRC officials to consider it a routine matter since the amounts discharged were well within the limits permitted by the NRC license, the feeling developed that neither the government nor the plant management had dealt candidly with the public. The incident led to a requirement that TMI would not discharge any radioactivity to the river, a limitation that remains in effect to this writing.

Management of contaminated water was the first major problem that had to be faced in the long-term program for decontaminating TMI-2. Initial measurements of water from the reactor building sump showed that there were 400,000 gal of water contaminated to a level of 270 μCi/ml. Wastewater stored in various tanks contained a comparable quantity of water contaminated to only a slightly smaller extent. Ion exchange systems that were designed after the accident to process this water performed well, and decontamination soon progressed to the point where tritium was the only remaining contaminant and the water could be discharged to the river subject to the limitations normally imposed on operating reactors. However, as noted above, the plant was prohibited from discharging *any* radioactivity to the river, and how this matter will be resolved remains to be seen.

Entry into the reactor building to begin the process of fact-finding and decontamination was prevented for many months because of the presence of about 57,000 Ci of krypton-85 in the building atmosphere. The various options for disposing of this gas were examined, and it was decided that venting to the atmosphere under controlled conditions would be the safest method. A study by the National Council on Radiation Protection and Measurements (NCRP, 1980b) concluded that exposure to the nearby residents would be trivial. The principal risk of exposure to krypton-85 is skin cancer, but the venting would result in a risk equivalent to the risk of about 20 min of exposure to sunlight.[2]

EXPOSURE OF THE NEARBY POPULATION

The task of estimating dose to the surrounding population was assigned to an Ad Hoc Population Dose Estimate Group (Battist *et al.*, 1979), which included representatives from all of the federal agencies involved in the accident investigation. It was concluded that the highest doses were received by a few people within a 2-mile radius who received doses between 20 and 70 mrem (Gerusky, 1981). Four different methods of estimating population dose gave estimates ranging from 1600 to 5300 person–rem. The most credible estimate was thought to be 3300 person–rem. Using the risk coefficients discussed in Chapter 2, this would imply that the health effect of the gamma exposure was somewhat less than one cancer. Beta radiation from the noble gases would have resulted in a skin dose as much as four times the gamma dose, but this would have been attenuated to an unknown extent by clothing.

The dose from ingested radionuclides was less than that from external radiation. Thousands of samples of air, milk, water, produce, and other environmental media were taken and analyzed by the various agencies.

---

[2]The Union of Concerned Scientists, an antinuclear group which conducted a parallel study, reached similar conclusions, but took note of the existence of psychological stress in the surrounding population, and recommended that to provide further assurance of safety, the top of the 250-ft stack should be connected to a plastic tube held aloft by a tethered balloon to carry the gases even higher!

PSYCHOLOGICAL EFFECTS ON THE SURROUNDING COMMUNITY

The sociopolitical ramifications of the accident's aftermath were in every respect as complicated as the technical problems. Public concern increased markedly on the third day of the accident because of apparent disagreements among the experts about the danger of the hydrogen bubble. Later surveys showed that 52% of the people living within 20 miles of TMI left the area, most of them on Friday, March 30 (Dohrenwend *et al.*, 1981). The Governor's advisory to pregnant women and preschool children was not lifted until April 9, and the schools within 5 miles of TMI did not open until April 11 (Flynn, 1981). Great distrust developed of the credibility of the utility, the Nuclear Regulatory Commission, and other agencies of state and federal government. The tension was exacerbated by the presence of hundreds of reporters and other media representatives who descended on the surrounding communities. It is reported that 1 year later, when the press was asked at a conference to comment on the possibility that the performance of key people involved in the accident may have been affected by the enormous pressure from the media, the press responded "that is one of the prices we must pay for the public's right to know" (Trunk and Trunk, 1981). It is estimated that 300–400 reporters were present in the community of Middletown.

In the course of time, it became apparent that the only health effects of the reactor accident would be those due to the psychological trauma (New York Academy of Sciences, 1981). The environs of TMI will be fertile laboratories for psychosocial scientists for years to come.

Despite the small doses that people received, there were many claims for damages. Diseases in farm animals and children that would normally have been accepted as due to the vicissitudes of living were ascribed to the accident, notwithstanding the absence of any evidence that the frequency of such occurrences was greater than normal after the accident. Claims for damages against the utility resulted in major awards by the courts.

Evidence for a Major Nuclear Accident in the Soviet Union (1957–1958)

In 1976 Zhores Medvedev, a Soviet geneticist living in exile, first published accounts of a major release of radioactive material that he claimed took place in the Urals in 1957 or 1958 (Medvedev, 1976, 1977, 1979). Until this writing, there has been no confirmation by the Soviet Union of such an occurrence, but the Medvedev claims have received support from U.S.

investigators, using many unclassified reports from Soviet radioecologists published in the open literature (Trabalka *et al.*, 1980).

According to Medvedev, extensive contamination resulted from an explosion at a military site in Cheliabinsk Province during the winter of 1957–1958. Hundreds of civilian casualties were said to have resulted from the contamination, which covered several thousand square miles.

Trabalka and associates (1980) undertook a comprehensive review of Soviet radioecological publications and found a pattern of anomalous features consistent with the claims made by Medvedev. Among these were the following:

1.   Results of radioecological research were published in which single applications were made of mixed fission products in nitrate solution, whereas it was customary for Soviet scientists to use chloride solutions. This is significant because high-level wastes are often stored in liquid form as nitrates.

2.   The research was done with single applications of millicuries per square meter, much higher than is necessary for convenient radiochemical analyses.

3.   The locations of the research were not given, which is very unusual for radioecological studies.

4.   The methods of field application were not given.

5.   The first reports of the studies were not published until an unusually long time period had passed (6–14 years).

Evidence was found, from published maps, that population in an area of 100–1000 km^2 had been resettled.

The contaminated zone was estimated to contain from 10^5 to 10^6 Ci ^{90}Sr. Thus far unexplained is the unusually low ratio ^{137}Cs : ^{90}Sr. Although this ratio is close to 1 for reactor-produced fission products, the studies of Trabalka *et al.* (1980) concluded that the ratio was about 0.01 for the episode in the Urals.

The most credible explanation is that an explosion occurred in stored high-level wastes, possibly because of steam generated by failure of the required cooling system. Trabalka and associates (1980) have also postulated that if the nitrate wastes were in dry form, detonation of NH_4NO_3 might have occurred. These and other aspects of the matter must remain in the realm of speculation until a statement is issued by the Soviet Union. The Soviets have learned much from published reports of nuclear accidents in Western countries, but the flow of information has been only in one direction.

The Reactor Accident at the Chernobyl Nuclear Power Station in the U.S.S.R., April 26, 1986[3]

A water-cooled, pressure tube, graphite-moderated, 1000 MWe reactor at the Chernobyl power station located on the Pripyat River in the U.S.S.R., about 90 km north of Kiev, was destroyed by an accident that occurred in the early morning of April 26, 1986. The damaged reactor, Unit No. 4 at the Chernobyl station, had operated since 1983, was the newest of four similar units at that station, and was a popular design for which the Russian acronym is RBMK. This accident was by far the worst that had ever occurred to any reactor, and it was the first power reactor accident to result in radiation casualties, the first to result in extensive contamination of the environs, and the first to require evacuation of nearby residents. It was unquestionably the most costly industrial accident in history. Although the consequences of the accident would have resulted in worldwide concern in any case, the global reaction was exacerbated by East–West political and cultural differences and the fact that many days passed before the U.S.S.R. admitted that the accident had occurred.

DESCRIPTION OF THE REACTOR

The Chernobyl nuclear power station comprised four 1000 MWe units. The core of the RBMK type reactor is a graphite cylinder 12 m in diameter and 7 m high, which is penetrated vertically by 1661 fuel channels and 222 control rod channels (Fig. 14-5). The fuel channels are housed within zirconium alloy pressure tubes 8.8 cm in diameter. Mounted within each of the tubes are subassemblies of 18 zircalloy clad fuel pins, each of which is 13.5 mm in diameter. The fuel was 2% enriched uranium dioxide (U.S.S.R., 1986), but it is reported elsewhere that metallic uranium can also be used (Amer. Nuc. Soc., 1986a) to facilitate plutonium production. The reactor utilizes a complex flow path for the coolant. Light water passes vertically through each of the fuel channels, and the water, with a void (steam) fraction of 14.5%, then passes, via an extraordinarily complex system of piping, to steam separators from which the dry steam passes directly to the turbine. The Soviets have constructed 14 reactors of this type. An

---

[3]This account of the Chernobyl accident relies almost entirely on a report entitled "The Accident at the Chernobyl Nuclear Power Plant and its Consequences," prepared by the U.S.S.R. State Committee on the Utilization of Atomic Energy and presented to a meeting of experts convened by the IAEA in Vienna, August 25–29, 1986. Unless otherwise indicated, the information presented in this section was obtained from the report, from oral presentations by U.S.S.R. experts, or from the notes of the author who attended the conference and benefited from many private discussions.

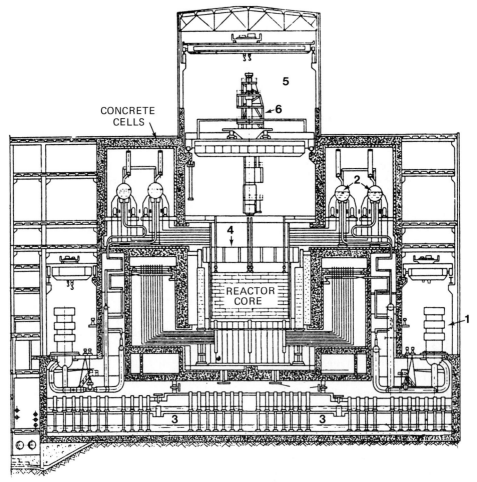

CONCRETE
CELLS

REACTOR
CORE

Fig. 14–5. Cross-section of the RBMK nuclear power plant. The reactor core is shown at the center. The main circulating pumps (1) send the coolant through a complex system of piping into the fuel channels (not shown) that pass vertically through the reactor core. Water, with a steam (void) content of 14%, passes from the top of the reactor to the steam separators (2), from which the dry steam passes to the turbines. The coolant system is located within massive concrete cells designed to vent steam to vapor suppression pools (3) in the event of a loss-of-coolant accident. Above the core is a massive concrete pad (4) assembled from removable blocks through which the coolant passes, and penetrated by the fuel channels. The high bay (5) above the reactor is of ordinary industrial construction, and houses the fuel handling machine (6).

important feature of the design is that the reactor can be refueled while operating at full power. This avoids the need to shut down for periodic refueling and also allows the fuel burnup to be optimized for plutonium production if so desired.

A serious defect in the RBMK design is its positive temperature coefficient of reactivity. This is caused by an increase in the void fraction that reduces neutron absorption by the moderator, thereby adding reactivity to the system. The reverse is true in light water reactors in which reactivity is reduced by the presence of voids (p. 195). Another deficiency in design is the absence of a full containment structure. Power reactors in Western countries are housed in buildings that are designed to withstand the pressures that would be generated during a loss of coolant accident. For many years, reactor experts who visited the Soviet Union have commented on the fact that the RBMK reactors were not fully surrounded by containment buildings of the type considered necessary in the West. The RBMK reactors are too large to be enclosed in this way. Instead, partial containment is provided by massive concrete walls that enclose most of the coolant system, but the region above the reactor floor is not so enclosed and the roof of the reactor building is of ordinary industrial construction. (This can be seen in Fig. 14-5.) Another basic difference is that the RBMK reactor core is not contained within steel pressure vessels such as are used in LWRs. Instead, the top of the reactor is a working area in which there are about 1900 closure plugs for the fuel channels and control rods. Above this floor is a high bay of conventional industrial construction that houses a tall fuel loading machine. The roof of the building is not designed to contain a pressure surge, but the lateral components of a surge can probably be absorbed or deflected by the massive concrete shields. Beneath the reactor is a vapor suppression pool similar in principle to the one described for the BWR on p. 228. The pool would be useful in the event of sudden depressurization of the coolant circulating system, but it is not designed so that steam produced in the core can vent through it, and it is not designed to relieve pressure surges that vent through the working floor of the reactor.

THE CIRCUMSTANCES OF THE ACCIDENT

A routine shutdown of the reactor was scheduled for April 25, prior to which it was intended to conduct an experiment concerned with the ability of the momentum of the turbines to continue production of electricity for start-up of emergency equipment in the event of an interruption in the steam supply. During the test the core contained 1659 fuel channels, with an average burnup of 1000 MW/ton. Similar tests had been performed previously without incident. This particular experiment was designed to test a method of compensating for a reduction in the busbar voltage that had occurred in previous tests before the rotor inertia had been fully expended.

Reactor power was gradually reduced from the full power level of 3200 MWt beginning at 0100 on April 25 in preparation for the test, which was to have been performed at 700–1000 MWt. By 1400 the emergency core cooling system was disconnected in accordance with the approved test procedure. However, because a need for power from Unit 4 developed, the load dispatcher ordered the shutdown to be delayed for 9 hours. The reactor continued to operate at a level of 1600 MWt without the ECCS, which was in violation of operating rules. The automatic system for adjusting reactivity was also disconnected. During the 9-hour delay, core reactivity was greatly reduced because of the buildup of xenon-135, which acts as a reactor "poison." The operator found the reactor difficult to control in the absence of the automatic controls, and at one point the reactor power fell to 30 MWt. It then stabilized at 200 MWt but could not be increased further because of insufficient reactivity reserve. The latter rate was below the reserve at which the rules required shutdown, an action that would have avoided the accident. Instead, at a few minutes after 0100 on the morning of April 26, the operator took a number of actions, the net effect of which was to increase the void fraction and cause the power level to increase so rapidly that two explosions occurred at 0124. The steam released by the failure of one or more pressurized tubes had reacted exothermally with the zirconium, producing hydrogen that exploded. Parts of the core were scattered about the building as well as on the roofs of the reactor, turbine, and auxiliary buildings.

Thirty fires were started, and firefighting units based at Chernobyl and Pripyat, especially trained for nuclear plant emergencies, were dispatched within minutes. The building fires were extinguished about 4 hours after the accident, but the graphite continued to burn much longer, abetted by residual heat from the inventory of fission products. A source of radioactive emissions to the atmosphere was thus established that persisted for about 2 weeks.

No announcement that the accident had occurred was made by the U.S.S.R. at the time. However, other countries soon learned of it from routine measurements made at nuclear power stations in Finland and Sweden where monitoring equipment detected fission products on the clothes of workers entering the plants, and also on the filters of air monitors. Consideration of the regional meteorology and radiochemical measurements quickly made it clear that the radioactivity originated from the U.S.S.R. Inquiries made through diplomatic channels for several days, however, elicited no information from the Soviet government. Finally, in response to worldwide pressure for information, an announcement that an accident had occurred was made on Soviet television on May 14, by Mikail S. Gorbachev, nearly three weeks after the accident took place. No quantitative information about the consequences were provided by the U.S.S.R. until late August when a comprehensive report of the cause of the accident and its consequences was presented to a conference at the Vienna headquarters of the International Atomic Energy Agency.

The failure of the Soviet government to make a prompt announcement of the disaster led to unfortunate rumors and speculation. A wire service dispatch from an Eastern European country stated that 2000 people in the Chernobyl area died from radiation exposure on the first day. The fact that such rapid death from radiation exposure could only result from injury to the central nervous system at doses in excess of about 5000 rad led to speculation that a very much larger number of deaths would result from the acute radiation injury in subsequent weeks among those exposed to smaller doses. It later turned out that there were only two deaths on the first day, both of which were the result of traumatic injuries. In all, a total of 31 deaths resulted in the weeks following the accident.

Most of the radioactive cloud from Chernobyl first drifted toward Scandinavia, and its lower portion headed in a southeastern direction toward Poland and other Eastern European countries. Many countries of Europe soon detected the cloud, and it was eventually detected in the U.S. and Japan. The fallout in parts of Europe was relatively heavy in areas over which the cloud passage coincided with rainfall. Worldwide excitement was created by reports of contaminated air and food.

In the meantime, while the world focused on the drifting radioactive cloud and speculated about its health consequences, a drama was unfolding at Chernobyl. There were 176 personnel on duty at the four power plants which were operating at full power when the accident occurred. In addition, 268 construction workers were on duty at the site. An attempt was first made to extinguish the fire and cool the core by flooding it with water. This proved unsuccessful, whereupon it was decided to cover the active part of the reactor with sand, clay, lead, dolomite, and a boron compound. These materials were selected because of specific properties that proved useful. The lead held a number of fission products in solution and also absorbed heat in the process of melting. The dolomite, a magnesium calcium carbonate, also served as a heat absorbant in the process of calcining to the oxide. Between April 27 and May 10 nearly 5000 tons of these materials were dropped on the core from helicopters, which succeeded in reducing substantially the emission of aerosols. Cooling of the core was also assisted by pumping nitrogen into it.

Unit Nos. 1 and 2 were shut down about one day after the accident because the buildings became contaminated by the radioactive air entering via the ventilation systems. By June 10, after decontamination by swabbing with chemical solutions, the levels in those units were reduced to less than 10 mr/hr, which was due mainly to direct radiation from Unit 4. The first two units, however, as well as Unit 3, were shut down for an indefinite period. Thus an additional 3000 MWe of generating capacity was removed from the grid.

Soviet investigators estimated that 81 MCi were discharged to the atmosphere by May 6, by which time the emissions had practically ceased. This quantity was about 3.5% of the reactor inventory at the time of the accident. The quantities of the various radionuclides released are given in Table 14-6, which shows that the releases varied from 100% of the inventory of unreactive noble gases to 2–3% of the nonvolatile elements such as cerium and the actinide elements. Intermediate were 20% for I-131 and 10–13% for the cesium nuclides. The consistency of the percentage releases of the actinide elements and cerium, which are relatively nonvolatile, suggested that about 3.5% of the fuel was dispersed in fine enough particulate form to behave as an aerosol. The quantity of radioactivity released was enormous and was capable of doing great harm over considerable distances. The amount of Cs-137 discharged, about 1.0 MCi, was about 3% of that released to the environment by all weapons tests conducted in the atmosphere. In parts of the U.S.S.R. and Western Europe where the cloud passage coincided with rainfall, the ground deposits of Cs-137 were much greater than that from all of the weapons tests. The quantity of Pu-239 released, about 700 Ci, was very much less than the 350,000 Ci distributed globally by the weapons tests. However, as in the case of cesium, there are undoubtedly places in the U.S.S.R. and western Europe where the deposition of plutonium from Chernobyl is greater than from the weapons tests. The emissions took place mainly over a period of about 10 days, much of it consisting of finely divided fuel of sufficiently small particle size to be transportable over considerable distances.

Steps taken to reduce the emissions were only partially effective during the period of April 26 until May 2, during which time the fuel particles, enriched in iodine, cesium, and tellurium, were convected from the core by hot air and gases generated by the burning graphite. The rate of emissions reduced in part by the cover of materials dropped from helicopters, had declined from rates measured in megacuries per day between April 6 and May 5 to less then one curie per day by the end of May.

During May and June, a massive cooling system was installed in a concrete slab poured under the foundation of the building to provide protection in the event of a total collapse of the reactor. Fortunately this did not occur. Soon after the accident a decision was made to entomb the damaged reactor in a massive concrete monolith.

TABLE 14–6

Estimated Quantities of Radioactive Materials Released
from the Chernobyl Reactor[a]

Nuclide	Releases (megacuries)		% of inventory released as of May 6, 1986
	April 26, 1986	By May 6, 1986	
XE-133	5	45	100
Kr-85m	0.15	—	100
Kr-95	—	0.9	10
I-131	4.5	7.3	20
Tc-132	4.0	1.3	15
Cs-134	0.15	0.5	10
Cs-137	0.3	1.0	13
Mo-99	0.45	3.0	2.3
Zr-95	0.45	3.8	3.2
Ru-103	0.6	3.2	2.9
Ru-106	0.2	1.6	2.9
Be-140	0.5	4.3	5.6
Ce-141	0.4	2.8	2.3
Ce-144	0.45	2.4	2.8
Sr-89	0.25	2.2	4.0
Sr-90	0.15	0.22	4.0
Pu-238	0.1×10^{-3}	0.8×10^{-3}	3.0
Pu-239	0.1×10^{-3}	0.7×10^{-3}	3.0
Pu-240	0.2×10^{-3}	1.1×10^{-3}	3.0
Pu-241	0.02	0.14	3.0
Pu-242	0.3×10^{-6}	2.1×10^{-6}	3.0
Cm-242	0.3×10^{-2}	2.1×10^{-2}	3.0
Np-234	2.7	1.2	3.2
Approx. total	20	81	

[a]U.S.S.R., 1986.

Radiological Consequences

At the time of the accident, the surface winds were weak and variable in direction, but at a height of 700 m to 1500 m the flow was to the northwest and north at velocities of 5–10 m/sec. Later in the day, the surface flow was in the westerly and northwesterly direction, crossing the frontier with Poland on April 26–27. The surface flow of the ground layer (to a height of 200 m) continued in the NNW direction through April 29. Information about the effective height of the releases during the first days have not been reported, but the emphasis given by the Soviet reports to meteorological conditions below 1.5 km suggests that the plume was initially contained below this level. Turbulent mixing would be expected to have eventually resulted in diffusion to somewhat higher altitudes.

Fallout from the plume was soon detected in many parts of the Soviet Union, several Eastern bloc countries, and Western Europe. The amounts of fallout were highly variable and were largely influenced by precipitation. The doses received in these "hot spots" were, in many cases, more than an order of magnitude larger than those received during the years of nuclear weapons testing in the atmosphere.

The radiological effects of the accident can be discussed under four headings: a) on site, b) the environs within 30 km, c) the European portion of the Soviet Union beyond 30 km from the reactor, and d) European countries outside the U.S.S.R. Fallout was also detected in North America and Asia, but only in traces. Since the release was confined to the troposphere, transport to the southern hemisphere did not occur.

On-Site Effects

One worker who was close to the explosion was never found and another died of extensive burns within a few hours. A team of specialists arrived within 12 hours and during the following day examined about 350 persons. Eighty-four persons, all of whom were either on-site at the time of the accident or participated in firefighting or other emergency activities, were diagnosed as having acute radiation sickness. This number was eventually increased to 203 cases, but none involved members of the general public. All acute cases were hospitalized either in Moscow (129 persons) or in Kiev (72 persons). The more severe cases were sent to Moscow, which has a clinic especially equipped for cases of radiation injury.

The exposures were to both external and internal sources. Neutron exposure was quickly ruled out by the absence of neutron-induced nuclides in blood samples. No information is available on the types of personal dosimeters worn by the employees. For the most part, the Soviet scientists relied on biological indicators of dose based apparently on unpublished earlier experiences with radiation injury in the U.S.S.R. Early signs of radiation injury such as the time of onset of nausea, vomiting, hematological changes, and chromosomal aberrations were used to estimate the dose received, the prognosis, and the methods of treatment. A large number of biochemical parameters were investigated to determine if some of them could be used as indicators of dose.

The Soviet physicians attributed major significance to the effects of beta and gamma radiation of the skin due to clothing that became drenched with highly radioactive water. These effects were first manifest within one or two days as a transient erythema that reappeared in greater severity in two or three weeks and sometimes resulted in blisters and severe ulceration. The skin exposure resulted from contact with clothing that had been soaked by radioactive water. Eight persons with damage to 60–100% of the skin died during the first few weeks. Although severe hematopoietic and intestinal injury was also present in all these cases, it was concluded that it was the skin damage that was sometimes incompatible with survival. Thus, in one case in which recovery of the bone marrow was occurring, death resulted on the 48th day from endogenous intoxication arising from extensive skin damage. In some cases the radiation effects were complicated by the presence of thermal burns.

There were 31 deaths among the 203 persons originally diagnosed as having radiation sickness. These patients had sustained whole body doses as high as 1600 rad. The on-site personnel also sustained significant internal exposure from inhaled and ingested radionuclides, mainly I-131. The thyroid burden among 171 persons who were monitored for I-131 three days postaccident ranged from about 20 microcuries to 200 microcuries. Potassium iodide had been administered to on-site personnel (250 mg twice daily), but not until about 90 min postaccident, too late to be fully effective. For two patients, it was found that urine contained 0.5 and 0.2 microcuries per liter I-131 and 0.1 and 0.07 microcuries per liter Cs-134/137. The whole body effective dose equivalent from internal emitters was estimated to be 400 rem and 150 rem.

Some of the firefighters inhaled massive amounts of radioactive smoke and sustained severe beta burns to the nasopharyngeal region.

Although much media attention was given to transplants of bone marrow (13 cases) and fetal liver (6 cases) in the treatment of those most severely affected, the Soviet scientists have reported that the treatments were not successful.

Effects within 30 Kilometers

The main population center within a radius of 30 km was the town of Pripyat in which 45,000 persons resided. Ambient gamma radiation measurements, which began about 3 hours postaccident, remained under 100 mr/hr during the first day and then climbed to more than 1000 mr/hr in some parts of the town. Evacuation of the village began about 36 hours after the accident, at which time the external gamma exposures ranged from 500 to about 1500 mr/hr. In the meantime, residents were cautioned to remain indoors. Distribution of KI was begun on a house-to-house basis beginning about 10 p.m. on the day of the accident. This was too late to be fully effective, but evidently did reduce the thryroid dose somewhat. The evacuation of Pripyat, which was completed in less than 3 hours, required the mobilization of 1100 buses. Evacuation of an additional 90,000 persons who resided within 30 km of the accident was completed by about the tenth day postaccident, bringing the total number of evacuees to 135,000.

Among the many logistical problems that had to be dealt with was the presence of many cattle that could not be cared for in the absence of their evacuated owners. These were slaughtered, butchered, and sent to cold storage facilities. It has been reported that the only significant contaminant of the cattle meat was I-131, which would not present a problem because of its short half-life. Accumulation of significant quantities of radioactive cesium had not yet occurred.

Based on data derived in several ways, Soviet scientists concluded that the dose to children's thyroids ranged from 25 to 250 rad, and that the collective effective dose equivalent received by the 135,000 evacuees was about 1.6 million person-rem. The effective dose equivalent received by individuals within the 30 km zone ranged from 0.4 rem to 300 rem. It is not yet known how long the area within 30 km must remain evacuated or what decontamination procedure will be needed before the residents can be allowed to return. Two months after the accident, it was estimated that between 8 and 14 MCi remained on the ground within the 30 km zone.

Among the many problems that must be dealt with in the immediate future is how to create a low radiation area in which the employees of the Chernobyl Nuclear Power Station can be housed so that the three undamaged units can be serviced, decontaminated, and placed back in operation.

Effects in European Russia beyond 30 Kilometers

It is estimated that about 14 MCi (as of May 6) were deposited in the U.S.S.R. beyond 30 km from the accident. The Soviet republics of the Ukraine and Byelorussia were the most affected, with external whole body gamma radiation levels averaging 0.44 mr/hr for the first 15 days postaccident at Kiev, one of the largest cities in the U.S.S.R., located about 90 km south of the Chernobyl station. The external gamma dose to the average rural residents of Kiev was estimated to be 0.74 rem for 1986, with a 50-year dose commitment of 2.5 rem.

Iodine levels in milk commonly reached 1 μCi/1 in southern Byelorussia. Leafy vegetables reached 10 μCi/kg. Standards for I-131 were based on the criteria that children should not receive thyroid doses in excess of 30 rem, which was met by limiting the I-131 content of milk to 0.1 μCi/liter, the same value used following the Windscale accident in 1957 (p. 352). Equivalent standards were established for other foods. The cesium standard was based on the criteria that the whole body dose commitment should not exceed 5 rem. The established standards were exceeded in many communities by a substantial fraction of the agricultural produce.

About 75 million people reside in the Ukraine and Byelorussia, and the food they produce is shipped throughout the U.S.S.R. The task of evaluating the dose commitments from foods grown in the region required massive effort. The resources of about 7000 laboratories in the Soviet Union were mobilized. About 100,000 children were scanned for thyroidal I-131. Thousands of whole body measurements for radiocesium were performed, and tens of thousands of measurements were made on food samples.

The projected dose commitments from Cs-137 proved to be particularly difficult to estimate because of peculiarities of the soils of the region. This had been recognized previously from studies of fallout from nuclear testing. The dose commitment from Cs-137 deposited on agricultural land is due to two fractions: that due to external radiation and that from ingestion of contaminated foods. In most areas of the world the ratio of the external and internal fractions is about 3, but Soviet investigators found, from studies of Cs-137 in weapons fallout, that the ratio was reversed, about 0.3. This has been attributed to the low organic and clay contents of the soil. Because of this, it is estimated that the dose commitment by ingestion from a given deposit of Cs-137 is eight to ten times higher in the southern U.S.S.R. than in most other places. By using models of the type discussed in Chapter 13, corrected for the anomalous uptake of cesium by growing plants in the U.S.S.R., the 50-year collective dose commitment from food grown in the area is estimated to be about 21 million person-rem. (The Soviet scientists originally estimated the collective dose to be 210 million person-rem, but this was revised downward in the course of the Vienna meeting.) To this estimate should be added an additional 29 million person-rem from external radiation originating from Cs-137 deposited on the ground. The total dose commitment from Cs-137 is thus estimated to be about 40 million person-rem, or about 0.53 rem per capita. The collective dose for 1986 is estimated to have been 8.6 person-rem, or about 0.1 rem per capita.

The dose for 1986 was thus about equivalent to that received annually from natural sources, and the 50-year dose commitment will be about 10% of the dose from nature. For the 75 million people of Byelorussia and the Ukraine, any increase in mortality from cancer would not be detectable, for the reasons given in Chapter 2. However, if we apply the linear hypotheses to a population that has received 50 million person-rem, it can be estimated that as many as 5000 deaths from radiation-induced cancer may result, which would be an increase of 0.05% in the normal mortality from cancer.

The U.S.S.R. scientists have also estimated the excess mortality to be expected from I-131 exposure and have concluded that about 1500 additional deaths from thyroid cancer can be expected in the population. This would be about a 1% increase over the 150,000 deaths from thyroid cancer that will occur normally in the population.

As with all estimates such as these, it should be remembered that the ICRP and BEIR III reports from which the risk coefficients are obtained caution that they provide upper bounds of risk and that the actual number of deaths may be much lower.

Effects in Western Europe

In contrast to the relatively few countries that made measurements of radioactive fallout during the period of atmospheric weapons testing, there were data obtained in many countries following the Chernobyl accident. This was due largely to the existence of nuclear power stations and the monitoring activities associated with them. In addition, many governments maintain emergency monitoring networks, systems for processing large volumes of data, and the organizations required to implement protective action guides. The World Health Organization played an important coordinating role in the weeks following the accident and issued reports that summarized the data being accumulated by the various countries and the protective actions taken (WHO 1986a,b). Interpretation of the data can be expected to be presented in future reports from the United Nations Scientific Committee on the Effects of Atomic Radiation.

There were large variations in both the amounts of fallout and the levels of radioactivity in food. This was the result of variations in meteorological conditions following the accident and, in particular, to the pattern of rainfall. The estimated effective dose equivalent for the first postaccident year for adult residents of seven western European countries ranged from 130 mrem in Switzerland to about 2 mrem in southern England. The effective dose equivalent for children ranged from 280 mrem in Switzerland to 5 mrem in southern England. The highest doses outside of the U.S.S.R. were reported from Poland, 95 mrem for adults and 660 mrem for children, in the eastern section of the country.

SOME GENERAL OBSERVATIONS

The Chernobyl accident provided the first good opportunity to test a number of models that have been developed in recent years to predict doses from a point source of radioactivity on a global scale. The experience should eventually result in improvements in the models and in a better understanding of both their effectiveness and limitations. The Chernobyl experience also served to identify many practical problems in the administration of protective action guides (PAGs). These have been established country by country and vary greatly. In North America, the PAG for I-131 in milk is ten times higher in the U.S. than in neighboring Canada. In Europe, the spread (for protection of the adult thyroid) varies about 30 fold among the various countries. There are comparable differences in the PAGs for Cs-137 in food.

The relatively long time it takes to obtain reliable information from radiation measurements is a fundamental difficulty in the application of PAGs. The problem is particularly difficult with respect to I-131, which has a short half-life and, because of its rapid uptake by grazing cows, can contaminate a milk supply in one day. This may be less time than it takes to collect representative samples of air, grass, or milk; transport them to a laboratory; complete the counting procedures; and report the results. The same problem exists with respect to cesium, but time is not as important a factor because of the long half-lives of the two cesium nuclides. Data on Sr-90 are likely to be totally lacking in the first few days after an accident because it is a beta emitter and can be measured only after lengthy radio-chemical procedures are completed.

The time factor is even more important with respect to the use of thyroid-blocking agents. To be maximally effective, KI should be administered prior to exposure, which will not be possible if the decision to administer KI is to depend on measurements. These difficulties will often make it necessary that implementation of PAGs depend on the use of models that can use estimates of the source strength and on-line meteorological information to predict the times and places where protective action must be initiated. This, of course, requires the availability of reliable models and the means to estimate the amount of radioactivity released. Unfortunately, as noted in Chapter 4, models are in a relatively primitive state of development and few of them have been adequately verified.

Finally, application of PAGs requires discipline on the part of local government, elected officials, scientists, and the media. The value of advance planning is greatly diminished when protective actions are taken (well meaning or otherwise) at lower levels of exposure than in the plan announced in advance.

MAJOR CONCLUSIONS THAT CAN BE DRAWN FROM
THE CHERNOBYL EXPERIENCE

The most urgent questions that arose in Western countries following the Chernobyl accident related to the implications of the accident in regard to the safety of existing civilian nuclear power plants, in particular the light water reactors that are predominant in the U.S. and elsewhere.

We have seen earlier that the RBMK design suffers from several inherent weaknesses, any one of which would be sufficient to prevent the reactor from being licensed in the United States or other Western countries. First is the positive void coefficient of reactivity, a factor that was of primary importance in the Chernobyl accident. Second is the large mass of the graphite moderator which, when ignited, adds enormously to the opportunity for extensive failure of the zirconium cladding, release of steam, and the generation of explosive hydrogen and carbon monoxide. Third is the absence of at least two of the defenses in depth inherent to LWRs, the massive pressure vessel in which the reactor core is located and a containment building. The Soviet investigators placed major emphasis on errors made by the reactor operators, and it is true that these were the immediate cause of the accident. However, the accident would not have occurred, or would have had far less serious consequences, if the inherent weaknesses in design did not exist.

There is also an institutional difference that may be most basic of all. The system of regulating nuclear power plants in the U.S. and other Western countries requires that all information about reactor design, construction, and operation be available to the public, and provision is made for participation by intervenors in the licensing procedures. The intervention is frequently costly and time-consuming, but it is an effective way of probing the safety of the reactor design and operation. The complex system of licensing and the very threat of intervention by members of the public is an effective way of assuring the best efforts on the part of the reactor designers, builders, and operators. The major lesson of Chernobyl may be that it was a convincing demonstration of the importance of a rigorous system of nuclear power reactor regulation, with unobstructed but reasonable opportunities for public participation in the licensing procedures.

Accidents Involving Military Aircraft Carrying Nuclear Weapons

There have been a number of crashes of military aircraft carrying nuclear weapons, but the exact number of accidents or their consequences are not known. In two reported accidents involving U.S. aircraft, the bombs were destroyed and plutonium was dispersed, but no nuclear reactions took place. There are no reports of such accidents from other countries.

A B-52 bomber of the U.S. Air Force crashed near Thule Air Base in northern Greenland, resulting in detonation of the chemical explosives in four nuclear weapons being carried aboard the aircraft (Aarkrog, 1971a). An unspecified quantity of ^{239}Pu and ^{241}Am was dispersed on the sea ice and, with its breakup in the spring, was transferred to the sea and sea bottom. Radioecological studies during the subsequent 11 years have investigated the behavior of these elements in the sediments and benthic organisms (Aarkrog *et al.*, 1984). No Pu or Am was found in plants collected from surface water or in fish or sea mammals.

Chapter 15

Radiation Exposure and Risk: Some Contemporary Social Aspects

It is understandable that when the World War II atomic energy program was being organized in 1942, there was apprehension about the effects of the program on the health of both the workers and the general public. The phenomenon of radioactivity had already been understood for nearly half a century, but radioactive substances were still a novelty: only about 1400 g of radium had been extracted from the earth's crust, but its uses and misuses had been responsible for more than 100 deaths. A much larger number of injuries and deaths had resulted among the physicians, physicists, and technicians who had been working with X-ray equipment. It was known that the World War II atom bomb project and the contemplated programs of civilian uses of atomic energy that would follow the wartime program would produce radioactive materials in enormous quantities, and there was concern as to whether the program could be managed safely.

Four decades have passed since the end of World War II, and the atomic energy industries in many countries of the world are full-grown. In the United States, more than 150,000 employees staff the atomic energy production and research facilities of the U.S. government (U.S. DOE, 1984c). Another 180,000 are employed in the civilian power industry (Atomic Industrial Forum, 1985), and an unknown but certainly equally large number of persons work with radioactive materials and X rays, either in the country's research laboratories or in clinics. Despite the enormous potential risks, the safety record of the nuclear energy industry has been exemplary and compares favorably with the safety record of industry in general. How did it happen that after such a bad start during the first 40 years after the discovery of X ray and radium, a great new industry could develop with an excellent safety record despite the enormous potential risks of using radioactive materials?

The basic reason was that the ionizing radiations were known to be so hazardous that early steps were taken to develop safe procedures for their use. As discussed in Chapters 2 and 3, this resulted in the establishment of the International Commission on Radiological Protection in 1928 and the formation of its U.S. counterpart, the National Council on Radiation Protection and Measurements, in 1929. By 1941, two basic standards were established that, unknown to anyone at the time, would prove to be of critical importance to establishment of safe practices when the wartime atomic energy program began 1 or 2 years later.

A second important factor was that the organizers of the wartime atomic energy program, and the postwar governmental agencies established to develop and regulate the atomic energy industry, recognized the need for large-scale research support. Major government-owned laboratories were established during World War II that have developed worldwide reputations for their research on the effects of ionizing radiations.[1] Additional funds have been supplied to support scholarly research in university laboratories, and fellowship programs were set up to provide opportunities for graduate education in the fields of health physics and environmental hygiene. Finally, a system of strict federal regulation over the use of radioactive materials were developed. As a result of all of these activities, more is known about the sources of ionizing radiation exposure and their effects on human health than any other of the many physical and chemical agents of disease that are known to be present in the environment.

The Safety Record of the Atomic Energy Industry

OCCUPATIONAL EXPOSURE

This book has been concerned until now with the external environment and has dealt only in passing with the subject of on-site health and safety. However, some discussion of that subject is appropriate in this concluding

---

[1]The techniques that were developed in these laboratories to study the effects of radioactive materials in the environment also proved useful for studying the effects of toxic, carcinogenic, and mutagenic chemicals when that subject began to attract belated attention in the mid-1960s.

chapter because of its relevance to the basic question of whether the nuclear energy program has been conducted in a responsible manner, with proper concern for the health of the workers. It is axiomatic that accidents in the nuclear industry are more likely to result in exposure of the workers than the public. An analysis of the record of occupational safety should thus serve as an indicator of the prudence with which the industry discharges its responsibilities for protection of health and safety more generally.

The Department of Energy and its predecessor agencies have reported regularly on the radiation exposure of their employees in reports that also include statistics on industrial accidents of all kinds, both those that involve radiation exposure and those that do not. The accident frequency among DOE employees has been consistently less than one-third of the average for all U.S. industry. This is all the more an outstanding accomplishment when one considers that the DOE program involves heavy construction, chemical processing, and other types of industrial activities that are potentially more hazardous than those in average industry. It is particularly notable that among all the industries in the United States for which the National Safety Council has consistently published accident statistics, none has had a lower accident rate than the DOE plants and laboratories (National Safety Council, 1984).

Despite the relative excellence of this 44-year-old program, there have been 360 work-related fatal accidents among government atomic energy workers. Most of these resulted from falls or falling objects, motor vehicle mishaps, electric shock, and various other traditional causes of injury and death among industrial workers. Six of the deaths resulted from radiation exposure in the governmental programs (U.S. AEC, 1975). Three of the six deaths resulted from the SL-1 accident (which was described in Chapter 14), and the other three were associated with relatively hazardous laboratory work in a weapons research laboratory early in the program. A seventh death, which resulted from a radiation accident in a privately operated industrial plant in 1964 (Karas and Stanbury, 1965), has been the only fatality to occur in nongovernmental facilities.

Thus, although radiation is potentially the most serious of all occupational hazards in the atomic energy program, it is significant that during the 41 years from 1943 to 1985 the fatalities caused by radiation accidents accounted for a small fraction of all deaths due to accidents in the workplace. It is also significant that the fatal accidents were associated primarily with experimental programs and that, as of this writing, the last of the fatal radiation accidents in the government program occurred more than 25 years ago at the SL-1 in 1961, and that 22 years have elapsed since the only fatal radiation accident in the commercial sector of the nuclear energy program.

In addition to the seven fatalities, radiation accidents in the U.S. governmental and privately operated nuclear energy programs have resulted in 25 cases of clinically observable radiation injuries. These included β burns of the skin (7 cases), excision of tissue in which plutonium or americium had lodged mechanically (3 cases), and symptoms of whole-body or partial-body overexposure (14 cases). Four of the latter 14 cases were from radiation sources such as X-ray spectrometers that are used in industry generally as research tools.

So much for the results of accidental massive exposure to radiation; there remains the important question of the frequency with which injuries or deaths have occurred in the past or will occur in the future because of the delayed effects of repeated small exposures. Except for the mining industry, there have been no known injuries from the delayed effects of radiation in the atomic energy industry. Ironically, uranium mining is the one part of the atomic energy program that had reason to take meticulous precautions, based on experience in the European mines, as discussed in Chapters 2 and 8. Had the uranium mine atmospheres been controlled to meet the standard established in 1941 to control the radon hazard in another industry, the tragic epidemic of lung cancer among uranium miners of the southwestern United States would have been avoided. Regrettably, regulation of the mines was not preempted by the federal government, as was the rest of the atomic energy program, but was left to the states, which lacked either the means or the will to deal with the problem in an effective manner.

The absence of clearly identifiable cases of delayed injury from ionizing radiation exposure in the atomic energy industry (other than in mining) is somewhat reassuring. The industry is more than 40 years of age, which should be long enough for such cases to become manifest if they are to occur in significant numbers from the practices adopted early in the program.

It is, of course, possible that some injuries or deaths from such exposure have occurred, but that a few cases here and there might not be recognized as such against the background of normal morbidity and mortality. This is an extremely important question at the present time. The hundreds of thousands of atomic energy workers as well as the military veterans of nuclear weapons tests conducted in the atmosphere have been aging over the years and have reached the stage of life where they have become more likely to develop cancer. About one person in five in the United States is likely to develop life-threatening forms of cancer from causes unrelated to any radiation exposure to which they may have been subjected. The atomic energy workers have entered the cancer-prone years of life at a time of unprecedented litigious activity in the United States, as a result of which lawsuits in considerable numbers are being brought against the U.S. government and private industry to seek compensation for alleged radiation injuries. A similar problem exists among members of the public who resided in the vicinity of the testing sites in Nevada. By 1985 there were about 4000 claims for radiation injury pending against the U.S. government (Jose, 1985). The vast majority of these claims have been filed by persons who lived or worked near the Nevada Test Site, or who were soldiers who claim to have been exposed in the course of military maneuvers (NAS–NRC, 1985b). In these cases, radiation exposure is being claimed to be responsible despite the fact that the doses received rarely exceeded a few rem and were usually less than 1 rem. The legal dilemma is posed by the fact that the linear no-threshold hypothesis makes it impossible to prove that even the smallest dose *did not* cause the cancer. As discussed in Chapter 2, it is hoped that the question of causality can be judged in probabilistic terms, with the use of tables that take into consideration the many factors that must be considered (NAS–NRC, 1984; U.S. DHHS, 1985; Bond, 1982).

The Nuclear Regulatory Commission publishes annual summaries of the radiation exposure of reactor workers. In 1982 there were 84 civilian nuclear power plants in operation in the United States, in which 84,000 workers were employed. The average dose received by these workers during the 5-year period 1978–1982 was 0.7 rem, which represented a collective dose of about 59,000 person–rem (U.S. NRC, 1983), and if one assumes a cancer risk coefficient of 10^{-4} cases per person-rem it can be estimated that the collective dose may result in about six fatal cancers among the 84,000 reactor workers in their lifetime, from 1 year of exposure. If the average worker continues to be so exposed for 20 years, which is probably an overestimate for the average worker, the lifetime expectation of radiation-induced cancer among the entire population of workers will be $20 \times 6 = 120$ cases, compared to an expected 21,000 cancer deaths that would occur in the absence of radiation exposure. Put another way, the probability that one one of the cancers was radiation-induced would be $120 \div 21,120 = 0.006$, which serves to illustrate the basis of the legal dilemma that exists.

It is clear from the above how important it is that the utility industry maintain complete medical records and radiation data on all employees so that an actuarial base can be developed that may eventually answer the question of whether the incidence of cancer among reactor workers is greater than expected.

GENERAL POPULATION EXPOSURE

The various sources of environmental radioactivity to which the public is exposed have been discussed in the preceding chapters. These included natural radioactivity, the nuclear power industry (from mining to waste

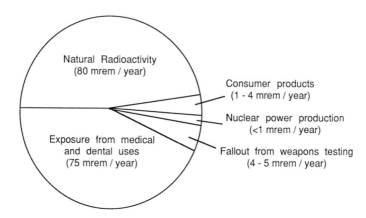

Fig. 15–1. Estimates of per capita radiation dose from all sources of exposure. (From Eisenbud, 1984.)

disposal), fallout from nuclear weapons testing, and a number of miscellaneous sources including the use of radiopharmaceuticals in the practice of medicine. However, one important source was not included—the use of X rays in the healing arts. It was estimated in 1980 that more than 300,000 X-ray units were being used for medical diagnosis and therapy. About 170,000 of these units were being used in the practice of dentistry and the balance by physicians, chiropractors, and podiatrists. The last comprehensive survey of the dose to the general public from these sources, which was published in 1977 (Shleien *et al.*, 1977), estimated that the per capita dose rate to the bone marrow in the adult U.S. population, which was 83 mrem/year in 1964, had increased to 103 mrem/year in 1970. It was estimated that about 70% of this dose rate originated from medical radiographic procedures, with fluoroscopic and dental examinations accounting for 20% and 3%, respectively. The per capita dose increases with age, from 52 mrem/year in adolescents and young adults to 151 mrem/year in persons more than 65 years old. The dose per examination varies with the type of procedure, from 9.4 mrem for a set of dental X rays to 595 mrem for a pelvic examination. There were wide variations from the average dose, depending on the equipment and techniques.

From all the foregoing, the sources of exposure to the general public can be approximated as shown in Fig. 15–1. The figures given are taken from various sources and should be used only to understand roughly how exposure of the general public is apportioned. Ninety-four percent of the total is contributed by two sources, natural radioactivity and exposure from medical and dental uses. The estimate for natural radioactivity is probably more reliable than the estimate for medical and dental uses.

The dose received by the bronchial epithelium from inhalation of the radon products is very much higher than the whole-body dose shown in Fig. 15–1. This was discussed in Chapters 2 and 7.

An interesting insight into the impact of the atomic energy industry on human health is provided by an Environmental Protection Agency report (U.S. EPA, 1979), the purpose of which was to estimate the number of cancers caused by the emission of radionuclides into the atmosphere. The study included radium mining, milling, the radiopharmaceutical industry, test reactors, plutonium fabrication facilities, and 27 major Department of Energy research and production facilities. The industry surveyed was thus a large one, employing about 170,000 persons. The estimated impact on the health of the U.S. population from all of these facilities was estimated by EPA to be less than one cancer per year. The study did not include licensed power-producing reactors. The NRC estimated that in 1981, when 73 commercial nuclear power plants were in operation, the collective dose received by 98 million people living within 80 km of the 48 different sites was 160 person–rem (Baker and Peloquin, 1985). This would, in round numbers, result in about 0.02 fatal cancers per year. The effect of the radioactive materials emitted by 250 coal-fired power plants in the United States was estimated by EPA to be 1.5 cancers per year. In short, if the current models for estimating risk from exposure to airborne radioactivity are applied to the large complex of atomic energy facilities in the United States, the total impact is less than that of the radioactive emissions from coal-fired power stations.

The public is understandably also concerned about catastrophic events which have a low probability of occurrence but can involve severe consequences. The Rasmussen report (U.S. NRC, 1975) places the probability of reactor accidents with fatal consequences orders of magnitude lower than that of other catastrophes caused by human activities, such as dam failures, explosions, or air crashes. Fatalities from reactor accidents are also far less probable than fatalities from most natural causes of disaster. The Three Mile Island accident, which was by far the most severe accident to a civilian nuclear power plant in the U.S., involved no overexposure of either the employees or members of the public, but enormous economic consequences did result from destruction of the nuclear power plant.

Many people live near many operating reactors without undue concern. Fear of a catastrophic event appears to be greatest near reactors that have not yet started operation and are in the licensing process. An example is the Shoreham plant of the Long Island Lighting Company, which has been ready to operate for several years but has been denied an operating license because the local government would not cooperate in developing the evacuation plans that would be needed in the event of an emergency. Fear of the catastrophic accident will hopefully diminish if it is confirmed that the amounts of radioactivity that would be released in the event of a loss-of-coolant accident have been greatly overestimated (Chapter 9).

The Chernobyl accident is certain to increase the fears of the public about the dangers of catastrophic reactor accidents. Unfortunately, the designs of the Russian graphite-moderated reactors did not provide all of the basic safety features that have been incorporated routinely into the construction of civilian power plants in the United States and other Western countries.

The Disparity between Actual and Perceived Risk

If the nuclear industry has had as safe a record as is indicated by the facts, why is there such widespread public opposition to the development of nuclear energy? The disparity between the actual and perceived risk from nuclear power is illustrated by a study conducted in Eugene, Oregon, in which four groups of people were asked to rate 30 different sources of risks according to the risk of death from each (Slovic et al., 1980). The four groups included college students, members of the League of Women Voters, 25 business and professional individuals, and an "expert group" of 15 persons selected nationwide for their professional involvement in risk assessment. From the responses received, it was found that the experts rated nuclear power 20th on the list of 30 sources of risk, whereas both the members of the League of Women Voters and the college students put nuclear power first, ahead of motor vehicles, handguns, and smoking! The fact that the disparity exists between actual and perceived risk, not only for nuclear power but also for food additives, pharmaceutical side effects, and chemical pollution, has been noted by many authors (Slovic et al., 1980; Fischhoff et al., 1981; Starr, 1969; Otway and Thomas, 1982).

There can be no doubt that the public attitude has been greatly influenced by the association of radioactivity and nuclear weapons. Everyone knows that nuclear bombs can cause great destruction and produce lethal amounts of radioactivity, and this knowledge influences the thinking of the public when it is confronted with decisions involving the more benign applications of nuclear energy.

A contributing factor may be the enormous range of quantities that are involved. In this book, for example, the quantities of radioactivity discussed have been as small as femtocuries (10^{-15} Ci) and as large as hundreds of megacuries (10^8 Ci). This is a range of 23 orders of magnitude, a spread of values totally without precedent insofar as the public and most scientists are concerned. It is not a simple matter to place risk in perspective when the magnitude of risk must be related to such a huge range of exposures.

The fears of the public are understandably affected by knowledge that the ionizing radiations are capable of producing both cancer and genetic effects, but very few people are able to place the probabilities of occurrence in a frame of reference. Few people have a "feel" for probabilities. The fact that radioactivity produces cancer is easily comprehended, but for many people the perception of risk will not be influenced by an added statement that the additional risk is "only" one in a million. To many people, the fact that radiation can produce cancer is much more important than whether the probability of occurrence is 10^{-2} or 10^{-8}.

There is also insufficient awareness that more is known about the effects of ionizing radiation than about any other of the many physical and chemical agents of disease that are known to be present in the environment. The U.S. government has spent $2 billion on research concerned with the health effects of radiation (General Accounting Office, 1981). All too often, the public is influenced by statements that nothing is known about the effects of low levels of radiation. Strictly speaking, we do know very little about the effects of doses below a few rem per year. This is because if the effects occur at all, they occur so infrequently that they cannot be measured. Estimates of effects at such low doses are made by extrapolation of risk coefficients obtained at higher doses and, usually, at higher dose rates. Most scientists believe that extrapolations made in this way overestimate the risk coefficients. Caveats in this regard are usually given in scientific reports, but are not known to the general public.

Many of the risk assessments made during the past several years have shown that the risk to individuals from many components of the nuclear energy industry is exceedingly small, but that a finite number of cancers can be predicted if the small dose is applied to a sufficiently large population. One example, discussed in Chapter 8, is that the risk to an individual of developing lung cancer from the radon released from mill tailings is about 3×10^{-10}. However, if the populations of the United States, Mexico, and Canada are taken into consideration, and if it is estimated that the population will reach 600 million persons in the next century, it can be estimated that about six cases of cancer will develop per year. Thus, the risk to the individual is certainly negligible by any criteria, but consideration must be given to the question of whether action should be taken to reduce or eliminate the small number of cases that may develop in the general population. This involves a political decision that must take many factors into consideration, including cost. If the government is to be required to spend large sums of money to reduce the risk, the cost of doing so must be assessed against other demands on the government to protect the public health. The point is that there is a difference between individual and collective risk that has not been explained to the public. There are times when both the individual and collective risk may be finite but the risks are trivial.

Although the scale of risks may vary over many orders of magnitude, the reaction of the public to a newly proposed source of radiation exposure is apt to be purely qualitative and dependent only on the question of whether radioactivity is or is not present. This was seen in New York City, where a large medical center proposed that it routinely incinerate low-level biological waste containing nanocurie amounts of relatively innocuous radionuclides. This was permissible under NRC regulations, and the procedure had the strong support of the New York Academies of both Medicine and Sciences, as well as the local medical and health physics community. The procedure would have led to insignificant exposures by any standard (Eisenbud, 1980; NYAM, 1983). The application to build the incinerator was nevertheless rejected by the local City Council on the advice of the City Health Department.

The disproportionate amount of concern for radiation effects is seen in the unusually large expenditures to reduce the levels of risk (Cohen, 1980). The cost to society of saving a life on the highways has been estimated to range from $3600 by the use of seat belts to $148,000 for mandatory air bags plus full adherence to the 55-mph speed limit. If smoke detectors were made mandatory in all rooms used for sleeping, the cost would be $40,000 per life saved (Graham and Vaupel, 1981). In contrast, the NRC policy of justifying expenditures of $1000 per person–rem avoided is equivalent to a cost of $10 million per cancer prevented, based on a risk coefficient of 10^{-4} cancers per rem. One proposal of EPA to reduce the atmospheric emissions of ^{210}Po from five producers of elemental phosphorus would have cost about $200 million to avert one cancer (U.S. EPA, 1984).

While many people find it repugnant to place a dollar value on life, the fact remains that in the administration of public health programs it is customary and necessary to consider the costs of proposed programs in relation to the benefits that will result from them.

There is a principle in law, *de minimis non curat lex*—"the law does not concern itself with trifles." This maxim should be incorporated into regulatory procedures, not only for radioactive substances but for toxic materials as well. Modern methods of environmental modeling now make it possible to predict levels of exposure far into the future and at concentrations that are beyond the capabilities of even our most sophisticated instrumentation. In an environmental impact statement prepared by the Department of Energy in connection with a proposed high-level waste storage site it was stated that a person living near the proposed site "would receive a 1.1×10^{-8} rem dose to the bone over a 50-year period," and the expected health effects were accordingly estimated (U.S. DOE, 1979a). There is no longer any lower limit to the health effects that can be estimated with risk coefficients based on the linear hypothesis in conjunction with computer programs that are capable of predicting the time and place of decay of the very last of the radioactive atoms discharged into a waste stream. That calculations are carried to such extremes is not a criticism of the good work done by the health physicists and environmental scientists who are asked by regulatory officials to make the calculations. It would be much more logical if the regulatory establishment would define a level of *de minimis* dose—the lower limit of regulatory concern (Eisenbud, 1981b). A *de minimis* dose of 10 mrem/year could be easily defended on the grounds that it is well within the limits of the variations in whole-body radiation exposure from natural sources. There are, of course, those who would point out that a dose of 10 mrem/year received by 240 million people would result in a collective dose of 2.4×10^6 person–rem, which, with a risk coefficient of about 10^{-4} cancers per rem, would result in 240 deaths per year, which is not an inconsequential effect. The fallacy in this argument is that it is unlikely that the sum of the doses from the nuclear power industry and other sources of radioactivity could produce an average dose of 10 mrem to 240 million people. Emissions from reactors, incinerators, or waste storage sites would result in measurable exposures to only a small fraction of the population. One exception would be with the military use of nuclear weapons, as was the case during the period of atmospheric testing. Another exception might be accumulation of long-lived nuclides such as ^{14}C, ^{129}I, or ^{85}Kr in the atmosphere, resulting in exposure on a global scale. However, these emissions would originate from a relatively few fuel-reprocessing plants for which controls could be installed for a very low cost.

The vast majority of sources of radioactive emissions could be excluded from regulatory concern if 10 mrem/year were considered the *de minimis* dose. Ten millirem per year would produce a lifetime dose of about 0.7 rem, which, with a risk coefficient of 10^{-4} cancers per rem, would increase the lifetime probability of developing cancer from roughly 0.2 to 0.20007. This would be a small incremental risk, in exchange for risks that could be eliminated if nuclear power could replace power generated from fossil fuels.

Public perception about radioactivity is also influenced by the widely publicized long lives of some of the waste products of the fission process. This is particularly true of the transuranic actinide elements such as plutonium. This is a specious argument which overlooks the fact that the stable toxic elements such as lead, arsenic, and mercury have infinite half-lives. While it is true that gram for gram the long-lived nuclides are more hazardous than the stable elements, it was pointed out in Chapter 1 that the latter exist in far greater quantities than the radionuclides. The hazardous characteristics of a radioactive waste are related not so much to the fact that it has a long half-life, but rather to the manner in which it is being stored and the geochemical and hydrological mechanisms that are available to transport the material into the biosphere.

How do these misperceptions originate and why are they perpetuated? This question has attracted a good deal of interest in recent years on the part of social scientists. The association between nuclear power and war, the inability of many people to comprehend risk when expressed in probabilistic terms, the association of nuclear energy with cancer and genetic effects, and other factors, some of which were mentioned above, are good reasons for the public to feel some initial concern. But why, after more than 40 years, and despite the excellent safety record, has there been an erosion of public confidence, not only in nuclear power but also concerning such relatively trivial matters such as the disposal of low-level radioactive wastes from hospitals?

The average person has no base of personal experience on which to form a judgment. His or her perspective must be influenced by external sources of information. It cannot be otherwise. Few people are familiar with the methods by which the primary sources of information can be reached in libraries. Moreover, there would rarely be the time or the technical training that is required. Because the public needs information that can be obtained conveniently and can be comprehended readily, the printed and electronic media become the primary sources of information. A secondary source which, in turn, undoubtedly exerts some influence on the media are the publications of the public interest groups, many of which have strong antinuclear orientations.

A series of illuminating surveys of the attitudes of various leadership groups in the United States have been published by Rothman and Lichter (1985). In response to the question "Are nuclear plants safe?" only 6.4% of the leadership of public interest organizations answered in the affirmative. This compares with 60.2% for a random sample of 929 scientists selected from "American Men and Women of Science." Among a sample of 279 scientists selected from energy-related fields, 76% answered "yes," as did 98.7% of the nuclear energy experts. In contrast, only 14% of the directors and producers of motion pictures answered affirmatively, as did only 13% of the producers and directors of prime-time television programs. The surveys have further shown that 50% of the journalists underestimated the extent of scientific support for nuclear energy, as did 85% of the directors, writers, and producers of television programs. This may explain why Ralph Nader, a nonscientist with well-known antinuclear views, was reported to be the most frequently interviewed "expert" on the subject of nuclear energy on prime-time television during the 1970s (Media Institute, 1979).

What about the Future?

Technological development is unpredictable and inventions have radically changed the way of life during this century. People live longer, have more leisure, are relatively wealthy by the standards of the last century, and their productive years are interrupted by far fewer episodes of illness. A basic reason for the vast improvement in our well-being has been the availability of convenient and inexpensive sources of electric energy in the home and workplace.

The demand for electricity was growing at an annual rate of about 6% for several decades, but this was reduced during the mid-1970s as a result of the unforeseen increases that took place in the price of oil and the worldwide economic depression that followed. At this writing, recovery from the global recession is occurring, but the demand for electricity is not likely to return to the former 6% annual growth of electrical energy production. Many economists now predict that future growth will be about 2–2.5% per year. While this is much less than the growth experienced in earlier decades, even a 2% annual growth will require that the nation's installed generating capacity be doubled every 35 years or less. New power plants will be required in large numbers in the years ahead.

Solar and other renewable sources of energy are not likely to have a significant impact until at least the middle years of the 21st century, if even then. The same can be said about energy from nuclear fusion. Thus, for the foreseeable future, the nation will be required to depend on coal, oil, and nuclear power, but the cost of oil and our dependence on foreign sources make it a highly questionable choice of fuel for electrical generation. Coal is a possible candidate to meet future needs, but there are obvious occupational safety and environmental problems associated with its use, of which the most intractable is likely to be atmospheric effects of carbon dioxide produced by the burning of coal (NAS–NRC, 1983b). The likelihood that the increase in atmospheric CO_2 will eventually result in a global warming trend seems more plausible with each passing decade, and the social and economic consequences of this "greenhouse effect" are bound to receive increasing attention from energy policy-makers. It could be a major reason for a resumption in the demand for nuclear power in the decades ahead.

The requirement that CO_2 emissions be reduced may also create a demand for electric-powered automobiles as well as a greater dependence on electricity for heating homes and other buildings.

There is every reason to believe that the nuclear energy industry of the future can continue to develop while at the same time fulfilling its responsibilities for protection of the safety of both its employees and the public. The demands for energy that can be expected to develop in the next century will be enormous in comparison with present needs, and the real danger is not that we may be using nuclear energy, but that if we do not it will not be possible to maintain the standard of living that society requires in a modern world.

When one considers the benefits that will accrue to mankind from civilian utilization of atomic energy on an extensive scale, the risks due to environmental radioactivity, though finite, are minuscule. However, when one contemplates the effects of radioactivity from the military uses of nuclear energy and, more particularly, the interaction of the effects of radioactivity with the effects due to blast and fire, the cost to humanity of military use of the atom in war would be so great as to make war unacceptable in relation to any benefits that could possibly accrue to any one country or to the world. More than 40 years ago, on June 14, 1946, U.S. Representative Bernard M. Baruch presented a program for international control of atomic energy to the infant United Nations (U.S. Department of State, 1946). It was then only 10 months since Hiroshima and only 5 months since the preceding January, when the first General Assembly of the United Nations convened in London. Mr. Baruch's address was delivered toward the end of a long and productive life intimately involved with the history of his times. His introductory words, among the first ever spoken at the United Nations on the need for nuclear sanity, provide an appropriate closing for this volume, in which we have dealt with the environmental atom in both war and peace.

"My Fellow Citizens of the World:

We are here to make a choice between the quick and the dead. . . ."
"Behind the black portent of the new atomic age lies a hope which, seized upon with faith, can work our salvation. If we fail, then we have damned every man to be the slave of Fear. Let us not deceive ourselves: We must elect World Peace or World Destruction. . . ."

That choice remains to be made.

Appendix

The Properties of Certain Radionuclides

The following pages summarize some of the important properties of the nuclides of principal interest. The biological and physical constants are subject to revision from time to time and the reader should be prepared to consult the primary sources of such information.

Explanations of the subheadings used, and the principal citations, are as follows:

Half-Life:
  Physical (Lederer *et al.*, 1978).
  Biological (ICRP, 1979).
Specific Activity (elemental): For naturally occurring radionuclides, the elemental specific activity is given in units of picocuries (pCi) per gram of element.
Principal Human Metabolic and Dosimetric Parameters: (as defined for adult radiation workers; ICRP, 1979):

  f_1: fractional transfer coefficient from the gastrointestinal tract to the blood.
  Inhalation Class (IC): describes the clearance rate of radioactive materials from the lungs. Classifications are: D (pulmonary half-times of <10 days), W (pulmonary half-times of 10–100 days), and Y (pulmonary half-times of >100 days).

Annual Limit on Intake (ALI): Secondary limit on the annual intake (in microcurries) of a radionuclide by either inhalation or ingestion such that the 50-year
committed effective dose equivalent attributable to intake during that year
is ≤5 rem for limiting stochastic effects and ≤50 rem for limiting nonstochastic effects. All limits given are for stochastic effects unless otherwise noted.

Derived air concentration (DAC): That concentration microcurries per cubic
meter of a radionuclide aerosol (1 µ AMAD) in air which if breathed by
Reference Man for a working year of 2000 hr (40 hr/week) would result in
one ALI by inhalation.

Systemic Transfer Fraction: That fraction of the inhaled or ingested radionuclide
leaving the transfer compartment (e.g., blood) and deposited within a specific organ or tissue.

Miscellaneous numerical values: For documentation, unless given, see elsewhere
in this text or the more recent reports of the United Nations Scientific Committee on the Effects of Atomic Radiation (1977, 1982).

Tritium (^{3}H)

Half-Life:
  Physical: 12.3 years.
  Biological: ~10 days (range: 4–18) total body for HTO.
Natural Levels: 6–24 pCi/liter of H_2O in surface prior to advent of bomb testing.
Sources: Cosmic-ray interactions with N and O; ternary fission; spallation from
    cosmic rays, $^6Li(n, \alpha)^3H$. World inventory of naturally produced tritium is
    approximately 26 MCi. Approximately 3600 MCi have been added to the North-
    ern Hemisphere from weapons testing.
Principal Modes and Energies of Decay (MeV): β^- 0.018.
Special Chemical and Biological Characteristics: Not selectively concentrated in
    any organ. Metabolized as H_2O.
Principal Organ: Total body.
Amount of Element in Body: 7×10^3 g.
Principal Human Metabolic and Dosimetric Parameters:
  $f_1 = 1.0$.
  ALI (µCi) = 8.1×10^4 (HTO)
  DAC (µCi/cm³) = 2.2×10^{-7} (HTO).
  Retention in body described by single exponential function: $R(t) = \exp(-0.693t/10)$, where t is the elapsed time in days.
Additional Dosimetric Considerations (rem/µCi): Total body (6.3×10^{-5} oral or
    inhalation). Annual dose equivalent rate from natural ^{3}H ~10^{-3} mrem/year
    and ~1.9 mrem/year from fallout ^{3}H (Northern Hemisphere).
Other: Production rates, 6.6 MCi/megaton of thermonuclear bombs; in light-water
    reactors, produced primarily by ternary fission and secondarily by neutron
    interactions with light elements. 5–10 Ci/MWe year for PWR; 0.3–0.9 CI/MWe
    year for BWR.

Carbon-14

Half-Life:
   Physical: 5730 years.
   Biological: Total body, 40 days.
Specific Activity of Living Carbon and Atmospheric CO_2 Prior to Era of Nuclear Testing: 6.1 pCi/g. Lower values in urban areas owing to fossil fuel combustion.
Sources: Cosmic-ray neutron activation, $^{14}N(n, p)^{14}C$; production rate at earth's surface, 2.28 atoms/cm² sec or 0.038 MCi/year: nuclear weapons testing 3.4 × 10⁴ Ci/megaton.
Principal Modes and Energies of Decay (MeV): β⁻ 0.155.
Special Characteristics: Metabolism varies considerably with the chemical compound of which it is a part.
Normal Daily Intake of Stable Element: 300 g.
Average Natural Tissue and/or body Burden: ~0.1 μCi.
Amount of Stable Element in Body: 1.6 × 10⁴ g.
Principal Human Metabolic and Dosimetric Parameters:
   f_1 = 1.0.
   ALI (μCi) = 2.2 × 10⁵ (inhalation of $^{14}CO_2$).
   DAC (μCi/cm³) = 8.1 × 10⁻⁵ ($^{14}CO_2$).
Additional Dosimetric Considerations (rem/μCi): Whole body (2.4 × 10⁻⁵ inhalation). From natural sources: gonads, 0.5 mrem/year; whole body, 1.3 mrem/year; endosteal cells, 2.0 mrem/year; bone marrow, 2.2 mrem/year (UNSCEAR, 1977).

Potassium-40

Half-Life:
   Physical: 1.28 × 10⁹ years.
   Biological: Whole body, 30 days.
Specific Activity: 853 pCi/g K.
Sources: Naturally occurring primordial nuclide.
Principal Modes and Energies of Decay (MeV): β⁻ 1.31 (89%); EC (11%); γ 1.46 (11%).
Special Chemical and Biological Characteristics: The element is distributed throughout the body, mainly in muscle.
Daily Intake: 3.3 g.
Typical Body Burden: 0.12 μCi (70-kg man).
Amount of Element in Body: 140 g (total body).
Principal Human Metabolic and Dosimetric Parameters:
   f_1 = 1.0.
   ALI (μCi) = 2.7 × 10² (oral or inhalation).
   Inhalation Class (IC): D.
   DAC (μCi/cm³) = 1.6 × 10⁻⁷.

Additional Dosimetric Considerations (rem/μCi): Whole body (2×10^{-2} oral, 1.3 $\times$ 10^{-2} inhalation). Dose equivalent rate from natural body burdens: 17 mrem/year whole body; 15 mrem/year endosteal cells; 27 mrem/year bone marrow.

Special Aspects: Body burden decreases with age and in muscle-wasting diseases.

Miscellaneous Information: Stable K constitutes 2.59% of earth's crust and 380 ppm in seawater; ^{40}K concentration in soils, 1–30 pCi/g. Potassium-40 is the predominant radioactive component in normal foods and human tissue.

Manganese-54

Half-Life:
  Physical: 312 days.
  Biological: Bone, 40 days; liver, 40 days.
Sources: Nuclear weapons testing; activation product—light-water reactors ^{54}Cr(p, n)^{54}Mn, ^{54}Fe(n, p)^{54}Mn.
Principal Modes and Energies of Decay (MeV): EC; γ or X rays, 0.835 (100%).
Special Characteristics: Stable manganese is an essential trace element for both plants and animals. In animals, bone and brain retain Mn more avidly than do most other organs.
Normal Daily Intake of Stable Element: 3.7×10^{-3} g/day.
Amount of Element in Body: 1.2×10^{-2} g.
Principal Human Metabolic and Dosimetric Parameters:
  $f_1 = 0.1$ (all compounds).
  ALI (μCi) $= 1.9 \times 10^3$ (oral),
        $= 8.1 \times 10^2$ (inhalation).
  Inhalation Class (IC): W (oxides, hydroxides, halides, and nitrates),
                D (all other compounds).
  DAC (μCi/cm^3) $= 8.1 \times 10^{-4}$ (for IC-D or -W).
  Metabolic Transfer Fractions: Bone (0.35), liver (0.15), kidney (0.1), other organs (0.2), other tissues (0.2).
Additional Dosimetric Considerations (rem/μCi): Lower large intestine wall (8.1 $\times 10^{-3}$ oral); liver (1.7×10^{-2} inhalation, IC-D, 9.3×10^{-3} inhalation, IC-W); lungs (2.5×10^{-2} inhalation, IC-W).

Iron-55

Half-Life:
  Physical: 657 days.
  Biological: Whole body, 2000 days.
Sources: Nuclear weapons testing, 1.7×10^7 Ci/megaton; produced in light-water reactors by ^{54}Fe(n, γ)^{55}Fe and ^{56}Fe(n, 2n)^{55}Fe.
Principal Modes and Energies of Decay: EC 6.0 keV, Auger electron 5.0 keV.

Special Biological Characteristics: 70% of total body iron is bound in hemoglobin with the remainder stored in other constituents of blood.

Normal Daily Intake of Stable Element: 0.016 g/day.

Amount of Stable Element in Body: 4.2 g.

Principal Human Metabolic and Dosimetric Parameters:

f_1 = 0.1.

ALI (μCi) = 8.1×10^3 (oral),

= 1.9×10^3 (inhalation, IC-D)

= 5.4×10^3 (inhalation, IC-W).

Inhalation Class (IC): W (oxides, hydroxides and halides), D (all other compounds).

DAC (μCi/cm^3) = 1.6×10^{-6} (IC-W),

= 8.1×10^{-7} (IC-D).

Systemic Transfer Fractions: Liver (0.08), spleen (0.013), and remaining tissues and organs (0.91).

Additional Dosimetric Considerations (rem/μCi): Spleen (2.1×10^{-3} oral; 1.0×10^{-2} inhalation, IC-D; 3.5×10^{-3} inhalation, IC-W); liver (1.3×10^{-3} oral; 6.3×10^{-3} inhalation, IC-D; 2.1×10^{-3} inhalation, IC-W).

Special Ecological Aspects: Direct foliar deposition is the most important pathway for terrestrial plant contamination. Readily taken up into marine food chains due to the low stable iron content of ocean waters. Concentration in the blood of Lapps and Eskimos high owing to air–lichen–reindeer (caribou)–human food chain. Also higher in consumers of fish diets.

Iron-59

Half-Life:

Physical: 44.56 days.

Biological: Whole body, 2,000 days.

Sources: Nuclear weapons testing, 2.2×10^6 Ci/megaton; activation produc'
^{58}Fe(n, γ)^{59}Fe.

Principal Modes and Energies of Decay (MeV): β^- 0.27 (49%), 0.4 (5%), γ 0.19 (2.7%), 1.10 (56%), 1.29 (44%).

Normal Daily Intake: See ^{55}Fe.

Principal Human Metabolic and Dosimetric Parameters:

f_1 = 0.1.

ALI (μCi) = 8.1×10^2 (oral),

= 2.7×10^2 (inhalation, IC-D),

= 5.4×10^2 (inhalation, IC-W).

Inhalation Class (IC): W (oxides, hydroxides and halides),

D (all other compounds).

DAC (μCi/cm^3) = 2.2×10^{-7} (IC-W),

= 1.4×10^{-7} (IC-D).

System Retention Fractions: See ^{55}Fe.

Additional Dosimetric Considerations (rem/μCi): Lower large intestine wall (3.1 $\times$ 10^{-2} oral; 1.8 $\times$ 10^{-2} inhalation, IC-D; 1.7 $\times$ 10^{-2} inhalation, IC-W); lungs (1.3 $\times$ 10^{-2} inhalation, IC-D; 5.2 $\times$ 10^{-2} inhalation, IC-W); spleen (1.1 $\times$ 10^{-2} inhalation, IC-W).

Special Ecological Aspects: See ^{55}Fe.

Cobalt-60

Half-Life:
   Physical: 5.27 years.
   Biological: Approximately 7 days with additional long-term components of 60 and 800 days.
Sources: Activation product ^{59}Co(n, γ)^{60}Co; ^{60}Ni(n, p)^{60}Co. Produced in nuclear explosions and in reactors.
Principal Modes and Energies of Decay (MeV): β^- 0.315 (99+%); γ 1.173 (100%), 1.332 (100%).
Normal Intake of Element: 3 $\times$ 10^{-4} g/day.
Amount of Element in Body (grams): 5 $\times$ 10^{-3}.
Principal Human Metabolic and Dosimetric Parameters:
   f_1 = 0.3 (organically complexed and for most inorganic compounds).
   f_1 = 0.05 (oxides and hydroxides in the presence of carrier material).
   ALI (μCi) = 1.9 $\times$ 10^2 (f_1 = 0.3, oral),
            = 5.4 $\times$ 10^2 (f_1 = 0.05, oral),
            = 1.6 $\times$ 10^2 (f_1 = 0.05, inhalation, IC-W),
            = 2.7 $\times$ 10^1 (f_1 = 0.05, inhalation, IC-Y).
   Inhalation Class (IC): Y (oxides, hydroxides, halides, and nitrates),
                        W (all other compounds).
   DAC (μCi/cm^3) = 1.4 $\times$ 10^{-8} (IC-Y),
                  = 8.1 $\times$ 10^{-8} (IC-W).
   Retention in body for systemic Co, given by the expression: $R(t)$ = 0.5 exp($-0.693t/0.5$) + 0.3 exp($-0.693t/6$) + 0.1 exp($-0.693t/60$) + 0.1 exp($-0.693t/800$), where t is in days.
Additional Dosimetric Considerations (rem/μCi): Upper large intestine wall (2.1 $\times$ 10^{-2} oral f_1 = 0.05, 3.6 $\times$ 10^{-2} oral f_1 = 0.3); lungs (1.3 $\times$ 10^{-1} inhalation, IC-W; 1.3 inhalation, IC-Y).

Zinc-65

Half-Life:
   Physical: 244 days.
   Biological: Skeleton, 400 days; other soft tissues (two components: 20 and 400 days).
Sources: Nuclear weapons testing, activation product ^{64}Zn(n, γ)^{65}Zn, ^{63}Cu(^{2}H, γ)^{65}Zn.

Principal Modes and Energies of Decay (MeV): EC (98.5%); β⁺ 0.325 (1.5%); γ 0.51
from β⁺ (3%), 1.11 (51%).
Normal Daily Intake of Element: 0.013 g/day.
Amount of Stable Isotopes in Body: 2.3 g.
Special Biological Considerations: Uptake from gastrointestinal tract (f_1) depen-
dent to some extent on fasting state and possibly dietary Zn level.
Principal Human Metabolic and Dosimetric Parameters:
f_1 = 0.5.
ALI (μCi) = 2.7×10^2 (oral or inhalation).
Inhalation Class (IC): Y (oxides, hydroxides, nitrates and phosphates).
DAC (μCi/cm³) = 1.1×10^{-7}.
Systemic Transfer Fractions: 0.2 (skeleton), 0.8 (remainder of organs and
tissues).
Additional Dosimetric Considerations (rem/μCi): Total body (1.8×10^{-2} oral, 2.2
$\times 10^{-2}$ inhalation); lungs (7.8×10^{-2} inhalation).

Krypton-85

Half-Life:
Physical: 10.70 years
Sources: Produced primarily from fission at ~400 Ci/MWe year. Atmospheric bur-
den mainly attributable to nuclear fuel reprocessing. A minor product of cos-
mic-ray interactions.
Principal Modes and Energies of Decay (MeV): β⁻ 0.672.
Special Chemical Characteristics and Biological Characteristics: Inert noble gas.
Diffusion is sole mechanism of absorption into body.
Principal Human Metabolic and Dosimetric Parameters: Exposure in a semi-
infinite cloud of radioactive noble gas is determined by external irradiation,
and the DAC is based solely on this consideration.
DAC (μCi/cm³) = 1.4×10^{-4} (limited by effects on skin).
Additional Dosimetric Considerations (rem/hr per μCi/cm³): Dose equivalent
rate from immersion in ⁸⁵Kr cloud of 1 μCi/cm³: skin (1.7×10^2); gonads (1.9);
lens of eye (2.2).

Strontium-89

Half-Life:
Physical: 50.5 days.
Biological: Retention in body best represented by a complex exponential plus
power function equation (see ICRP, 1973a).
Source: Fission product.
Principal Modes and Energies of Decay (MeV): β⁻ 1.49 (100%).

Special Chemical Characteristics and Biological Characteristics: Alkaline earth
    element, with tendency to concentrate uniformly throughout the volume of
    mineral bone.
Normal Daily Intake: 1.9×10^{-3} g/day.
Amount of Stable Isotopes in Body: 0.32 g.
Principal Human Metabolic and Dosimetric Parameters:
    $f_1 = 0.3$ (all compounds except $SrTiO_3$).
    ALI (μCi) = 5.4×10^2 (oral),
               = 8.1×10^2 (inhalation, IC-D).
    Inhalation Class (IC): D.
    DAC (μCi/cm^3) = 2.7×10^{-7}.
Additional Dosimetric Considerations (rem/μCi): Bone surfaces (1.8×10^{-2} oral
    $f_1 = 0.3$, 3.1×10^{-2} inhalation, IC-D); lower large intestine wall (7.8×10^{-2}
    oral $f_1 = 0.3$, 1.3×10^{-2} inhalation IC-D); lungs (3.1×10^{-1} inhalation, IC-Y).

Strontium-90

Half-Life:
    Physical: 28.8 years.
    Biological: See [89]Sr.
Source: Fission product.
Principal Modes and Energies of Decay (MeV): β^- 0.54 (100%).
Special Chemical and Biological Characteristics: See [89]Sr.
Amount of Stable Isotopes in Body: See [89]Sr.
Principal Human Metabolic and Dosimetric Parameters:
    $f_1 = 0.3$ (all compounds except $SrTiO_3$).
    ALI (μCi) = 2.7×10^1 (oral, for limiting nonstochastic effects to bone surfaces)
               = 1.9×10^1 (inhalation, IC-D; see above limitation).
    Inhalation Class (IC): D.
    DAC (μCi/cm^3) = 8.1×10^{-9}.
Additional Dosimetric Considerations (rem/μCi): Bone surfaces (1.6 oral $f_1 = 0.3$,
    2.7 inhalation, IC-D); lungs (1.1×10^1 inhalation, IC-Y).
Special Ecological Aspects: Incorporates into calcium pool of the biosphere. Prin-
    cipal ecological pathway is grass → cow's milk → human.

Technetium-99

Half-Life:
    Physical: 2.1×10^5 years.
    Biological: Whole body, ~2 days; thyroid, 0.5 days.
Sources: Fission product released to the environment mainly from weapons testing
    and recycling of nuclear fuels.
Principal Modes and Energies of Decay (MeV): β^- 0.292.

Special Chemical Characteristics: Most stable as the very soluble pertechnetate anion (TcO_4^-).

Principal Human Metabolic and Dosimetric Parameters:

$f_1 = 0.8.$

ALI (μCi) = 2.7×10^3 (oral),

   = 5.4×10^3 (inhalation, IC-D, for limiting effects on the stomach wall),

   = 5.4×10^2 (inhalation, IC-W).

Inhalation Class (IC): W (oxides, hydroxides, halides, and nitrates),

   D (all other compounds).

DAC (μCi/cm³) = 2.2×10^{-6} (IC-D),

   = 2.7×10^{-7} (IC-W).

Systemic Transfer Fractions: Thyroid (0.04), stomach wall (0.1), liver (0.03), and all other tissues and organs (0.83).

Retention in whole body given by the expression: $R(t) = 0.76 \exp(-0.693t/1.6) + 0.19 \exp(-0.693t/3.7) + 0.043 \exp(-0.693t/22)$, where t is in days.

Additional Dosimetric Considerations (rem/μCi): Thyroid (5.9×10^{-3} oral, 4.4×10^{-3} inhalation, IC-D); stomach wall (1.3×10^{-2} oral, 9.3×10^{-3} inhalation, IC-D).

Special Ecological Aspects: TcO_4^- readily available for plant root uptake (plant/soil concentration ratio ~20). High mobility in soils and apparent short half-times in vegetation may limit dosimetric consequences from plant ingestion (Hoffman *et al.*, 1982).

Iodine-129

Half-Life:

Physical: 1.57×10^7 years.

Biological: See [131]I.

Sources: Naturally produced in the upper atmosphere from interactions of high-energy particles with xenon and, to a lesser extent, in the lithosphere, from spontaneous fission of [238]U. Also produced in nuclear explosions at a rate of 0.05 Ci/megaton and in light-water reactors at a rate of 1.3 μCi/MWd.

Principal Modes and Energies of Decay (MeV): β^- 0.15 (100%), γ 0.04 (7.5%).

Special Biological Characteristics: See [131]I.

Normal Intake of Stable Element: See [131]I.

Amount of Stable Isotopes in Body: See [131]I.

Principal Human Metabolic and Dosimetric Parameters:

$f_1 = 1.0.$

ALI (μCi) = 5.4 (oral, for limiting nonstochastic effects to the thyroid),

   = 8.1 (inhalation, see above limitation).

Inhalation Class (IC): D.

DAC (μCi/cm³) = 2.7×10^{-9}.

Systemic Transfer Fractions: See [131]I.

Additional Dosimetric Considerations: The specific activity of [129]I is too low (0.17 μCi/mg) to permit significant dose to individuals.

Special Ecological Aspects: Prior to weapons testing, ~30 Ci resided in the hydrosphere and 10 Ci in the lithosphere from natural sources. Weapons testing has approximately doubled the natural inventory (NCRP, 1983). Also, see [131]I.

Iodine-131

Half-Life:
   Physical: 8.04 days.
   Biological: 12 days (organic I in whole body), 120 days (thyroid).
Sources: A fission product produced by nuclear weapons at a rate of 125 MCi/-megaton. Attains equilibrium of 26 kCi/MWt of steady-state reactor operation.
Principal Modes and Energies of Decay (MeV): β^- 0.34 (13%), 0.61 (86%); γ 0.28 (6%), 0.36 (81%), 0.64 (7%), 0.72 (2%).
Special Biological Characteristics: Soluble form readily absorbed through the skin, lung, and alimentary tract, and concentrates in the thyroid.
Normal Intake of Stable Element: 2×10^{-4} g/day.
Amount of Stable Isotopes in Body: 1.1×10^{-3} g (1.0×10^{-3} g in thyroid).
Principal Human Metabolic and Dosimetric Parameters:
   $f_1 = 1.0$.
   ALI (μCi) = 1.7×10^1 (oral, for limiting nonstochastic effects to the thyroid),
            = 5.4×10^1 (inhalation, see above limitation).
   Inhalation Class (IC): D.
   DAC μCi/cm³) = 1.9×10^{-8}.
   Systemic Transfer Fractions: Thyroid (0.3).
Additional Dosimetric Considerations (rem/μCi): Adult thyroid (1.8 oral, 1.1 inhalation). Dose commitments ~10 times greater for a 1-year-old child.
Special Ecological Aspects: Principal route of human absorption is via fresh milk: grass → cow → milk → human chain, with rapid distribution of milk a primary consideration. Milk content of radioiodine reaches a peak 3 days after deposition on forage (Soldat, 1965). Effective half-time of removal from grass is about 5 days. 9.1×10^{-2} μCi ^{131}I/liter milk can be expected per μCi ^{131}I deposited on 1 m² of grass. The ratio pCi/kg grass:pCi/m³ air = 4200. The ratio pCi/liter milk:pCi/kg grass = 0.15 (Soldat, 1963).

Cesium-134

Half-Life:
   Physical: 2.06 years.
   Biological: Total body adults, 50–150 days; total body children, 44 days.
Sources: An activation product produced in nuclear reactors at a rate of 3 Ci/MWd, [134]Cs/[137]Cs = 0.4–0.6. Low-level discharges (generally <0.5 Ci/year) to liquid effluent from commercial nuclear reactors.

Principal Modes and Energies of Decay (MeV): β^- 0.089 (27%), 0.66 (70%), γ 0.605 (98%), 0.795 (85%).

Special Chemical Characteristics: Alkali metal with properties similar to K and Rb; most salts are soluble.

Biological Characteristics: Metabolism resembles that of potassium—distributed uniformly throughout body. Convenient to express concentration in biological material relative to K content.

Normal Intake of Stable Element: 1×10^{-5} g/day.

Amount of Stable Element in Body: 1.5×10^{-3} g.

Principal Human Metabolic and Dosimetric Parameters:

$f_1 = 1.0$.

ALI (μCi) = 8.1×10^1 (oral),

= 1.1×10^2 (inhalation).

Inhalation Class (IC): D.

DAC (μCi/cm^3) = 5.4×10^{-8}.

Retention in body given by the function: $R(t) = 0.1 \exp(-0.693t/2) + 0.9 \exp(-0.693t/110)$.

Additional Dosimetric Considerations (rem/μCi): Gonads (7.8×10^{-2} oral, 4.8×10^{-2} inhalation), whole body (8.5×10^{-2} oral, 5.6×10^{-2} inhalation).

Cesium-137

Half-Life:

Physical: 30.2 years.

Biological: See ^{134}Cs.

Sources: Nuclear weapons testing 0.17 MCi ^{137}Cs/megaton fission. Low-level discharges (generally <0.5 Ci/year) to liquid effluents from commercial nuclear reactors.

Principal Modes and Energies of Decay (MeV): β^- 0.51 (94%), 1.18 (6%), γ 0.66 (85%).

Special Chemical Characteristics and Biological Characteristics: See ^{134}Cs.

Amount of Stable Isotopes in Body: See ^{134}Cs.

Principal Human Metabolic and Dosimetric Parameters:

$f_1 = 1.0$.

ALI (μCi) = 1.1×10^2 (oral),

= 1.6×10^2 (inhalation).

Inhalation Class (IC): D.

DAC (μCi/cm^3) = 5.4×10^{-8}.

Retention function in body: See ^{134}Cs.

Additional Dosimetric Considerations (rem/μCi): Gonads (5.2×10^{-2} oral, 3.3×10^{-2} inhalation), whole body (4.8×10^{-2} oral).

Special Ecological Aspects: Higher body burdens in Lapps and Eskimos due to air $\rightarrow$ lichen $\rightarrow$ reindeer (caribou) $\rightarrow$ human food chain; strong retention to clay soils limits plant uptake; solubility much greater in marine than fresh waters; uptake by fish controlled by [K$^+$] content of water.

Lead-210 (Ra D)

Half-Life:
  Physical: 22.3 years.
  Biological: Several half-times have been proposed for different compartments in
    bone and various organs (see ICRP, 1979).
Sources: Naturally occurring from U decay chain.
Principal Modes and Energies of Decay (MeV): β^- 0.015 (81%), 0.061 (19%); γ
  0.0465 (3.86%); Bi X-rays—0.013 av. (23.35%).
Special Biological Characteristics: Lead can be substituted for Ca in apatite (min-
  eral bone) and, accordingly, is uniformly distributed throughout the bone
  volume.
Normal Ingested Intake of Stable Element: 4×10^{-4} g/day (1–2 pCi/day).
Average Natural Tissue and/or Body Burden: 160–860 pCi in skeleton (mean 380
  pCi; NCRP, 1984b); (5.4×10^{-2} pCi/g fresh bone).
Amount of Stable Element in Body: 0.12 g.
Principal Human Metabolic and Dosimetric Parameters:
  $f_1 = 0.2$.
  ALI (μCi) = 5.4×10^{-1} (oral, limited by nonstochastic effects to bone surfaces),
       = 2.4×10^{-1} (inhalation, see above limitation).
  Inhalation Class (IC): D.
  DAC (μCi/cm^3) = 1.1×10^{-10}.
  Systemic Transfer Fractions: Bone (0.55), liver (0.25), kidneys (0.02), and re-
    maining organs and tissues (0.18), uniform distribution).
Additional Dosimetric Considerations (rem/μCi): Bone surfaces (8.1×10^1 oral,
  2.0×10^2 inhalation); liver (2.3×10^1 oral, 5.6×10^1 inhalation). Dose equiv-
  alent rate from naturally present ^{210}Pb in equilibrium with ^{210}Po $\sim$50
  mrem/year to bone surfaces.

Polonium-210

Half-Life:
  Physical: 138.4 days.
  Biological: Whole body, 50 days.
Sources: Naturally occurring from uranium decay chain and ^{209}Bi(n, γ)^{210}Bi β^-
  ^{210}Po.
Principal Modes and Energies of Decay (MeV): α 5.3 (100%).
Special Chemical Characteristics: Tendency for radiocolloid formation.
Range of Normal Intake: 1–5 pCi/day.
Average Natural Tissue and/or Body Burden: 160–800 pCi, mean 360 pCi.
Principal Human Metabolic and Dosimetric Parameters:
  $f_1 = 0.1$.
  ALI (μCi) = 2.7 (oral),
       = 5.4×10^{-1} (inhalation, IC-D or -W).

Inhalation Class (IC): W (oxides, hydroxides, and nitrates),
> D (all other compounds).

DAC (μCi/m³) = 2.7 × 10^{-10} (IC-D or -W).

Systemic Transfer Fractions: Liver (0.1), kidney (0.1), spleen (0.1), and all other tissues and organs (0.7).

Retention in the body given by the expression $R(t) = \exp(-0.693t/50)$, where t is in days.

Additional Dosimetric Considerations (rem/μCi): Kidneys (9.3 oral; 4.4 × 10^1 inhalation, IC-D; 1.4 × 10^1 inhalation, IC-W); spleen (1.6 × 10^1 oral; 8.1 × 10^1 inhalation, IC-D; 2.5 × 10^1 inhalation, IC-W). Dose rate from natural body burden is 50 mrem/year to bone surfaces (assumes equilibrium with ^{210}Pb). Builds in from ^{210}Pb, especially in mineral bone.

Special Ecological Aspects: Air → lichen → reindeer → human pathway results in abnormally high body burdens, as do high fish and shellfish diets; also present in tobacco smoke.

Radon-222

Half-Life:
  Physical: 3.82 days.

Sources: Naturally occurring from uranium decay chain. Equilibrium global inventory ~40 MCi.

Principal Modes and Energies of Decay (MeV): α 5.49 (100%).

Special Chemical Characteristics: Inert noble gas, somewhat soluble in body fat. Under most conditions, the principal does is from α emissions by short-lived daughter products inhaled as attachments to inert dust normally present in the atmosphere.

Normal Range of Atmospheric Radon Content: 0.1–0.5 pCi/liter (mean 0.15 pCi/liter). Indoor concentrations quite variable depending on source strength, indoor location, and ventilation.

Dosimetric Considerations: A series of short-lived daughter products approach equilibrium with ^{222}Rn in a few hours. (See Chapter 7 and Fig. 7–3.) A special "unit," developed for use in U.S. uranium mines, is the working level (WL), defined as any combination of radon daughters in 1 liter of air that will produce 1.3 × 10^5 MeV of α energy. This is equivalent to a ^{222}Rn concentration of 100 pCi/liter, in equilibrium with its short-lived α-emitting progeny. For continuous environmental exposure (i.e., 730 hr/month) at 1 WL, one would accumulate 50 WLM in a year. The dose equivalent rate to the bronchial epithelium from a normal indoor ^{222}Rn daughter exposure level of 0.004 WL is about 3 rem/year (NCRP, 1984b), based on an average conversion factor for all ages of 0.7 rad/WLM (Harley and Pasternack, 1982) and a quality factor of 20.

Existing Exposure Limits:
  Occupational: 4.8 WLM/year (ICRP, 1981), 4.0 WLM/year (U.S. miners).
  Nonoccupational: Under consideration by EPA.

Other: When [226]Ra is deposited in the human skeleton, ~70% of the radon produced diffuses to the blood and is eliminated in the expired breath. One microcurie of [226]Ra in the skeleton results in a concentration of 13 pCi/liter of [222]Rn in exhaled air.

Radium-226

Half-Life:
   Physical: 1599 years.
   Biological: Whole-body retention is best represented by a complex exponential plus power function equation (see ICRP, 1973a).
Sources: Naturally occurring from [238]U decay chain.
Principal Modes and Energies of Decay (MeV): α 4.78 (94.4%), 4.60 (5.6%); γ 0.186 (3.3%).
Biological Characteristics: Deposits in bone with nonuniform distribution. Following decay of [226]Ra in bone, ~70% of [222]Rn diffuses to the blood and is exhaled.
Normal Intake: 2.3×10^{-12} g/day (2.3 pCi [226]Ra/day).
Average Natural Tissue and/or Body Burden: 3.1×10^{-11} g (~31 pCi).
Principal Human Metabolic and Dosimetric Parameters:
   $f_1 = 0.2$.
   ALI (μCi) = 1.9 (oral, for limiting nonstochastic effects to bone surfaces),
      = 5.4×10^{-1} (inhalation, see above limitations).
   Inhalation Class (IC): W.
   DAC (μCi/cm^3) = 2.7×10^{-10}.
   Systemic Transfer Fractions and Retention Expressions: See ICRP (1973a).
Additional Dosimetric Considerations (rem/μCi): Bone surfaces (2.5×10^1 oral, 2.8×10^1 inhalation), lungs (5.9×10^1 inhalation). (Dose commitments based on [226]Ra + daughters assuming ~30% [222]Rn retention.)

Radium-228

Half-Life:
   Physical: 5.8 years.
   Biological: See [226]Ra.
Sources: Occurs naturally as a decay product of [223]Th.
Biological Characteristics: See [226]Ra. No loss of [220]Rn occurs because of its short half-life (56 sec).
Normal Intake: No systematic measurements. Believed to be ¼ to ½ that of [226]Ra.
Normal Body Burden: Believed to be ¼ to ½ that of [226]Ra.
Principal Metabolic Parameters: See [226]Ra.

Thorium-230

Half-Life:
    Physical: 8.0×10^4 years.
    Biological: See [232]Th.
Specific Activity: 0.154 Ci/g [nat]Th.
Sources: Naturally occurring member of the [238]U series.
Principal Modes and Energies of Decay (MeV): α 4.688 (76.3%), 4.621 (23.4%), γ Ra
    L X-rays.
Special Chemical Characteristics: See [232]Th.
Biological Characteristics: See [232]Th.
Average Natural Tissue and/or Body Burden: ~3 pCi total body (Wrenn *et al.*,
    1981).
Principal Human Metabolic and Dosimetric Parameters:
    $f_1 = 2 \times 10^{-4}$.
    ALI (μCi) = 2.7 (oral, for limiting nonstochastic effects to bone surfaces),
               $= 5.4 \times 10^{-3}$ (inhalation, IC-W, see above limitation),
               $= 1.6 \times 10^{-2}$ (inhalation, IC-Y, see above limitation).
    Inhalation Class (IC): Y (oxides and hydroxides),
                        W (all other compounds).
    DAC (μCi/cm³) = 2.7×10^{-12} (IC-W),
                    $= 5.4 \times 10^{-12}$ (IC-Y).
    Systemic Transfer Fractions: See [232]Th.
Additional Dosimetric Considerations (rem/μCi): Bone surfaces (1.3×10^1 oral,
    3.2×10^3 inhalation, IC-Y); lungs (1.1×10^3 inhalation, IC-Y).
Special Ecological Aspects: [230]Th is found in elevated concentrations in uranium
    mill tailings and in certain phosphate fertilizers.

Thorium-232

Half-Life:
    Physical: 1.41×10^{10} years.
    Biological: Bone, 8×10^3 days; liver and remaining tissues, 700 days.
Specific Activity: 1.1×10^5 pCi/g Th.
Sources: Naturally occurring.
Principal Modes and Energies of Decay (MeV): α 4.01 (77%), 3.96 (23%), γ Ra L X-
    rays.
Special Chemical Characteristics: Hydroxides and oxides are insoluble; nitrates,
    sulfates, chlorides, and perchloride salts are readily soluble.
Biological Characteristics: Tendency to concentrate on bone surfaces.
Normal Intake: ~5×10^{-7} g/day (Linsalata *et al.*, 1986b).
Average Natural Tissue and/or Body Burden: ~12 μg or 1.3 pCi, total body
    (Wrenn *et al.*, 1981).

Principal Human Metabolic and Dosimetric Parameters:
$f_1 = 2 \times 10^{-4}$.
ALI (μCi) = 8.1×10^{-1} (oral, for limiting nonstochastic effects to bone
        surfaces),
        = 1.1×10^{-3} (inhalation, IC-W, see above limitation),
        = 2.7×10^{-3} (inhalation, IC-Y, see above limitation).
Inhalation Class (IC): Y (oxides and hydroxides);
                W (all other compounds).
DAC (μCi/cm^3) = 5.4×10^{-13} (IC-W);
           = 1.1×10^{-12} (IC-Y).
Systemic Transfer Fractions: Bone (0.7), liver (0.04), and all other tissues and
    organs (0.16).
Additional Dosimetric Considerations (rem/μCi): Bone surfaces (7.0×10^1 oral,
    1.9×10^4 inhalation, IC-Y); lungs (3.5×10^3 inhalation, IC-Y).
Special Ecological Aspects: Thorium-bearing minerals result in anomalously high
    natural radiation levels in certain areas in Brazil and India. Depending on the
    type of rock, the concentration of ^{232}Th in the earth's crust is 2–20 μg/g and
    about 10 μg/g in normal soils.

Uranium-235

Half-Life:
    Physical: 7.04×10^8 years.
    Biological: Kidney, 6 days; skeleton, 20 days with a longer component of 5000
        days.
Specific Activity (Elemental) (pCi/g): 1.54×10^4 pCi/g ^{nat}U.
Sources: Normal constituent (0.72%) of uranium in earth's crust. Enriched with
    respect to ^{238}U in fissionable materials for reactors and weapons.
Principal Modes and Energies of Decay (MeV): α 4.60 (84.6%), 4.56 (3.7%), 4.50
    (1.2%), 4.42 (4%), 4.40 (57%), 4.37 (18%); γ 0.144 (11%), 0.186 (54%), 0.205 (5%).
Normal Elemental Intake: 1.9×10^{-6} g/day.
Amount of Element in Body: 9×10^{-5} g ^{nat}U.
Principal Human Metabolic and Dosimetric Parameters:
    $f_1 = 0.05$ [water-soluble compounds, i.e., compounds of U(VI)].
    $f_1 = 2 \times 10^{-3}$ [compounds of U(IV), e.g., UO_2, UF_4, and U_3O_8].
    ALI (μCi) = 1.4×10^1 (oral, $f_1 = 0.05$),
             = 1.9×10^2 (oral, $f_1 = 2 \times 10^{-3}$),
             = 1.4 (inhalation, IC-D, $f_1 = 0.05$),
             = 8.1×10^{-1} (inhalation, IC-W, $f_1 = 0.05$),
             = 5.4×10^{-2} (inhalation, IC-Y, $f_1 = 2 \times 10^{-3}$).
    Inhalation Class (IC): D [U(VI) compounds], W (UO_3, UF_4, and UCl_4), Y (UO_2 and
        U_3O_8).
    DAC (μCi/cm^3) = 5.4×10^{-10} (IC-D)
                = 2.7×10^{-10} (IC-W)
                = 1.6×10^{-11} (IC-Y).

Retention in various organs and tissues given by the expressions:

$R_{bone}(t) = 0.2 \exp(-0.693t/20) + 0.023 \exp(-0.693t/5000)$ $^*R_{kidney}(t) = 0.12$ $\exp(-0.693t/6) + 5.2 \times 10^{-4} \exp(-0.693t/1500)$

Additional Dosimetric Considerations (rem/μCi): Bone surfaces (3.7 oral $f_1 =$ 0.05, 1.6×10^{-1} oral $f_1 = 2 \times 10^{-3}$, 3.7×10^1 inhalation, IC-D); lungs (5.6×10^1 inhalation, IC-W; 1.0×10^3 inhalation, IC-Y); kidneys (1.6 oral $f_1 = 0.05$, 1.6×10^1 inhalation, IC-D). Since ^{nat}U has a low specific activity, chemical damage to the kidney is likely to be more important than radiation damage. However, radiation injury to the lung or kidney must be considered if exposure is to enriched U.

Uranium-238

Half-Life:
   Physical: 4.5×10^9 years.
   Biological: See ^{235}U.
Specific Activity (Elemental) (pCi/g): 3.3×10^5 pCi/g ^{nat}U.
Sources: Naturally occurring in earth's crust. ^{238}U is present to extent of 99.28% by weight in natural uranium.
Principal Modes and Energies of Decay (MeV): α 4.20 (77%), 4.15 (23%).
Normal Annual Intake: See ^{235}U.
Principal Metabolic and Dosimetric Parameters:
   $f_1 = 0.05$ (see ^{235}U for additional information).
   $f_1 = 2 \times 10^{-3}$ (see ^{235}U for additional information).
   ALI (μCi) $= 1.4 \times 10^1$ (oral, $f_1 = 0.05$),
            $= 2.2 \times 10^2$ (oral, $f_1 = 2 \times 10^{-3}$),
            $= 1.4$ (inhalation, IC-D, $f_1 = 0.05$),
            $= 8.1 \times 10^{-1}$ (inhalation, IC-W, $f_1 = 0.05$),
            $= 5.4 \times 10^{-2}$ (inhalation, IC-Y), $f_1 = 2 \times 10^{-3}$).
   Inhalation Class (IC): D, W, and Y (see ^{235}U).
   DAC (μCi/cm³) $= 5.4 \times 10^{-10}$ (IC-D),
                 $= 2.7 \times 10^{-10}$ (IC-W),
                 $= 1.9 \times 10^{-11}$ (IC-Y).
   Retention Expressions: See ^{235}U.
Additional Dosimetric Considerations (rem/μCi): Bone surfaces (3.7 oral $f_1 =$ 0.05, 1.5×10^{-1} oral $f_1 = 2 \times 10^{-3}$, 3.6×10^1 inhalation, IC-D); lungs (5.2×10^1 inhalation, IC-W; 1×10^3 inhalation, IC-Y); kidneys (1.5 oral $f_1 = 0.05$, 1.5×10^1 inhalation, IC-D).

Neptunium-237

Half-Life:
   Physical: 2.1×10^6 years.

---

*Also applies to soft tissues.

Biological: Bone, 100 years; liver, 40 years. (*Author's note:* biological half-times
will probably be reevaluated toward much lower values based on research
in progress.)

Sources: Produced by neutron capture reactions with ^{238}U in fission reactors and
by decay of ^{241}Am.

Principal Modes and Energies of Decay: α 4.82 (1.5%), 4.80 (1.6%), 4.79 (51%), 4.77
(36%); γ 0.086 (12.6%).

Special Chemical Characteristics: Relatively soluble as the NpO_2^+ ion, which is
stabilized against hydrolysis.

Special Biological Characteristics: f_1 value given below from ICRP Rep. 30 (1980b)
recommendations is likely to overestimate true gut absorption by a factor of 10.
Similarly, systemic transfer fractions to bone and liver may be better represented by transfer fractions of 0.60 and 0.15, respectively (Thompson, 1982).
Uncertainty exists with respect to the uniformity of bone deposition.

Principal Human Metabolic and Dosimetric Parameters:

$f_1 = 0.01$.

ALI (μCi) $= 8.1 \times 10^{-2}$ (oral, for limiting nonstochastic
effects to bone surfaces),
$= 5.4 \times 10^{-3}$ (inhalation, see above limitation).

Inhalation Class (IC): W.

DAC (μCi/cm³) $= 2.4 \times 10^{-12}$ (IC-W).

Systemic Transfer Fractions: Skeleton (0.45) and liver (0.45).

Additional Dosimetric Considerations (rem/μCi): Bone surfaces (7×10^2 oral, 9
$\times 10^3$ inhalation); liver (1.5×10^2 oral, 2×10^3 inhalation).

Plutonium-238

Half-Life:

Physical: 87.7 years.

Biological: Bone, 100 years; liver, 40 years.

Sources: Minor constituent of nuclear weapons fallout (3.5% $^{239,240}Pu$ by activity).
Produced for power sources (SNAP) by $^{237}Np(n, \beta)^{238}Pu$.

Principal Modes and Energies of Decay (MeV): α 5.50 (71%), 5.46 (29%); γ 0.039
(11%).

Special Chemical Characteristics: Generally forms insoluble fluorides, hydroxides, and oxides. Solubility in water dependent on redox, pH, and organic
ligands present.

Biological Characteristics: Deposits mainly on endosteal surfaces of mineral
bone.

Principal Human Metabolic and Dosimetric Parameters:

$f_1 = 1 \times 10^{-4}$ (compounds other than oxides and hydroxides), $f_1 = 1 \times 10^{-5}$
(oxides and hydroxides).

ALI (μCi) $= 8.1$ (oral, $f_1 = 10^{-4}$),
$= 8.1 \times 10^1$ (oral, $f_1 = 10^{-5}$),
$= 5.4 \times 10^{-3}$ (inhalation, IC-W, $f_1 = 10^{-4}$),
$= 1.6 \times 10^{-2}$ (inhalation, IC-Y, $f_1 = 10^{-5}$).

Inhalation Class (IC): Y (oxides and hydroxides), W (all other compounds).
DAC (μCi/cm^3) = 2.4 $\times$ 10^{-12} (IC-W),
$\qquad\qquad\quad$ = 8.1 $\times$ 10^{-12} (IC-Y).
Systemic Transfer Fractions: Skeleton (0.45) and liver (0.45).
Additional Dosimetric Considerations (rem/μCi): Bone surfaces (6.7 oral f_1 = 10^{-4}; 6.7 $\times$ 10^{-1} oral f_1 = 10^{-5}; 8.1 $\times$ 10^3 inhalation, IC-W; 3.1 $\times$ 10^3 inhalation, IC-Y); liver (1.5 oral f_1 = 10^{-4}; 1.5 $\times$ 10^{-1} oral f_1 = 10^{-5}; 1.8 $\times$ 10^3 inhalation, IC-W; 6.7 $\times$ 10^2 inhalation, IC-Y); lungs (1.2 $\times$ 10^3 inhalation, IC-Y).

Plutonium-239

Half-Life:
$\quad$ *Physical:* 2.41 $\times$ 10^4 years.
$\quad$ *Biological:* See ^{238}Pu.
Sources: Produced in thermal reactors by neutron irradiation of ^{238}U. Used in nuclear weapons and as fuel for fast reactors. Approximately 325 kCi of ^{239,240}Pu have been distributed worldwide from nuclear weapons testing (Hardy *et al.*, 1973).
Principal Modes and Energies of Decay (MeV): α 5.10 (11%), 5.14 (15%), 5.15 (73%); L$_\alpha$ x-ray 0.014 (1.5%), x-ray 0.014 (10%), L$_\beta$ 0.018 (0.1%).
Special Chemical Characteristics and Biological Characteristics: See ^{238}Pu.
Principal Human Metabolic and Dosimetric Parameters:
$\quad$ f_1 = 1 $\times$ 10^{-4} (compounds other than oxides and hydroxides).
$\quad$ f_1 = 1 $\times$ 10^{-5} (oxides and hydroxides).
$\quad$ ALI (μCi) = 5.4 (oral, f_1 = 10^{-4}),
$\qquad\qquad\quad$ = 5.4 $\times$ 10^1 (oral, f_1 = 10^{-5}),
$\qquad\qquad\quad$ = 5.4 $\times$ 10^{-3} (inhalation, IC-W, f_1 = 10^{-4},
$\qquad\qquad\quad$ = 1.3 $\times$ 10^{-2} (inhalation, IC-Y, f_1 = 10^{-5}).
$\quad$ Inhalation Class (IC): Y (oxides and hydroxides),
$\qquad\qquad\qquad\qquad\quad$ W (all other compounds).
$\quad$ DAC (μCi/cm^3) = 2.2 $\times$ 10^{-12} (IC-W),
$\qquad\qquad\qquad\quad$ = 5.4 $\times$ 10^{-12} (IC-Y).
Systemic Retention Fractions: See ^{238}Pu.
Additional Dosimetric Considerations (rem/μCi): Bone surfaces (7.8 oral f_1 = 10^{-4}; 7.8 $\times$ 10^{-1} oral f_1 = 10^{-5}; 9.3 $\times$ 10^3 inhalation, IC-W; 3.5 $\times$ 10^3 inhalation, IC-Y); liver (1.6 oral, f_1 = 10^{-4}; 1.6 $\times$ 10^{-1} oral, f_1 10^{-5}; 2.0 $\times$ 10^3 inhalation, IC-W; 7.8 $\times$ 10^2 inhalation, IC-Y); lungs (1.2 $\times$ 10^3 inhalation, IC-Y).

Americium-241

Half-Life:
$\quad$ *Physical:* 432 years.
$\quad$ *Biological:* Bone, 100 years; liver, 40 years.
Sources: Decay product of ^{241}Pu.

Principal Modes and Energies of Decay (MeV): α 5.49 (84%), 5.44 (13%), 5.39 (1%); γ 0.059 (35.7%).

Special Chemical Characteristics: Oxides and hydroxides of Am(III) are relatively insoluble.

Special Biological Characteristics: Deposits mainly on endosteal but also on periosteal surfaces of bone.

Principal Human Metabolic and Dosimetric Parameters:

$f_1 = 5 \times 10^{-4}$.

ALI (μCi) = 1.4 (oral, for limiting nonstochastic effects to bone surfaces),

= 5.4×10^{-3} (inhalation, IC-W, see above limitation).

Inhalation Class (IC): W.

DAC (μCi/cm^3) = 2.2×10^{-12}.

Systemic Transfer Fractions: Skeleton (0.45) and liver (0.45).

Additional Dosimetric Considerations (rem/μCi): Bone surfaces (4×10^1 oral, 9×10^3 inhalation); liver (8.5 oral, 2×10^3 inhalation).

References

Aarkrog, A. (1971a). Radioecological investigations of plutonium in an arctic marine environment. *Health Phys.* **20**, 31–47.

Aarkrog, A. (1971b). Prediction models for ^{90}Sr and ^{137}Cs levels in the human food chain. *Health Phys.* **20**, 297–312.

Aarkrog, A., Dahlgaard, H., and Nilsson, K. (1984). Further studies of plutonium and americium at Thule, Greenland. *Health Phys.* **46**, 29–44.

Adler, H. L., and Weinberg, A. M. (1978). An approach to setting radiation standards. *Health Phys.* **34**, 719–720.

Adriano, D. C., Pinder, J. E., III, McLeod, K. W., Corey, J. C., and Boni, A. L. (1982). Plutonium contents and fluxes in a soybean crop ecosystem near a nuclear fuel chemical separation plant. *J. Environ. Qual.* **11**, 506–511.

Agricola, G. (1556). "De Re Metallica." (Reprinted: Transl. by H. Hoover and L. Hoover. Dover, New York, 1950.)

Albenesius, E. L. (1959). Tritium as a product of fission. *Phys. Rev. Lett.* **3**, 274.

Albert, R. E. (1966). "Thorium—Its Industrial Hygiene Aspects." Academic Press, New York.

Albert, R. E., and Altshuler, B. (1973). Considerations relating to the formulation of limits for unavoidable population exposures to environmental carcinogens. *In* "Radionuclide Carcinogenesis" (J. E. Ballon *et al.*, eds.), AEC Symp. Ser., CONF-72050, pp. 233–253. NTIS, Springfield, Virginia.

Alberts, J. J., Bobula, C. M., and Farras, D. T. (1980). A comparison of the distribution of industrially released ^{238}Pu and fallout ^{239,240}Pu in temperate, northern United states soils. *J. Environ. Qual.* **9**, 592–596.

Alexander, L. T. (1959). Strontium-90 distribution as determined by the analysis of soils. *In*

"Fallout from Nuclear Weapons Tests," Hearings before Joint Committee on Atomic Energy, Special Subcommittee on Radiation, Vol. I, pp. 278–371. USGPO, Washington, D.C.

Alexander, L. T. (1967). "Depth of Penetration of the Radioisotopes Strontium-90 and Cesium-137," Fallout Program Quarterly Summary, Rep. HASL-183. USAEC, New York.

Alexander, L. T., Hardy, E. P., and Hollister, H. L. (1960). Radioisotopes in soils: Particularly with reference to strontium-90. *In* "Radioisotopes in the Biosphere" (R. S. Caldecott and L. A. Snyder, eds.), pp. 3–22. Univ. of Minnesota Press, Minneapolis.

Altshuler, B., Nelson, N., and Kuschner, M. (1964). Estimation of the lung tissue dose from the inhalation of radon and daughters. *Health Phys.* **10,** 1137–1162.

American Nuclear Society (1971). Symposium for reactor containment spray system technology. *Nucl. Technol.* **10,** 400.

American Nuclear Society (1984). "Report of Special Committee on Source Terms." ANS, La Grange Park, Illinois.

American Nuclear Society (1986a). The Chernobyl Accident. *Nuclear News,* June 1986, pp. 87–94.

American Nuclear Society (1986b). Chernobyl: The Soviet Report. *Nuclear News,* September 1986, pp. 1–8.

Anderson, E. C. (1953). The production and distribution of natural radiocarbon. *Annu. Rev. Nucl. Sci.* **2,** 63.

Anderson, W., and Turner, R. C. (1956). Radon content of the atmosphere. *Nature (London)* **178,** 203.

Anderson, W., Mayneord, W. V., and Turner, R. C. (1954). The radon content of the atmosphere. *Nature (London)* **174,** 424.

Angelovic, J. W., White, J. C., and Davis, E. M. (1969). Interaction of ionizing radiation, salinity, and temperature on the estuarine fish *Fundulus heteroclitus. Proc. Natl. Symp. Radioecol., 2nd* CONF-670503, pp. 131–141. USAEC, Washington, D.C.

Anspaugh, L. R., Shinn, J. H., Phelps, P. L., and Kennedy, N. C. (1975). Resuspension and redistribution of plutonium in soils. *Health Phys.* **29,** 571–582.

Archer, V. E. (1981). Health concerns in uranium mining and milling. *J. Occup. Med.* **23,** 502–505.

Arthur, W. J., III, and Markham, O. D. (1984). Polonium-210 in the environment around a radioactive waste disposal area and phosphate ore processing plant. *Health Phys.* **46,** 793–799.

Atomic Industrial Forum (1961). *Memo* p. 27. Detroit.

Atomic Industrial Forum (1985). *INFO Data* Apr.

Auerbach, S. I., Nelson, D. J., Kaye, S. V., Reichle, D. E., and Coutant, C. C. (1971). Ecological considerations in reactor power plant siting. *Proc. Environ. Aspects Nucl. Power Stn.* IAEA, Vienna.

Aulenbach, D. B., and Davis, R. E. (1976). Long-term consumption of mineral water containing natural Ra-226. *Proc. Midyear Top. Symp., 10th, Health Phys. Soc., Rensselaer Polytech. Inst., Troy, New York,* 154–166.

Auxier, J. A., and Dickson, H. W. (1983). Concern over recent use of the ALARA philosophy. *Health Phys.* **44,** 595–600.

Axelson, O., and Sundell, L. (1978). Mining lung cancer and smoking. *Scand. J. Work Environ. Health* **4,** 46.

Baker, D. A., and Peloquin, R. A. (1985). "Population Dose Commitments Due to Radioactive Releases from Nuclear Power Plant Sites in 1981," Rep. NUREG/CR-2850, PNL-4221, Vol. 3. USNRC, Washington, D.C.

Barcinski, M. A., Abreu, M. C. A., de Almeida, J. C. C., Naya, J. M., Fonseca, L. G., and Castro, L.

E. (1975). Cytogenetic investigation in a Brazilian population living in an area of high natural radioactivity. *Am. J. Hum. Genet.* **27**, 802–806.

Barnaby, F. (1982). The effect of a global war: the arsenals. *Ambio* **11**, 76.

Barreira, F. (1961). Concentration of atmospheric radon and wind directions. *Nature (London)* **190**, 1092.

Bartlett, B. O., and Russell, R. S. (1966). Prediction of future levels of long-lived fission products in milk. *Nature (London)* **209**, 1062–1065.

Battist, L., Buchanan, J., Congel, F., Nelson, C., Nelson, M., Peterson, H., and Rosenstein, M. (1979). "Population Dose and Health Impact of the Accident at the Three Mile Island Nuclear Station," A preliminary assessment for the period March 28 through April 7, 1979, Ad Hoc Population Dose Assessment Group. USGPO, Washington, D.C.

Beasley, T. M., and Palmer, H. E. (1966). Lead-210 and polonium-210 in biological samples from Alaska. *Science* **152**, 1062–1063.

Beck, H. L. (1966). Environmental gamma radiation from deposited fission products, 1960–1964. *Health Phys.* **12**, 313–322.

Beck, H. L. (1972). The physics of environmental radiation fields. *In* "The Natural Radiation Environment II" (J. A. S. Adams, W. M. Lowder, and T. F. Gesell, eds.), USDOE CONF-720805-P1, pp. 101–133. NTIS, Springfield, Virginia.

Beck, H. L. (1980). "Factors for Radionuclides Deposited on the Ground," EML Rep. 378. USDOE, New York.

Beck, H. L., and dePlanque, G. (1968). "The Radiation Field in Air Due to Distributed Gamma-Ray Sources in the Ground," Rep. HASL-195. USAEC, New York.

Beck, H., and Krey, P. W. (1983). Radiation exposures in Utah from Nevada nuclear tests. *Science* **220**, 18–24.

Beck, H. L., Lowder, W. M., Bennett, B. G., and Condon, W. J. (1966). "Further Studies of External Environmental Radiation," Rep. HASL-170. USAEC, New York.

Beck, H. L., Gogolak, C. V., Miller, K. M., and Lowder, W. M. (1980). Perturbations on the natural radiation environment due to the utilization of coal as an energy source. *In* "Natural Radiation Environment III" (T. F. Gesell and W. M. Lowder, eds.), CONF-89-42, Vol. 2, pp. 1521–1558. USDOE, Washington, D.C.

Becquerel, H., and Curie, P. (1901). Action physiologique des rayons du radium. *C. R. Hebd. Seances Acad. Sci.* **132**, 1289–1291.

Bedrosian, P. H., Easterly, D. G., and Cummings, S. L. (1970). "Radiological Survey Around Power Plants Using Fossil Fuel," Rep. EERL-71-3. USEPA, Rockville, Maryland.

Beebe, G. W., Kato, H., and Land, C. E. (1978). Studies of the mortality of A-bomb survivors. 6. Mortality and radiation dose. *Radiat. Res.* **75**, 138–201.

Behounek, F. (1970). History of exposure of miners to radon. *Health Phys.* **19**, 56–57.

Bell, M. J. (1973). "ORIGEN," Oak Ridge Isotope and Depletion Code, ORNL Rep. 4628. Oak Ridge Natl. Lab., Oak Ridge, Tennessee.

Belyaev, V. E., Kolesnikov, A. G., and Nelepo, B. A. (1965). Estimate of the intensity of radioactive pollutants in the oceans on the basis of new data on decay processes. *Proc. Int. Conf. Peaceful Uses At. Energy, 3rd, Geneva, 1964* **14**, 83–86.

Bennett, B. G. (1972). "Estimation of ^{90}Sr levels in the Diet," HASL Rep. 246. USAEC, New York.

Bennett, B. G. (1978). "Environmental Aspects of Americium," EML Rep. 348. USDOE, New York.

Bennett, G. L. (1981). Overview of the U.S. flight safety process for space nuclear power. *Nucl. Saf.* **22**, 423–434.

Bennett, G. L., Lombardo, J. J., and Rock, B. J. (1984). U.S. radioisotope thermoelectric generator space operating experience (June 1961–December 1982). *Nucl. Eng.* **25**, 49–58.

Bensen, D. W., and Sparrow, A. H., eds. (1971). Survival of Food Crops and Livestock in the Event of Nuclear War. *Proc. Symp. Brookhaven Natl. Lab., 1970* CONF-700909. USAEC, Washington, D.C.

Bertini, H. W. (1980). "Descriptions of Selected Accidents that Have Occurred at Nuclear Reactor Facilities," Rep. ORNL/NSIC-176. Oak Ridge Natl. Lab., Oak Ridge, Tennessee.

Bjornerstedt, R., and Engstrom, A. (1960). Radioisotopes in the skeleton: Dosage implications based on microscopic distribution. *In* "Radioisotopes in the Biosphere," Chap. 27. Univ. of Minnesota Press, Minneapolis.

Black, D. (1984). "Investigation of the Possible Increased Incidence of Cancer in West Cumbria," Report of the Independent Advisory Group. HM Stationery Off., London.

Blanchard, R. L., and Holaday, D. A. (1960). Evaluation of radiation hazards created by thoron and thoron daughters. *Am. Ind. Hyg. Assoc. Q.* **21**, 201–206.

Blanchard, R. L., Fowler, T. W., Horton, T. R., and Smith, J. M. (1982). Potential health effects of radioactive emissions from active surface and underground uranium mines. *Nucl. Saf.* **23**, 439–450.

Blaylock, B. G. (1982). Radionuclide data bases available for bioaccumulation factors for freshwater biota. *Nucl. Saf.* **23**, 427–438.

Blaylock, B. G., and Trabalka, J. R. (1978). Evaluating the effects of ionizing radiation on aquatic organisms. *Adv. Radiat. Biol.* **7**, 103–151.

Blifford, I. H., Lockhart, L. B., *et al.* (1952). "On the Natural Radioactivity of the Air," Rep. 4036. Nav. Res. Lab., Washington, D.C.

Blomeke, J. O., and Harrington, F. E. (1968). Waste management at nuclear power stations. *Nucl. Saf.* **9**, 239–248.

Blum, M., and Eisenbud, M. (1967). Reduction of thyroid irradiation from ^{131}I by potassium iodide. *J. Am. Med. Assoc.* **200**, 1036–1040.

Boice, J. D., Jr. (1982). Risk estimates for breast. *In* "Critical Issues in Setting Radiation Dose Limits," Proc. 17th Annu. Meet. NCRP, pp. 164–181. Natl. Counc. Radiat. Prot. Meas., Bethesda, Maryland.

Bond, V. P. (1982). The conceptual basis for evaluating risk from low-level radiation exposure. *In* "Issues in Setting Radiation Standards," Proc. 17th Annu. Meet. NCRP, pp. 25–26. Natl. Counc. Radiat. Prot. Meas., Bethesda, Maryland.

Bondietti, E. A., and Francis, C. W. (1979). Geologic migration potentials or technetium-99 and neptunium-237. *Science* **203**, 1337–1340.

Bondietti, E. A., Trabalka, J. R., Garten, C. T., and Killough, G. G. (1979). Biogeochemistry of actinides: A nuclear fuel cycle perspective. *In* "Radioactive Waste in Geological Storage" (S. Fried, ed.), *Am. Chem. Soc. Symp. Ser.* No. 100, 241–266.

Boone, F. W., Kantelo, M. V., Mayer, P. G., and Palms, J. M. (1985). Residence half-times of [129]I in undisturbed surface soils based on measured soil concentration profiles. *Health Phys.* **48**, 401–413.

Borduin, L. C., and Taboas, A. L. (1981). DOE radioactive waste incineration technology. *Nucl. Saf.* **22**, 56–69.

Bowen, V. T., Noshkin, V. E., Livingston, H. D., and Volchok, H. L. (1980). Fallout radionuclides in the Pacific Ocean: Vertical and horizontal distributions, largely from Geosec stations. *Earth Planet. Sci. Lett.* **49**, 411–343.

Brazilian Academy of Sciences (Academia Brasileira de Ciencias) (1977). "Proceedings of International Symposium on Areas of High Natural Radioactivity." Rio de Janeiro, Brazil.

Brecker, R., and Brecker, E. (1969). "The Rays: A History of Radiology in the United States and Canada." Williams & Wilkins, Baltimore, Maryland.

Breslin, A. J., and Glauberman, H. (1970). Investigation of radioactive dust dispersed from uranium tailings piles. *In* "Environmental Surveillance in the Vicinity of Nuclear Facilities" (W. C. Reinig, ed.). Thomas, Springfield, Illinois.

Brewer, A. W. (1949). Evidence for a world circulation provided by the measurements of helium and water vapour distribution in the stratosphere. *Q. J. R. Meteorol. Soc.* **75**, 351–363.

Briggs, G. A. (1984). Plume rise and buoyancy effects. *In* "Atmospheric Science and Power Production" (D. Randerson, ed.), Rep. DOE/TIC-27601, pp. 327–366. USDOE, Washington, D.C.

Brinck, W. L., Schliekelman, R. J., Bennet, D. L., Bell, C., and Markwood, I. M. (1978). Radium removal efficiencies in water treatment processes. *J. Am. Water Works Assoc.* **70**, 31–43.

British Nuclear Fuels Limited (1984). "Annual Report on Radioactive Discharges and Monitoring of the Environment, 1978." Health Saf. Dir., Risley, Warrington, Cheshire, England.

Brittan, R. O., and Heap, J. C. (1958). Reactor containment. *Proc. U.N. Int. Conf. Peaceful Uses At. Energy, 2nd, Geneva* p. 437.

Broecker, W. S., Goddard, J., and Sarmiento, J. L. (1976). The distribution of ^{226}Ra in the Atlantic Ocean. *Earth Planet. Sci. Lett.* **32**, 220–235.

Broseus, R. W. (1970). Cesium-137/strontium-90 ratios in milk and grass from Jamaica. M.S. Thesis, New York Univ., New York.

Brown, S. H., and Smith, R. C. (1980). A model for determining the overall release rate and annual source term for a commercial *in situ* leach uranium facility. *In* "Uranium Resource Technology," pp. 794–800, Seminar III. Colorado Sch. Mines, Boulder.

Bruce, F. R. (1960). Origin and nature of radioactive wastes in the United States atomic energy programme. *Disposal Radioact. Wastes, Proc. Sci. Conf., Monaco, 1959.*

Buchanan, J. R. (1963). SL-1 final report. *Nucl. Saf.* **4**, 83–86.

Burnett, T. J. (1970). A derivation of the "Factor of 700" for ^{131}I. *Health Phys.* **18**, 73–75.

Burton, E. F. (1904). A radioactive gas from crude petroleum. *Philos. Mag.* **8**, 498–508.

Cantril, S. T., and Parker, H. M. (1945). "The Tolerance Dose," MDDC-1100. USAEC, Washington, D.C.

Catlin, R. J. (1980). "Assessment of the Surveillance Program of the High-Level Waste Storage Tanks at Hanford." Rep. to U.S. Dep. Energy, Asst. Sec. Environ., Washington, D.C.

Chamberlain, A. C. (1955). "Aspects of Travel and Deposition of Aerosol and Vapor Clouds," Rep. HP/R1261. U.K. At. Energy Auth., Res. Comp.

Chamberlain, A. C. (1960). Aspects of the deposition of radioactive and other gases and particles. *Int. J. Air Pollut.* **3**, 63–88.

Chamberlain, A. C. (1970). Interception and retention of radioactive aerosols by vegetation. *Atmos. Environ.* **4**, 57–78.

Chamberlain, A. C., and Chadwick, R. C. (1966). Transport of iodine from atmosphere to ground. *Tellus* **18**, 226–237.

Chang, T. Y., Cheng, W. L., and Weng, P. S. (1974). Potassium, uranium, and thorium content of building materials of Taiwan. *Health Phys.* **27**, 385–387.

Chepil, W. A. (1957). Erosion of soil by wind. *In* "The Yearbook of Agriculture," pp. 308–314. U.S. Dep. Agric., Washington, D.C.

Clapp, C. A. (1934). "Cataract." Lea & Febiger, Philadelphia, Pennsylvania.

Clark, H. M. (1954). The occurrence of an unusually high-level radioactive rainout in the area of Troy, N.Y. *Science* **119**, 619–622.

Clarke, R. H. (1974). An analysis of the Windscale accident using the WEERIE code. *Am. Nucl. Sci. Eng.* **1**, 73.

Claus, W. D. (1958). "Radiation Biology and Medicine." Addison-Wesley, Reading, Massachusetts.

Clayton, G. D., Arnold, J. R., and Patty, F. A. (1955). Determination of sources of particulate atmospheric carbon. *Science* **122,** 751–753.

Clegg, J. W., and Foley, D. D. (1958). "Uranium Ore Processing." Addison-Wesley, Reading, Massachusetts.

Clements, W., and Wilkening, M. (1974). Atmospheric pressure effects on radon transport across the earth–air interface. *J. Geophys. Res.* **79,** 5025–5029.

Codman, E. A. (1902). A study of the cases of accidental x-ray burns hitherto recorded. *Philadelphia Med. J.* pp. 438–442.

Cohen, B. L. (1980). Society's valuation of life saving in radiation protection and other contexts. *Health Phys.* **38,** 33–51.

Cohen, B. L. (1982). Effects of ICRP 30 and the 1980 BEIR report on hazard assessment of high level wastes. *Health Phys.* **42,** 133–143.

Cohen, B. S., Eisenbud, M., and Harley, N. H. (1980). Measurement of the α-radioactivity on the mucosal surface of the human bronchial tree. *Health Phys.* **39,** 619–632.

Colby, L. J. (1976). Fuel reprocessing in the U.S.: A review of problems and some solutions. *Nucl. News* (*La Grange Park, Ill.*) Jan.

Collins, J. T., Bell, M. J., and Hewitt, W. M. (1978). Radioactive waste source terms. *In* "Nuclear Power Waste Technology" (A. A. Moghissi, H. W. Godbee, M. S. Ozker, and M. W. Carter, eds.), pp. 167–199. Am. Soc. Mech. Eng., New York.

Colorado Committee for Environmental Information (1970). "Report on the Dow Rocky Flats Fire: Implications of Plutonium Releases to the Public Health and Safety," Subcommittee on Rocky Flats, Boulder, Colo., Jan. 13, 1970, Rep. HASL-235 (Ref. 2, p. 38). USAEC, New York.

Comar, C. L. (1979). Risk: a pragmatic *de minimis* approach. *Science* **203,** 133–143.

Comar, C. L., and Wasserman, R. H. (1960). Radioisotope absorption and methods of elimination: differential behavior of substances in metabolic pathways. *In* "Radioisotopes in the Biosphere" (R. S. Caldecott and L. A. Snyder, eds.), pp. 526–540. Univ. of Minnesota Press, Minneapolis.

Comar, C. L., Wasserman, R. H., and Nold, M. M. (1956). Strontium calcium discrimination factors in the rat. *Proc. Soc. Exp. Biol. Med.* **92,** 859–863.

Comar, C. L., Trum, B. F., Kuhn, U. S. G., III, Wasserman, R. H., Nold, M. M., and Schooley, J. C. (1957). Thyroid radioactivity after nuclear weapons tests. *Science* **126,** 16–18.

Committee on Armed Services (1981). "Hearings on Uranium Ore Residues: Potential Hazards and Disposition," June 24–25, 1981. USGPO, Washington, D.C.

Conard, R. A. (1984). Late radiation effects in Marshall Islanders exposed to fallout 28 years ago. *In* "Radiation Carcinogenesis: Epidemiology and Biological Significance" (J. D. Boice and J. F. Fraumeni, eds.), pp. 57–71. Raven, New York.

Conard, R. A., *et al.* (1960). "Medical Survey of Rongelap People Five and Six years After Exposure to Fallout," Rep. BNL-609. Brookhaven Natl. Lab., Upton, New York.

Conard, R. A., *et al.* (1970). "Medical Survey of the People of Rongelap and Utirik Islands Thirteen, Fourteen, and Fifteen Years After Exposure to Fallout Radiation (March 1967, March, 1968 and March, 1969)," Rep. BNL 50220 (T-562). Brookhaven Natl. Lab., Upton, New York.

Conard, R. A., Paglia, D. E., Larsen, P. R., Sutow, W. W., Dobyns, B. M., Robbins, J., Krotosky, W. A., Field, J. B., Rall, J. E., and Wolff, J. (1980). "Review of Medical Findings in a Marshallese Population Twenty-Six Years After Accident Exposure to Radioactive Fallout," Rep. BNL 52161. Brookhaven Natl. Lab., Upton, New York.

Coplan, B. V., and Baron, J. S. (1978). Treatment of liquid radwastes. *In* "Nuclear Power Waste Technology" (A. A. Moghissi, H. W. Godbee, M. S. Ozker, and M. W. Carter, eds.), pp. 233–272. Am. Soc. Mech. Eng., New York.

Cothern, C. R., and Lappenbusch, W., eds. (1985). Special issue. *Health Phys.* **48**(5), 529–712.

Cottrell, W. B. (1974). Control of radioactive wastes in operating nuclear facilities. *In* "Human and Ecologic Effects of Nuclear Power Plants" (L. A. Sagan, ed.), pp. 72–131. Thomas, Springfield, Illinois.

Court Brown, W. M., Doll, R., and Hill, A. B. (1960). Incidence of leukaemia after exposure to diagnostic radiation in utero. *Br. Med. J.* **ii**, 1539–1545.

Cowan, G. A. (1976). A natural fission reactor. *Sci. Am.* **235**, 36–47.

Crabtree, J. (1959). The travel and diffusion of the radioactive material emitted during the Windscale accident. *Q. J. R. Meteorol. Soc.* **85**, 362–370.

Crawford, T. V. (1978). Atmospheric transport of radionuclides: Report of working group on atmospheric dispersion, deposition and resuspension. *Proc. Workshop Eval. Models Used Environ. Assess. Radionuclide Releases, Gatlinburg, Tenn., 1977* ORNL Rep. CONF-770901, p. 5.

Crick, M. J., and Lindsley, G. S. (1983). Addendum to "An Assessment of the Radiological Impact of the Windscale Reactor Fire, October, 1957," Rep. R135. Nat. Radiat. Prot. Board.

Crick, M. J., and Lindsley, G. S. (1984). An Assessment of the Radiological Impact of the Windscale Reactor Fire, October, 1957. *Int. J. Radiat. Biol.* **46**, 479–506.

Crocker, G. R., O'Connor, J. D., and Freiling, E. C. (1966). Physical and radiochemical properties of fallout particles. *Health Phys.* **12**, 1099–1104.

Cronkite, E. P. (1961). Evidence for radiation and chemical as leukemogenic agents. *Environ. Health* **3**, 297.

Cross, F. T., Harley, N. H., and Hofmann, W. (1985). Health effects and risks from [222]Rn in drinking water. *Health Phys.* **48**, 649–670.

Cuddihy, R. G. (1982). Risks of radiation-induced lung cancer. *In* "Critical Issues in Setting Radiation Dose Limits," Proc. 17th Annu. Meet. NCRP, pp. 133–152. Natl. Counc. Radiat. Prot. Meas., Bethesda, Maryland.

Cuddihy, R. G., Clellan, R. O., and Griffith, W. C. (1979). Variability in target organ deposition among individuals exposed to toxic substances. *Toxicol. Appl. Pharmacol.* **49**, 179–187.

Cuthbert, F. L. (1958). "Thorium Production Technology." Addison-Wesley, Reading, Massachusetts.

Davis, J. J., Perkins, R. W., Palmer, R. F., Hanson, W. C., and Cline, J. F. (1958). Radioactive materials in aquatic and terrestrial organisms exposed to reactor effluent water. *Proc. U.N. Int. Conf. Peaceful Uses At. Energy, 2nd, Geneva.*

de Bartoli, M. C., and Gaglione, P. (1969). SNAP plutonium-238 fallout at Ispra, Italy. *Health Phys.* **16**, 197–204.

de Villiers, A. J., and Windish, J. P. (1964). Lung cancer in fluorspar mining community. I. Radiation, dust and mortality experience. *Br. J. Ind. Med.* **21**, 94–109.

Dickerson, M. H., Foster, T., and Gudiksen, P. H. (1984). Experimental and model transport and diffusion studies in complex terrain with emphasis on tracer studies. *Boundary-Layer Meteorol.* **30**, 333–350.

DiNunno, J. J., Anderson, F. D., Baker, R. E., and Waterfield, R. L. (1962). "Calculation of Distance Factors for Power and Test Reactor sites," Rep. TID-14844. USAEC, Washington, D.C.

Dobson, G. M. B. (1956). Origin and distribution of the polyatomic molecules in the atmosphere. *Proc. R. Soc. London, Ser. A* **236,** 187–193.

Dohrenwend, B. P., Dohrendwend, B. S., Warheit, G. J., Bartlett, G. S., Goldsteen, R. L., Goldsteen, K., and Martin, J. L. (1981). Stress in the community: a report to the President's Commission on the Accident at Three Mile Island. *In* "The Three Mile Island Nuclear Accident: Lessons and Implications" (T. H. Moss and D. L. Sills, eds.), *Ann. N.Y. Acad. Sci.* **365,** 159–174.

Dolphin, G. W. (1968). The risk of thyroid cancer following irradiation. *Health Phys.* **15,** 219–228.

Dolphin, G. W. (1971). Dietary intakes of iodine and thyroid dosimetry. *Health Phys.* **21,** 711–713.

Donaldson, L. R., Bonham, K., Eagleton, J. G., and Castle, P. (1969). "Effects of Chronic Irradiation on Chinook Salmon Research in Fisheries," Contrib. No. 300, pp. 52–53. Coll. Fish., Univ. of Washington, Seattle.

Dreicer, M., Hakonson, T. E., and White, G. C. (1984). Rainsplash as a mechanism for soil contamination of plant surfaces. *Health Phys.* **46,** 177–187.

Drew, R. T., and Eisenbud, M. (1966). The normal radiation dose to indigenous rodents on the Morro do Ferro, Brazil. *Health Phys.* **12,** 1267–1274.

Drinker, P., and Hatch, T. (1954). "Industrial Dust," 2nd Ed. McGraw-Hill, New York.

DuFrain, R. J., Littlefield, L. G., Joiner, E. J., and Frome, E. L. (1980). *In vitro* human cytogenetic dose-response systems. *In* "The Medical Basis for Radiation Accident Preparedness" pp. 358–374. Elsevier/North-Holland, New York.

Dunning, G. M. (1962). "Fallout from USSR 1961 Nuclear Tests," Rep. TID-14377. USAEC, Washington, D.C.

Dunster, H. J. (1958). The disposal of radioactive liquid wastes into coastal waters. *Proc. U.N. Int. Conf. Peaceful Uses At. Energy, 2nd, Geneva.*

Dunster, H. J. (1969). *United Kingdom studies on radioactive releases in the marine environment. In* "Biological Implications of the Nuclear Age." USAEC, Washington, D.C.

Dunster, H. J., Howells, H., and Templeton, W. L. (1958). District surveys following the Windscale Incident, October, 1957. *Proc. U.N. Int. Conf. Peaceful Uses At. Energy, 2nd, Geneva.*

Duursma, E. K., and Gross, M. C. (1971). Marine sediments and radioactivity. *In* "Radioactivity in the Marine Environment," pp. 147–160. Natl. Acad. Sci., Washington, D.C.

Eckert, J. A., Coogan, J. S., Mikkelsen, R. L., and Lem, P. N. (1970). Cesium-137 concentrations in Eskimos, Spring, 1968. *Radiol. Health Data Rep.* **11,** 219–225.

Ehrlich, P., and Sagan, C. (1984). "The Cold and the Dark." Norton, New York.

Eisenbud, M. (1957). Global distribution of strontium-90 from nuclear detonations. *Sci. Mon.* **84,** 237–244.

Eisenbud, M. (1959). Deposition of strontium-90 through October, 1958. *Science* **130,** 76–80.

Eisenbud, M. (1980). Radioactive wastes from biomedical institutions. (Editorial.) *Science* **207,** 1299.

Eisenbud, M. (1981a). The status of radioactive waste management: Needs for reassessment. *Health Phys.* **40,** 429–437.

Eisenbud, M. (1981b). The concept of *de minimis* dose. *In* "Quantitative Risk in Standard Setting," pp. 64–75. Natl. Counc. Radiat. Prot. Meas., Bethesda, Maryland.

Eisenbud, M. (1984). Sources of ionizing radiation exposure. *Environment* **26**(10), 1–33.

Eisenbud, M., and Harley, J. H. (1953). Radioactive dust from nuclear detonations. *Science* **117,** 141–147.

Eisenbud, M., and Harley, J. H. (1956). Radioactive fallout through September, 1955. *Science* **124**, 251–255.

Eisenbud, M., and Petrow, H. (1964). Radioactivity in the atmospheric effluents of nuclear power plants that use fossil fuels. *Science* **144**, 288–289.

Eisenbud, M., and Quigley, J. (1956). Industrial hygiene of uranium processing. *AMA Arch. Ind. Health* **14**, 12–22.

Eisenbud, M., Mochizuki, Y., Goldin, A. S., and Laurer, G. R. (1962). Iodine-131 dose from Soviet nuclear tests. *Science* **136**, 370–374.

Eisenbud, M., Mochizuki, Y., and Laurer, G. L. (1963a). I-131 dose to human thyroids in New York City from nuclear tests in 1962. *Health Phys.* **9**, 1291–1298.

Eisenbud, M., Pasternack, B., Laurer, G. R., Mochizuki, Y., Wrenn, M. E., Block, L., and Mowafy, R. (1963b). Estimation of the distribution of thyroid doses in a population exposed to I-131 from weapons tests. *Health Phys.* **9**, 1281–1290.

Eisenbud, M., Pasternack, B., Laurer, G. R., and Block, L. (1963c). Variability of the I-131 concentrations in the milk distribution system of a large city. *Health Phys.* **9**, 1303–1305.

Eisenbud, M., Petrow, H., Drew, R. T., Roser, F. X., Kegel, G., and Cullen, T. L. (1964). Naturally occurring radionuclides in foods and waters from the Brazilian areas of high radioactivity. *In* "The Natural Radiation Environment I" (J. A. S. Adams and W. M. Lowder, eds.), pp. 837–854. Univ. of Chicago Press, Chicago, Illinois.

Eisenbud, M., Krauskopf, K., Penna Franca, E., Lei, W., Ballad, R., Linsalata, P., and Fujimori, K. (1984). Natural analogues for the transuranic actinide elements: An investigation in Minas Gerais, Brazil. *Environ. Geol. Water Sci.* **6**, 1–9.

Electric Power Research Institute (1982). "Survey of Plume Models for Atmospheric Application," Rep. EA-2243. Energy Anal. Environ. Div., Palo Alto, California.

Emerson, E. L., and McClure, J. D. (1985). "Radioactive Material (RAM) Accident/Incident Data Analysis Program," Rep. NUREG/CR-3611. USNRC, Washington, D.C.

Etherington, H., ed. (1958). "Nuclear Engineering Handbook." McGraw-Hill, New York.

Evans, R. D. (1943). Protection of radium dial workers and radiologists from injury by radium. *J. Ind. Hyg. Toxicol.* **25**, 253–269.

Evans, R. D. (1966). The effect of skeletally deposited alpha-ray emitters in man. *Br. J. Radiol.* **39**, 881–895.

Evans, R. D. (1967). The radium standard for boneseekers—evaluation of the data on radium patients and dial painters. *Health Phys.* **13**, 267–278.

Evans, R. D., and Raitt, R. W. (1935). The radioactivity of the earth's crust and its influence on cosmic ray electroscope observations made near ground level. *Phys. Rev.* **48**, 171.

Evans, R. D., Keane, A. T., Kolenkow, R. J., Neal, W. R., and Shanaham, M. M. (1969). Radiogenic tumors in the radium and mesothorium cases studied at M.I.T. *In* "Delayed Effects of Bone-Seeking Radionuclides" (C. W. Mays *et al.*, eds.). Univ. of Utah Press, Salt Lake City.

Failla, H. P. (1932). Radium protection. *Radiology* **19**, 12–21.

Faul, H. (1954). "Nuclear Geology." Wiley, New York.

Federal Radiation Council (1960). "Background Material for the Development or Radiation Protection Standards," Rep. No. 1. USGPO, Washington, D.C.

Feely, H. W. (1960). Strontium-90 content of the stratosphere. *Science* **131**, 645–649.

Feely, H. W., Biscaye, P. E., and Lagarmarsino, R. J. (1965). Atmospheric circulation rates from measurements of nuclear debris. *CACR Symp. Atmos. Chem., Circ. Aerosols, Visby, Swed.* Cited in United Nations Scientific Committee on the Effects of Atomic Radiation (1966).

Ferlic, K. P. (1983). "Fallout: Its Characteristics and Management," Rep. AFRI TR83-5. Armed Forces Radiobiol. Res. Inst., Bethesda, Maryland.

Fischhoff, B., Lichtenstein, S., Slovic, P., Derby, S. L., and Kenney, R. L. (1981). "Acceptable Risk." Cambridge Univ. Press, London and New York.

Fisenne, I. M. (1982). Table of radionuclides. In "Environmental Report, November 2, 1982," EML Rep. 412, pp. V-1–V-2. USDOE, New York.

Fisenne, I. M., and Keller, H. W. (1970). "Radium-226 in the Diet of Two U.S. Cities," Rep. HASL-224. USAEC, New York.

Fisenne, I. M., Cohen, N., Neton, J. W., and Perry, P. (1980). Fallout plutonium in human tissues from New York City. Radiat. Res. 83, 162–168.

Fisenne, I. M.. Keller, H. W., and Harlev, N. H. (1981). Worldwide measurement of ^{226}Ra in human bone: Estimate of skeletal α dose. Health Phys. 40, 163–171.

Fisher, H. B., List, E. J.. Koh, R. C. Y., Imberger, I., and Brooks, N. H. (1979). "Mixing in Inland and Coastal Waters." Academic Press, New York.

Flynn, C. B. (1981). Local public opinion. In "The Three Mile Island Nuclear Accident: Lessons and Implications" (T. H. Moss and D. L. Sills, eds.), Ann. N.Y. Acad. Sci. 365, 146–158.

Folsom, T. R., and Vine, A. C. (1957). "Tagged Water Masses for Studying the Oceans," Rep. 551. Natl. Acad. Sci., Washington, D.C.

Food and Agriculture Organization (1960). "Radioactive Materials in Food and Agriculture." FAO, United Nations, Rome.

Foster, G. R., and Hakonson, T. E. (1986). Erosional losses of fallout plutonium. In "Environmental Research for Actinide Elements" (J. E. Pinder, III, ed.). NTIS, Springfield, Virginia. In press.

Foster, R. F. (1959). Distribution of reactor effluent in the Columbia River. In "Industrial Radioactive Waste Disposal," Hearings before Joint Committee on Atomic Energy.

Frederickson, L., et al. (1958). Studies of soil–plant–animal interrelationship with respect to fission products. Proc. U.N. Int. Conf. Peaceful Uses At. Energy, 2nd, Geneva p. 177.

Freiling, E. C., and Kay, M. A. (1965). "Radionuclide Fractionation in Air Burst Debris," Rep. USNRDL-TR-933. U.S. Nav. Radiol. Def. Lab., Washington, D.C.

Fresco, J., Jetter, E., and Harley, J. (1952). Radiometric properties of the thorium series. Nucleonics 10, 60.

Freudenthal, P. C. (1970a). "Aerosol Scavenging by Ocean Spray," Rep. HASL-232. USAEC, New York.

Freudenthal, P. C. (1970b). Strontium 90 concentrations in surface air: North America versus Atlantic Ocean from 1966 to 1969. J. Geophys. Res. 75, 4089–4096.

Fry, R. J. M., Powers-Risius, P., Alpen, E. L., and Ainsworth, E. J. (1985). High-LET carcinogenesis. Radiat. Res. 104, S180–S195.

Gahr, W. N. (1959). Uranium mill wastes. In "Industrial Radioactive Waste Disposal," Hearings before Joint Congressional Committee on Atomic Energy.

Gallaghar, R. G., and Saenger, E. L. (1957). Radium capsules and their associated hazards. Am. J. Roentgenol. Radium Ther. Nucl. Med. 77, 511–523.

Garner, R. J. (1960). An assessment of the quantities of fission products likely to be found in milk in the event of aerial contamination of agricultural land. Nature (London) 186, 1063.

Garten, C. T. (1978). A review of parameter values used to assess the transport of Pu, U and Th, in terrestrial food chains. Environ. Res. 17, 437–452.

Garten, C. T. (1980). Statistical uncertainties in predicting plutonium dose to lung and bone from contaminated soils. Health Phys. 39, 99–102.

General Accounting Office of the U.S. (1981). "Problems in Assessing the Cancer Risks of Low-Level Ionizing Radiation." Washington, D.C.

General Electric Company (1985). "Final Safety Analysis Report for the Galileo Mission and the Ulysses Mission." Rep. GESP 7200, Valley Forge, Pennsylvania.

German Federal Minister of Research and Technology (1979). "The German Risk Study: Summary." Ges. Reaktorsicherheit (GRS) (React. Saf. Co.), Cologne.

Gerusky, T. M. (1981). Three Mile Island: Assessment of radiation exposures and environmental contamination. *In* "The Three Mile Island Nuclear Accident: Lessons and Implications" (T. H. Moss and D. L. Sills, eds.), *Ann. N.Y. Acad. Sci.* **365**, 54–62.

Gesell, T. F. (1975). Some radiological health aspects of radon-222 in liquified petroleum gas. *In* "Noble Gases" (R. E. Stanley and A. A. Moghissi, eds.), Energy Research and Development Administration, TIC Rep. CONF-730915, pp. 612–629. Washington, D.C.

Gesell, T. F. (1983). Background atmospheric ^{222}Rn concentrations outdoors and indoors: A review. *Health Phys.* **45**, 289–302.

Gesell, T. F., and Prichard, H. M. (1975). The technologically enhanced natural radiation environment. *Health Phys.* **28**, 361–366.

Gesell, T. F., Johnson, R. H., and Bernhardt, D. E. (1975). "Assessment of Potential Radiological Population Health Effects from Radon in Liquified Petroleum Gas," Rep. EPA-520 1-75-002. USEPA, Washington, D.C.

Giardina, P. A., *et al.* (1977). "Summary Report on the Low-Level Radioactive Burial Site, West Valley, New York (1963–1975)." USEPA, Washington, D.C.

Gifford, F. A. (1968). An outline of theories of diffusion in the lower layers of the atmosphere. *In* "Meteorology and Atomic Energy—1968" (D. H. Slade, ed.), Rep. TID-24190, pp. 65–116. USAEC, Washington, D.C.

Gifford, F. A. (1974). Power reactor siting: A summary of U.S. practice. *In* "Human and Ecologic Effects of Nuclear Power Plants" (L. A. Sagan, ed.), pp. 46–71. Thomas, Springfield, Illinois.

Gilinsky, V. (1978–1979). Plutonium, proliferation, and the price of reprocessing. *Foreign Affairs* **57**, 374–386.

Glasstone, S. (1955). "Principles of Nuclear Reactor Engineering." Van Nostrand-Reinhold, New York.

Glasstone, S. (1962). "The effects of Nuclear Weapons." USAEC, Washington, D.C.

Glasstone, S., and Dolan, P. J. (1977). "The Effects of Nuclear Weapons." U.S. Dep. Def., Washington, D.C.

Glasstone, S., and Sesonske, A. (1980). "Nuclear Reactor Engineering," 2nd Ed. Van Nostrand-Reinhold, Princeton, New Jersey.

Gold, S., Barkhau, H. W., Shleien, B., and Kahn, B. (1964). Measurement of naturally occurring radionuclides in air. *In* "The Natural Radiation Environment I" (J. A. S. Adams and W. M. Lowder, eds.), pp. 369–382. Univ. of Chicago Press, Chicago, Illinois.

Goldman, M., and Yaniv, S. S. (1978). Naturally occurring radioactivity in ophthalmic glass. *In* "Radioactivity in Consumer Products" (A. A. Moghissi, P. Paras, M. W. Carter, and R. F. Barker, eds.), Rep. NUREG/CP-0001, pp. 227–240. USNRC, Washington, D.C.

Goldsmith, W. A. (1976). Radiological aspects of inactive uranium-milling sites: An overview. *Nucl. Saf.* **17**, 722–732.

Goodjohn, A. J., and Fortescue, P. (1971). Environmental aspects of high temperature gas-cooled reactors. *Proc. Am. Power Conf.*

Graham, J. D., and Vaupel, J. W. (1981). Value of a life: What difference does it make? *Risk Anal.* **1**, 89–95.

Groves, L. R. (1962). "Now It Can Be Told: The Story of the Manhattan Project." Harper, New York.

Grubbe, E. H. (1933). Priority in the therapeutic use of x-rays. *Radiology* **21,** 156.

Guimond, R. J. (1978). The radiological aspects of fertilizer utilization in radioactivity in consumers' products. *In* "Radioactivity in Consumer Products" (A. A. Moghissi, P. Paras, M. W. Carter, and R. F. Barker, eds), USNRC Rep. NUREG/CP0003, pp. 381–393. NTIS, Springfield, Virginia.

Guimond, R. J., and Windham, S. T. (1980). Radiological evaluation of structures constructed on phosphate-related land. *In* "Natural Radiation Environment III" (T. F. Gesell and W. M. Lowder, eds.), CONF-78042, Vol. 2, pp. 1457–1475. USDOE, Washington, D.C.

Gustafson, P. F., Nelson, D. M., Brar, S. S., and Muniak, S. E. (1970). "Recent Trends in Radioactive Fallout," Rep. ANL-7760, Part III, p. 246. Argonne Natl. Lab., Argonne, Illinois.

Gustafson, P. J. (1969). Cesium-137 in freshwater fish during 1954–1965. *Proc. Natl. Symp. Radioecol., 2nd* CONF-670503, pp. 249–257. USAEC, Washington, D.C.

Gwaltney, R. C. (1969). Missile generation and protection in light-water-cooled reactors. *Nucl. Saf.* **10,** 300–307.

Hairr, L. (1974). An investigation of mechanisms of radiocesium cycling in estuarine sediments. Ph.D. Thesis, New York Univ., New York.

Halitsky, J. (1968). Gas diffusion near buildings. *In* "Meteorology and Atomic Energy" (D. H. Slade, ed.), Rep. TID-24190, pp. 221–255. USAEC, Washington, D.C.

Hallden, N. A., Fisenne, I. M., Ong, L. D. Y., and Harley, J. H. (1961). "Radioactive Decay of Weapons Debris," Rep. HASL-117. USAEC, New York.

Hamilton, E. I. (1971). Relative radioactivity in building materials. *Am. Ind. Hyg. Assoc. J.* **32,** 398–403.

Hansen, W. G., *et al.* (1964). "Farming Practices and Concentrations of Fission Products in Milk," Publ. 999-R-6. U.S. Public Health Serv., Washington, D.C.

Hanson, W. C., ed. (1980). "Transuranic Elements in the Environment," Rep. DOE/TIC-22800. Tech. Inf. Cent./U.S. Dep. Energy, Washington, D.C.

Hardy, E. P. (1974a). "Depth Distributions of Global Fallout ^{90}Sr, ^{137}Cs and ^{239,240}Pu in Sandy Loam Soal," Fallout Program Quarterly Summary Rep. HASL-286, pp. I-2–I-10. USAEC, New York.

Hardy, E. P. (1974b). Worldwide distribution of plutonium. *In* "Plutonium and Other Transuranic Elements," Rep. WASH 1359, pp. 115–128. USAEC, Washington, D.C.

Hardy, E., and Alexander, L. T. (1962). Rainfall and deposition of strontium-90 in Challam County, Washington. *Science* **136,** 881–882.

Hardy, E. P., Krey, P. W., and Volchok, H. L. (1973). Global inventory and distribution of fallout plutonium. *Nature (London)* **241,** 444–445.

Harley, J. H. (1952). A study of the airborne daughter products of radon and thoron. Ph.D. Thesis, Rensselaer Polytech. Inst., Troy, New York.

Harley, J. H. (1969). Radionuclides in food. *In* "Biological Implications of the Nuclear Age," AEC Symp. Ser. No. 16. USAEC, Oak Ridge, Tennessee.

Harley, J. H. (1978). Radioactivity in building materials. *In* "Radioactivity in Consumer Products" (A. A. Moghissi, P. Paras, M. W. Carter, and R. F. Barker, eds.), Rep. NUREG/CP-0001, pp. 332–343. USNRC, Washington, D.C.

Harley, N. H. (1980). Comments on the proposed ICRP lung model as applied to occupational limits for short-lived radon daughters: A comparison with epidemiologic and dosimetry models. *Berlin Colloq.*

Harley, N. H. (1984). Comparing radon daughter dose: Environmental vs. underground exposure. *Radiat. Prot. Dosim.* **7,** 371–375.

Harley, N. H., and Cohen, B. (1980). Polonium-210 in tobacco. *In* "Radioactivity in Consumer Products" (A. A. Moghissi, P. Paras, M. W. Carter, and R. F. Barker, eds), Rep. NUREG/-CP0001, pp. 199–216. USNRC. Washington, D.C.

Harley, N. H., and Pasternack, B. S. (1982). Environmental radon daughter alpha dose factors in a five-lobed human lung. *Health Phys.* **42,** 789–799.

Harper, W. R. (1961). "Basic Principles of Fission Reactors." Wiley (Interscience), New York.

Harris, K. (1986). Personal Communication. Pittway Corp., Northbrook, Illinois.

Hartung, F. H., and Hesse, W. (1879). Die Lungenkrebs, die Bergkrankheit, in den Schneeberger Gruben. *Vierteljahresschr. Gerichtl. Med. Oeff. Gesundheitwess.* **30,** 296.

Harvey, R. S. (1970). Temperature effects on the sorption of radionuclides by freshwater algae. *Health Phys.* **19,** 293–297.

Hashizume, T., Maruyama, T., Kumamoto, Y., Kato, Y., and Kawamura, S. (1969). Estimation of gamma-ray dose from neutron-induced radioactivity in Hiroshima and Nagasaki. *Health Phys.* **17,** 761–771.

Hawkins, M. (1961). "Procedures for the Assessment and Control of the Shorter Term Hazards of Nuclear Warfare Fallout in Water Supply Systems." Res. Inst. Eng., Univ. of California, Berkeley.

Hawley, C. A., Jr., Sill, C. W., Voelz, G. L., and Islitzer, N. F. (1964). "Controlled Environmental Radioiodine Tests National Reactor Testing Station," IDO-12035. USAEC, Idaho Oper. Off., Idaho Falls, Idaho.

Healy, J. W. (1981). "Statement before Procurement and Military Systems Subcommittee," Committee on Armed Services, House of Representatives, June 24–25, 1981. USGPO, Washington, D.C.

Healy, J. W. (1984). Radioactive cloud dose calculations. *In* "Atmospheric Science and Power Production" (D. Randerson, ed.), Rep. DOE/TIC-27601, pp. 685–745. USDOE, Washington, D.C.

Heinemann, K., and Vogt, K. J. (1979). Messungen zur ablagerung and biologischen Halbwertszeit von jod auf Vegetation. *Proc. Annu. Symp. Ger.–Swiss Fachverb. Strahlenschultz, 12th, Norderney, FRG 1978.*

Hemplemann, L. H., Hall, W. J., Philips, M., *et al.* (1975). Neoplasms in persons treated with X rays in infancy. *J. Natl. Cancer Inst.* **55,** 519–530.

Henshaw, P. S. (1958). Whole-body irradiation syndrome. *In* "Radiation Biology and Medicine" (W. D. Claus, ed.), pp. 317–340. Addison-Wesley, Reading, Massachusetts.

Hess, C. T., Casparius, R. E., Norton, S. A., and Brutsaert, W. F. (1980). Investigations of natural levels of radon-222 in groundwater in Maine for assessment of related health effects. *In* "The Natural Radiation Environment III" (T. F. Gesell and W. M. Lowder, eds.), USDOE CONF-708422, Vol. 1, pp. 529–546. NTIS, Springfield, Virginia.

Hess, C. T., Norton, S. A., Brutsaert, W. F., Lowry, J. F., Weiffenbach, C. V., Casparius, R. E., Coombs, E. G., and Brandow, J. E. (1981). "Investigation of ^{222}Rn, ^{226}Ra, and U in Air and Groundwaters of Maine," Rep. B-017-Me. Land Water Resour. Cent., Univ. of Maine, Orono.

Hess, C. T., Weiffenbach, C. V., and Norton, S. A. (1983). Environmental radon and cancer correlations in Maine. *Health Phys.* **45,** 339–348.

Hess, C. T., Michel, J., Horton, T. R., Prichard, H. M., and Coniglio, W. A. (1985). The occurrence of radioactivity in public water supplies in the United States. *Health Phys.* **48,** 553–586.

Hewlett, R. C., and Anderson, O. E. (1962). "The New World," A History of the Atomic Energy Commission, Vol. 1. Pennsylvania State Univ. Press, University Park.

Hill, C. R. (1966). Polonium-210 content of human tissues in relation to dietary habit. *Science* **152,** 1261–1262.

Hillel, D. (1971). "Soil and Water: Physical Principles and Processes." Academic Press, New York.

Hilsmeier, W. F., and Gifford, F. A., Jr. (1962). "Graphs for Estimating Atmospheric Dispersion," Rep. ORO-545. USAEC, Washington, D.C.

Hodges, P. C. (1964). "The Life and Times of Emil H. Grubbe." Univ. of Chicago Press, Chicago, Illinois.

Hoffman, D. A. (1984). Late effects of I-131 therapy in the United States. *In* "Radiation Carcinogenesis: Epidemiology and Biological Significance" (J. D. Boice and J. F. Fraumeni, eds.), pp. 273–280. Raven, New York.

Hoffman, F. O., and Baes, C. F. (1979). "A Statistical Analysis of Selected Parameters for Predicting Food Chain Transport and Internal Dose of Radionuclides," Rep. NUREG/-CR-1004. Oak Ridge Nat. Lab., Oak Ridge, Tennessee.

Hoffman, F. O., Garten, C. T., Lucas, D. M., and Huckabee, J. W. (1982). Environmental behavior of technetium in soil and vegetation: Implications for radiological assessments. *Environ. Sci. Technol.* **16,** 214–217.

Holaday, D. A. (1959). The nature of wastes produced in the mining and milling of ores. *In* "Industrial Radioactive Waste Disposal," Hearings before Joint Committee on Atomic Energy. USGPO, Washington, D.C.

Holaday, D. A. (1969). History of the exposure of miners to radon. *Health Phys.* **16,** 547–552.

Holaday, D. A., Rushing, D. E., Coleman, R. D., Woolrich, P. F., Kusnetz, H. L., and Bale, W. F. (1957). "Control of Radon and Daughters in Uranium Mines and Calculations on Biologic Effects," Public Health Serv. Publ. No. 494. USGPO, Washington, D.C.

Holaday, D. A., Archer, V. E., and Lundin, F. (1968). A summary of United States exposure experiences in the uranium mining industry. *Diagn. Treat. Deposited Radionuclides, Proc. Symp., Richland, Wash., 1967* pp. 451–456.

Holcomb, W. F. (1980). Inventory (1962–1978) and projections (to 2000) of shallow land burial of radioactive wastes at commercial sites: An update. *Nucl. Saf.* **21,** 380–388.

Holcomb, W. F. (1982). A history of ocean disposal of low level radioactive waste. *Nucl. Saf.* **23,** 183–197.

Holland, J. Z. (1953). "A Meteorological Survey of the Oak Ridge Area," Rep. ORO-99. USAEC, Washington, D.C.

Holland, J. Z. (1959a). Stratospheric radioactivity data obtained by balloon sampling. *In* "Fallout from Nuclear Weapons Tests," Hearings before Congressional Joint Committee on Atomic Energy. USGPO, Washington, D.C.

Holland, J. Z. (1959b). "Summary of New Data on Atmospheric Fallout," Rep. TID-5554. USAEC, Washington, D.C.

Holland, J. Z. (1963). Physical origin and dispersion of radioiodine. *Health Phys.* **9,** 1095–1103.

Holleman, D. F., Luick, J. R., and Whicker, F. W. (1971). Tranfer of radiocesium from lichen to reindeer. *Health Phys.* **21,** 657–666.

Holm, W. M. (1978). Radium in consumer products: An historical perspective. *In* "Radioactivity in Consumer Products" (A. A. Moghissi, P. Paras, M. W. Carter, and R. F. Barker, eds.). Rep. NUREG/CP-0001, pp. 118–121. USNRC, Washington, D.C.

Holtzman, R. B. (1964). Lead-210 (RaD) and polonium-210 (RaF) in potable waters in Illinois. *In* "The Natural Radiation Environment I" (J. A. S. Adams and W. M. Lowder, eds.), pp. 227–237. Univ. of Chicago Press, Chicago, Illinois.

Holtzman, R. B. (1980). Normal dietary levels of radium-226, radium-228, lead-210, and polonium-210 for men. *In* "The Natural Radiation Environment III" (T. F. Gesell and W. M. Lowder, eds.), USAEC CONF-780422, Vol. 2, pp. 755–782. NTIS, Springfield, Virginia.

Holtzman, R. B., and Ilcewicz, F. H. (1966). Lead-210 and polonium-210 in tissues of cigarette smokers. *Science* **153,** 1259–1260.

Horan, J. R., and Gammill, W. P. (1963). The health physics aspects of the SL-1 accident. *Health Phys.* **9,** 177–186.

Hosker, R. P. (1982). "Methods of Estimating Wake Flow and Effluent dispersion Near Simple Blocklike Buildings," USNRC Rep. NUREG/CR-252. NTIS, Springfield, Virginia.

Howells, H. (1966). Discharges of low-activity, radioactive effluent from the Windscale works into the Irish Sea. *Disposal Radioact. Wastes Seas, Oceans Surf. Waters, Proc. Symp., Vienna* p. 769.

Hubner, K. F., and Fry, S. A., eds. (1980). "The Medical Basis for Radiation Accident Preparedness." Elsevier/North-Holland, New York.

Hueper, W. C. (1942). "Occupational Tumors and Allied Diseases." Thomas, Springfield, Illinois.

Hultqvist, B. (1956). Studies on naturally occurring ionizing radiation. *K Sven. Vetenskapsakad. Handl., Suppl.*

Hunkin, G. G. (1980). Solution mining economics. *In* "Uranium Resources Technology," pp. 153–172, Seminar III. Colorado Sch. Mines, Boulder.

Hunter, H. F., and Ballou, N. E. (1951). Fission product decay rates. *Nucleonics* **9**, C2.

International Air Transport Association (1982). "Restricted Articles Regulations." IATA, Montreal.

International Atomic Energy Agency (1973). "Regulations for the Safe Transportation of Radioactive Materials." IAEA, Vienna.

International Atomic Energy Agency (1975). "The Oklo Phenomenon." Rep. STI/PUB/405. IAEA, Vienna.

International Atomic Energy Association (1985). Press release. Mar. 6.

International Civil Aviation Organization (1983). "Technical Constructions for the Safe Transport of Dangerous Goods by Air." INTEREG, Chicago, Illinois.

International Commission on Radiation Units and Measurements (1971). "Radiation Quantities and Units," Rep. No. 19. ICRU Publ., Washington, D.C.

International Commission on Radiological Protection (1960). Report of Committee II on permissible dose for internal radiation (with bibliography for biological, mathematical and physical data). *Health Phys.* **3**, Spec. Vol., 380 pp.

International Commission on Radiological Protection (Task Group on Lung Dynamics) (1966a). Deposition and retention models for internal dosimetry of the human respiratory tract. *Health Phys.* **12**, 173–207.

International Commission on Radiological Protection (1966b). "Recommendations of the International Commission on Radiological Protection (Adopted September 17, 1965)," ICRP Publ. No. 9. Pergamon, Oxford.

International Commission on Radiological Protection (1973a). "Alkaline Earth Metabolism in Adult Man," ICRP Publ. No. 20. Pergamon, Oxford.

International Commission on Radiological Protection (1973b). "Implications of Commission Recommendations That Doses Be Kept as Low as Readily Achievable," ICRP Publ. No. 22. Pergamon, Oxford.

International Commission on Radiological Protection (1975). "Reference Man: Anatomical, Physiological and Metabolic Characteristics," ICRP Publ. No. 23. Pergamon, Oxford.

International Commission on Radiological Protection (1977). "Recommendations of the ICRP," ICRP Publ. No. 26. *Ann. ICRP* **1**(3).

International Commission on Radiological Protection (1979). "Limits for Intakes of Radionuclides by Workers," ICRP Publ. No. 30, Part 1. *Ann. ICRP* **2**(3/4).

International Commission on Radiological Protection (1980a). "Biological Effects of Inhaled Radionuclides," ICRP Publ. No. 31. *Ann. ICRP* **4**(1/2).

International Commission on Radiological Protection (1980b). "Limits for Intakes of Radionuclides by Workers," ICRP Publ. No. 30, Part 2. *Ann. ICRP* **4**(3/4).

International Commission on Radiological Protection (1981). "Limits of Inhalation of Radon Daughters by Workers," ICRP Publ. No. 32. *Ann. ICRP* **6**(1).

International Commission on Radiological Protection (1983). "Cost–Benefit Analysis in the Optimization of Radiation Protection," ICRP Publ. No. 37. *Ann. ICRP* **10**(2/3).

International Maritime Organization. "International Maritime Dangerous Goods Code." The most recent code should be consulted.

Ishihara, T., and Kumatori, T. (1983). Cytogenetic follow-up studies in Japanese fisherman exposed to fallout radiation. *In* "Radiation-Induced Chromosome Damage in Man," pp. 475–490. Alan R. Liss, New York.

Ishimaru, T., Ichimaru, M., Mikami, M., Yamada, Y., and Tomonaga, Y. (1982). "Distribution of Onset of Leukemia Among Atomic Bomb Survivors in the Leukemia Registry by Dose, Hiroshima and Nagasaki, 1946–75," Tech. Rep. RERF TR 12-81. Radiat. Eff. Res. Found.

Jablon, S., and Kato, H. (1970). Childhood cancer in relation to prenatal exposure to atomic-bomb radiation. *Lancet*, 1000–1003.

Jacobi, W. (1964). The dose to the human respiratory tract by inhalation of short-lived ^{222}Rn- and ^{220}Rn-decay products. *Health Phys.* **10**, 1163–1175.

Jacobs, D. G., Szluha, A. T., Gablin, K. A., and Croff, A. G. (1985). "A Brief Historical Perspective on the Definition of High-Level Nuclear Wastes," Rep. ORNL/Sub/84-13833/1. Oak Ridge Natl. Lab., Oak Ridge, Tennessee.

Jaffe, L. (1981). Technical aspects and chronology of the Three Mile Island accident. *In* "The Three Mile Island Nuclear Accident: Lessons and Implications" (T. H. Moss and D. L. Sills, eds.), *Ann. N.Y. Acad. Sci.* **365**, 37–47.

Japan Society for the Promotion of Science (1956). "Research in the Effects and Influences of the Nuclear Bomb Test Explosions." Jpn. Soc. Promotion Sci., Tokyo.

Jaworowski, Z. (1967). "Stable and Radioactive Lead in Environment and Human Body." Nucl. Energy Inf. Cent., Warsaw.

Jaworowski, Z., Bilkiewicz, J., Kownacka, L., and Wlodek, S. (1975). Artificial sources of natural radionuclides in environment. *In* "Natural Radiation Environment II" (J. A. S. Adams, W. M. Lowder, and T. F. Gesell, eds.), CONF-71, p. 809. USDOE, Washington, D.C.

Jenne, E. A. (1968). Controls on Mn, Fe, Co, Ni, Cu, and Zn concentrations in soils and water: The significant role of hydrous Mn and Fe oxides. *Adv. Chem. Ser.* No. 73, 337–389.

Jinks, S. M. (1975). An investigation of the factors influencing radiocesium concentrations of fish inhabiting natural aquatic ecosystems. Ph.D. Thesis, New York Univ., New York.

Jinks, S. M., and Wrenn, M. E. (1976). Radiocesium transport in the Hudson River estuary. *In* "Environmental Toxicity of Aquatic Radionuclides: Models and Mechanisms" (H. W. Miller and J. N. Stannard, eds.). Ann Arbor Sci. Publ., Ann Arbor, Michigan.

Jirka, G. H., Findikakis, A. N., Onishi, Y., and Ryan, P. J. (1983). Transport of radionuclides in surface waters. *In* "Radiological Assessment" (J. E. Till and H. R. Meyer, eds.), Rep. NUREG/CR-3332, Chap. 3. USNRC, Washington, D.C.

Johnson, J. E. (1978). Smoke detectors containing radioactive materials. *In* "Radioactivity in Consumer Products" (A. A. Moghissi, P. Paras, M. W. Carter, and R. F. Barker, eds.), Rep. NUREG/CP-0001, pp. 434–440. USNRC, Washington, D.C.

Johnson, R. H., Jr., Bernhardt, D. E., Nelson, N. S., and Calley, H. W., Jr. (1973). "Assessment of Potential Radiological Health Effects from radon in Natural Gas," Rep. EPA-520/1-73-004. USEPA, Washington, D.C.

Johnson, W. B., and Bailey, P. G. (1983). Study of Radon daughter concentrations in Polk and Hillsborough Counties. *Health Phys.* **45**, 432–434.

Jose, D. E. (1985). U.S. court practice concerning compensation for alleged radiation injuries. *Edison Elec. Inst.–Health Phys. Conf.*

Kaplan, S. I. (1971). HTGR safety. *Nucl. Saf.* **12**, 438–447.

Karas, J. S., and Stanbury, J. B. (1965). Fatal radiation syndrome from an accidental nuclear excursion. *N. Engl. J. Med.* **272**, 755–761.

Kauranen, P., and Miettinen, J. K. (1969). [210]Po and ^{210}Pb in the arctic food chain and the natural radiation exposure of Lapps. *Health Phys.* **16,** 287–296.

Kemeny, J. G. (1979). "President's Commission on the Accident at Three Mile Island," Report Task Force on Public Health and Safety. USGPO, Washington, D.C.

King, L. J., and McCarley, W. T. (1961). "Plutonium Release Incident of November 20," Rep. ORNL-2989. Oak Ridge Natl. Lab., Oak Ridge, Tennessee.

King, P. T., Michel, J., and Moore, W. S. (1982). Groundwater geochemistry of Ra-228, Ra-226 and Rn-222. *Geochim. Cosmochim. Acta* **46,** 1173–1182.

Kinne, O., ed. (1970). "Marine Ecology. Vol. I: Environmental Factors." Wiley (Interscience), New York.

Kirchmann, R., Darcheville, M., and Koch, G. (1980). Accumulation of radium-226 from phosphate fertilizers in cultivated soils and transfer to crops. *In* "Natural Radiation Environment III" (T. F. Gesell and W. M. Lowder, eds.), CONF-78042, Vol. 2, pp. 1667–1672. USDOE, Washington, D.C.

Kleinman, M. T. (1971). "The Stratospheric Inventory of Pu-238," Rep. HASL-245. USAEC, New York.

Klement, A. W. (1959). "A Review of Potential Radionuclides Produced in Weapons Detonations," Rep. WASH-1024. USAEC, Washington, D.C.

Klement, A. W. (1965). Radioactive fallout phenomena and mechanisms. *Health Phys.* **11,** 1265–1274.

Klevin, P. B., Weinstein, M. S., and Harris, W. B. (1956). Groundlevel contamination from stack effluents. *Am. Ind. Hyg. Assoc., Q.* **17,** 189–192.

Klingsberg, C., and Duguid, J. (1980). "Status of Technology for Isolating High-Level Radioactive Wastes in Geologic Repositories," DOE/TIC 11207 (Draft). USDOE, Washington, D.C.

Klusek, C. S. (1984a). "Sr-90 in the U.S. Diet, 1982," EML Rep. 429. USDOE, New York.

Klusek, C. S. (1984b). "Sr-90 in Human Bone in the U.S., 1982," EML Rep. 435. USDOE, New York.

Knapp, A. H. (1961). "The Effect of Deposition Rate and Cumulative Soil Level on the Concentration of Strontium-90 in U.S. Milk and Food Supplies," Rep. TID-13945. USAEC, Washington, D.C.

Knapp, H. A. (1963). "Iodine-131 in Fresh Milk and Human Thyroids Following a Single Deposition of Nuclear Test Fallout," TID-19266. USAEC, Washington, D.C.

Kneale, G. W., and Stewart, A. M. (1976). Mantel-Haenszel analysis of Oxford data. I. Independent effects of several birth factors including fetal irradiation. *J. Natl. Cancer Inst.* **56,** 879–883.

Knief, R. A. (1981). "Nuclear Energy Technology." McGraw-Hill, New York.

Kobayashi, R., and Nagai, I. (1956). Cooperation by the United States in the radiochemical analyses. *In* "Research in the Effects and Influences of the Nuclear Bomb Test Explosions II," pp. 1435–1445. Jpn. Soc. Promotion Sci., Tokyo.

Kocher, D. C. (1983). External dosimetry. *In* "Radiological Assessment" (J. E. Till and H. R. Meyer, eds.), pp. 8.1–8.52. USNRC, Washington, D.C.

Kocher, D. C., Sjoreen, A. L., and Bard, C. S. (1983). "Uncertainties in Geological Disposal of High Level Wastes: Ground Water Transport of Radionuclides and Radiological Consequences," Rep. 5838. Oak Ridge Natl. Lab., Oak Ridge, Tennessee.

Kochupillai, N., Verma, I. C., Grewal, M. S., and Ramalingaswami, V. (1976). Down's syndrome and related abnormalities in an area of high background radiation in coastal Kerala. *Nature (London)* **262,** 60–61.

Koczy, F. F. (1960). The distribution of elements in the sea. *Disposal Radioact. Wastes, Proc. Sci. Conf., Monaco, 1959.*

Kohman, T. (1959). *In* "Radiation Hygiene Handbook" (H. Blatz, ed.), Sect. 6. McGraw-Hill, New York.

Kohman, T., and Saito, N. (1954). Radioactivity in geology and cosmology. *Annu. Rev. Nucl. Sci.* **4.**

Kolb, W. (1974). Influence of building materials on the radiation dose to the population. *Kernenerg. Offentlicht.* **4,** 18–20.

Kolb, W., and Schmier, H. (1978). Building material induced radiation exposure of the population. *In* "Radioactivity in Consumer Products" (A. A. Moghissi, P. Paras, M. W. Carter, and R. F. Barker, eds.), Rep. NUREG/CP-0001, pp. 344–349. USNRC, Washington, D.C.

Kouts, H., and Long, J. (1973). Tritium production in nuclear reactors. *In* "Tritium" (A. A. Moghissi and M. W. Carter, eds.), pp. 38–45. Messenger Graphics, Phoenix, Arizona.

Kraner, H. W., Schroeder, G. L., and Evans, R. D. (1964). Measurements of the effects of atmospheric variables on radon-222 flux and soil–gas concentrations. *In* "The Natural Radiation Environment" (J. A. S. Adams and W. L. Lowder, eds.), pp. 191–215. Univ. of Chicago Press, Chicago, Illinois.

Krauskopf, K. B. (1986). Thorium as an analog for plutonium and rare-earth metals as analogs for heavier actinides. *Chem. Geol.* (in press).

Krey, P. W. (1967). Atmospheric burnup of a plutonium-238 generator. *Science* **158,** 769–771.

Krey, P. W., and Hardy, E. P. (1970). "Plutonium in Soil Around the Rocky Flats Plant," Rep. HASL-235. USAEC, New York.

Krey, P. W., and Krajewski, B. (1970). Comparison of atmospheric transport model calculations with observations of radioactive debris. *J. Geophys. Res.* **75,** 2901–2908.

Krey, P. W., Kleinman, M. T., and Krajewski, B. T. (1970). "Sr-90, Zr-95 and Pu-238 Stratospheric Inventories 1967–1969," Rep. HASL-227, pp. 39–69. USAEC, New York.

Krieger, H. L., and Burmann, F. J. (1969). Effective half-time of [85]Sr and [134]Cs for a contaminated pasture. *Health Phys.* **17,** 811–824.

Krishnaswami, S., Graunstein, W. C., Turekian, K. K., and Dowd, J. J. (1982). Radium, thorium, and radioactive lead isotopes in groundwater: Application to the *in situ* determination of adsorption–desorption rate constants and retardation factors. *J. Water Resour. Res.* **18,** 1633–1675.

Kumatori, T., Ishihara, T., Hirashima, K., Sugiyama, H., Ishii, S., and Miyoshi, K. (1980). Follow-up studies over a 25-year period on the Japanese fishermen exposed to radioactive fallout in 1954. *In* "The Medical Basis for Radiation Accident Preparedness" (K. F. Hubner and S. A. Fry, eds.), pp. 33–54. Elsevier/North-Holland, New York.

Kupferman, S. L., Livingston, H. D., and Bowen, V. T. (1979). A mass balance for ^{137}Cs and ^{90}Sr in the North Atlantic Ocean. *J. Mar. Res.* **37,** 157–159.

Kuroda, P. K., Hodges, H. L., Fry, L. M., and Moore, H. E. (1962). Stratospheric residence time of strontium-90. *Science* **137,** 15–17.

Lacy, W. J., and Stangler, M. J. (1962). The postnuclear attack water contamination problem. *Health Phys.* **8,** 423–427.

Lamarsh, J. R. (1966). "Nuclear Reactor Theory." Addison-Wesley, Reading, Massachusetts.

Lamarsh, J. R. (1975). "Introduction to Nuclear Engineering." Addison-Wesley, Reading, Massachusetts.

Lamont, L. (1965). "Day of Trinity." Atheneum, New York.

Land, C. E., McKay, F. W., and Machado, S. G. (1984). Childhood leukemia and fallout from the Nevada nuclear tests. *Science* **223,** 139–144.

Lappenbusch, W. L., and Cothern, C. R. (1985). Regulatory development of the interim and revised regulations for radioactivity in drinking water: Past and present issues and problems. *Health Phys.* **48,** 535–551.

Laurer, G. R., and Eisenbud, M. (1963). Low-level *in vivo* measurement of iodine-131 in humans. *Health Phys.* **9,** 401–405.

Lederer, C. M., and Shirley, V. S., eds. (1978). "Table of Isotopes," 7th Ed. Wiley, New York.

Lederer, C. M., Hollander, J. M., and Perlman, I. (1967). "Table of Isotopes," 6th Ed. Wiley, New York.

Lei, W., Linsalata, P., Penna Franca, E., and Eisenbud, M. (1986). Distribution and mobilization of cerium, lanthanum and neodymium in the Morro do Ferro basin, Brazil. *Chem. Geol.* (in press).

Leifer, R. Z., and Juzdan, Z. R. (1984). "Cosmos 1402," Environmental Measurements Laboratory Annual Report for 1984, p. 23. USDOE, New York.

Leifer, R. Z., Juzdan, Z. R., and Larsen, R. (1984). "The High Altitude Sampling Program: Radioactivity in the Stratosphere," EML Rep. 434. USDOE, New York.

Leipunsky, O. I. (1957). Radioactive hazards from clean hydrogen bomb and fission atomic bomb explosions. *At. Energ.* **3,** 530.

Lengemann, F. W. (1966). Predicting the total projected intake of radioiodine from milk by man. I. The situation where no counter measures are taken. *Health Phys.* **12,** 825–835.

Lentsch, J. W., Kneip, T. J., Wrenn, M. E., Howells, G. P., and Eisenbud, M. (1972). Stable manganese ad Mn-54 distributions in the physical and biological components of the Hudson River estuary. *Proc. Natl. Symp. Radioecol., 3rd* CONF-710501-P2, pp. 752–768. USAEC, Washington, D.C.

Lessard, E. T., Miltenberger, R. P., Cohn, S. H., Musolino, S. V., and Conard, R. A. (1984). Protracted exposure to fallout: The Rongelap and Utirik experience. *Health Phys.* **46,** 511–527.

Lewis, E. B. (1959). Statement. *In* "Fallout from Nuclear Weapons Tests," Hearings before Joint Committee on Atomic Energy, Special Subcommittee on Radiation. Vol. II, pp. 1552–1554. USGPO, Washington, D.C.

Lewis, H. W., Budnitz, R. J., Kouts, H. J. C., Loewenstein, W. B., Rowe, W. D., von Hippel, F., and Zachariasen, F. (1979). "Risk Assessment Review Group Report to the U.S. Nuclear Regulatory Commission," Ad Hoc Risk Assessment Review Group, Rep. NUREG/CR-0400. NTIS, Springfield, Virginia.

Lewis, W. L. (1955). "Arthritis and Radioactivity." Christopher Publ. House, Boston, Massachusetts.

Libby, W. F. (1952). "Radiocarbon Dating." Univ. of Chicago Press, Chicago, Illinois.

Libby, W. F. (1956). *Proc. Natl. Acad. Sci. U.S.A.* **42,** 365,945.

Libby, W. F. (1958). *Swiss Acad. Med., Lausanne.*

Libby, W. F. (1959). Remarks at University of Washington, Mar. 13, 1959. *In* "Fallout from Nuclear Weapons Tests," Hearings before Joint Committee on Atomic Energy, Special Subcommittee on Radiation, Vol. III, pp. 2227–2259. USGPO, Washington, D.C.

Linsalata, P. (1984). Sources, distribution and mobility of plutonium and radiocesium in soils, sediments and water of the Hudson River and watershed. Ph.D. Thesis, New York Univ., New York.

Linsalata, P., Hickman, D., and Cohen, N. (1985a). Dosimetry of radiocesium within the Hudson River estuary based on long-term environmental measurements. *Proc. Midyear Top. Symp. Health Phys. Soc.—Environ. Radiat., 18th, Colorado Springs, Colo.* pp. 353–542.

Linsalata, P., Simpson, H. J., Olsen, C. R., Cohen, N., and Trier, R. M. (1985b). Plutonium and radiocesium in the water column of the Hudson River estuary. *Environ. Geol. Water Sci.* **7,** 193–204.

Linsalata, P., Eisenbud, M., and Penna Franca, E. (1986a). "Studies of Transport Pathways of Th, U, Rare Earths, ^{228}Ra, and ^{226}Ra from Soil to Farm Animals," Third Progress Report to U.S. Department of Energy, Office of Health and Environmental Research (Agreement No. DE-AK-02-83ER60134). New York Univ. Med. Cent., New York.

Linsalata, P., Eisenbud, M., and Penna Franca, E. (1986b). Ingestion estimates of Th and the light rare earth elements based on measurements of human feces. *Health Phys.* **50,** 163–167.

Linsalata, P., Penna Franca, E., Campos, M. J., Lobão, N., Ballad, R., Lei, W., Ford, H., Morse, R. S., and Eisenbud, M. (1986c). Radium, thorium and the light rare earth elements in soils and vegetables grown in an area of high natural radioactivity. *In* "Environmental Research for Actinide Elements" (J. E. Pinder III, ed.). NTIS, Springfield, Virginia. In press.

Linsalata, P., Cohen, N., and Wrenn, M. E. (1986d). Sources, behavior and transport of ^{137}Cs in sediments and water of the Hudson River estuary. Unpublished manuscript. New York Univ. Med. Cent., New York.

Little, C. A. (1983). Development of computer codes for radiological assessments. *In* "Radiological Assessment" (J. E. Till and H. R. Meyer, eds.), Rep. NUREG/CR-3332, Chap. 13. USNRC, Washington, D.C.

Little, C. A., and Miller, C. W. (1979). "The Uncertainty Associated with Selected Environmental Transport Models," Rep. ORNL-5528. Oak Ridge Nat. Lab., Oak Ridge, Tennessee.

Little, J. B., Radford, E. P., McCombs, H. L., and Hunt, V. R. (1965). Distribution of polonium in pulmonary tissues of cigarette smokers. *N. Engl. J. Med.* **173,** 1343.

Lockhart, L. B. (1958). "Atmospheric Radioactivity studies at U.S. Naval Research Laboratory," Rep. 5249. Nav. Res. Lab., Washington, D.C.

Lockhart, L. B. (1964). Radioactivity of the radon-222 and radon-220 series in the air at ground level. *In* "The Natural Radiation Environment I" (J. A. S. Adams and W. M. Lowder, eds.), pp. 331–344. Univ. of Chicago Press, Chicago, Illinois.

Lodge, J. P., Bien, G. S., and Suess, H. E. (1960). The carbon-14 content of urban airborne particulate matter. *Int. J. Air Pollut.* **2,** 309–312.

Logsdon, J. E., and Chissler, R. I. (1970). "Radioactive Waste Discharges to the Environment from Nuclear Power Facilities," Rep. BRH/DER 70-2. Bur. Radiol. Health, Environ. Radiat. Div., Rockville, Maryland.

Look Magazine (1960). A sequel to atomic tragedy. *Look Mag.* Apr. 12.

Lorenz, E. (1944). Radioactivity and lung cancer; a critical review in miners of Schneeberg and Joachimstahl. *J. Natl. Cancer Inst.* **5,** 1–15.

Lough, S. A., Hamada, G. H., and Comar, C. L. (1960). Secretion of dietary strontium-90 and calcium in human milk. *Proc. Soc. Exp. Biol. Med.* **104,** 194–198.

Lowder, W. M., and Solon, L. R. (1956). "Background Radiation," Rep. NYO-4712. USAEC, Washington, D.C.

Lowman, F. G., Rice, T. R., and Richards, F. A. (1971). Accumulation and redistribution of radionuclides by marine organisms. *In* "Radioactivity in the Marine Environment," pp. 161–199. Natl. Acad. Sci., Washington, D.C.

Loysen, P. (1965). Particle size distribution of stratospheric aerosols at 110,000 ft. *In* "Radioactive Fallout from Nuclear Weapons Tests," AEC Symp. Ser. No. 5. USAEC, Washington, D.C.

Loysen, P., Breslin, A. J., and DiGiovanni, H. J. (1956). "Experimental Collection, Efficiency of a Stratospheric Air Sampler," Rep. NYO-4708. USAEC, Washington, D.C.

Lucas, H. F., Jr. (1982). Ra-226 and Ra-228 in drinking water. *Annu. Meet. Health Phys. Soc., 27th, Las Vegas, Nev.*

Luetzelschwab, J. W., and Googins, S. W. (1984). Radioactivity released from burning gas lantern mantles. *Health Phys.* **46,** 873–881.

Lushbaugh, C. A., Fry, S. A., Hubner, K. F., and Ricks, R. D. (1980). Total-body irradiation: A historical review and follow-up. *In* "The Medical Basis for Radiation Accident Preparedness" (K. F. Hubner and S. A. Fry, eds.), pp. 3–15. Elsevier/North-Holland, New York.

Lyon, J. L., Klauber, M. R., Gardner, J. W., and Udall, K. S. (1979). Childhood leukemias associated with fallout from nuclear testing. *N. Engl. J. Med.* **300,** 397–402.

Machta, L. (1959). "Hearings on Biological and Environmental Effects of Nuclear War." Joint Committee on Atomic Energy. USGPO, Washington, D.C.

Machta, L., and List, R. J. (1959). Analysis of stratospheric strontium-90 measurements. *In* "Fallout from Nuclear Weapons Tests" Hearings before Joint Committee on Atomic Energy, Special Subcommittee on Radiation, Vol. I, pp. 741–762. USGPO, Washington, D.C.

Machta, L., List, R. J., and Hubert, L. F. (1956). World-wide travel of atomic debris. *Science* **124,** 474–477.

MacMahon, B. (1962). Prenatal x-ray exposure and childhood cancer. *J. Natl. Cancer Inst.* **28,** 1173–1191.

Magno, P. J., Groulx, P. R., and Apidianakis, J. C. (1970). Lead-210 in air and total diets in U.S. *Health Phys.* **18,** 383–388.

Manning, R. (1985). The future of nuclear power. *Environment* **27**(4), 12–38.

Marley, W. G., and Fry, T. M. (1956). Radiological hazards from an escape of fission products and the implications in power reactor location. *Proc. Int. Conf. Peaceful Uses At. Energy, 1st, Geneva, 1955.*

Marsden, E., and Collins, M. A. (1963). Particle activity and free radicals from tobacco. *Nature (London)* **198,** 962.

Marshall, N. (1983). "Nuclear Power Technology," Vol. 1. Oxford Univ. Press (Clarendon), London and New York.

Martell, E. A. (1959). Atmospheric aspects of strontium-90 fallout. *Science* **129,** 1197–1206.

Martell, E. A. (1974). Radioactivity of tobacco trichomes and insoluble cigarette smoke particles. *Nature (London)* **249,** 215–217.

Martland, H. S. (1925). Some unrecognized dangers in the use and handling of radioactive substances. *Proc. N.Y. Pathol. Soc.* **25,** Nos. 6–8.

Martland, H. S. (1951). "Collection of Reprints on Radium Poisoning, 1925–1939." USAEC, Oak Ridge, Tennessee.

Mason, B. (1960). "Principles of Geochemistry." Wiley, New York.

Mason, B. (1982). "Principles of Geochemistry," 4th Ed. Wiley, New York.

Mauchline, J., and Templeton, W. L. (1964). Artificial and natural radioisotopes in the marine environment. *Annu. Rev. Oceanogr. Mar. Biol.* **2,** 229–279.

Mayneord, W. V., Radley, J. M., and Turner, R. C. (1958). The alpha-ray activity of humans and their environment. *Proc. U.N. Int. Conf. Peaceful Uses At. Energy, 2nd. Geneva.*

Mayneord, W. V., Turner, R. C., and Radley, J. M. (1960). Alpha activity of certain botanical materials. *Nature (London)* **187,** 208.

Mays, C. W. (1983). *In* Discussion. *Health Phys.* **44,** Suppl. 1, 194–195.

McDowell-Boyer, L. M., and O'Donnell, F. R. (1978). "Radiation Dose Estimates from Timepieces Containing Tritium or ^{147}Pm in Radioluminescent Paints," Rep. ORNL/-NUREG/TM-150. Oak Ridge Nat. Lab., Oak Ridge, Tennessee.

McDowell-Boyer, L. M., Watson, A. P., and Travis, C. C. (1980). A review of parameters describing food-chain transport of lead-210 and radium-226. *Nucl. Saf.* **21,** 486–495.

McEachern, P., Myers, W. G., and White, F. A. (1971). Uranium concentrations in surface air at rural and urban localities within New York State. *Environ. Sci. Technol.* **5,** 700–703.

McLeod, K. W., Adriano, D. C., Boni, A. L., Corey, J. C., Horton, J. H., Paine, D., and Pinder, J. E., III (1980). Influence of a nuclear fuel chemical separation facility on the plutonium contents of a wheat crop. *J. Environ. Qual.* **9,** 306–315.

Means, J. L., Crerar, D. A., Borcsik, M. P., and Duguid, J. O. (1978). Adsorption of Co and selected actinides by Mn and Fe oxides in soils and sediments. *Geochim. Cosmochim. Acta* **42,** 1763–1773.

Media Institute (1979). "Television Evening News Covers Nuclear Energy: A Ten-Year Perspective." Washington, D.C.

Medical Research Council (1956). "The Hazards to Man of Nuclear and Allied Radiations." HM Stationery Off., London.

Medvedev, Z. A. (1976). Two decades of dissidence. *New Sci.* **72,** 264.

Medvedev, Z. A. (1977). Facts behind the Soviet nuclear disaster. *New Sci.* **74,** 761–764.

Medvedev, Z. A. (1979). "Nuclear Disaster in the Urals." Norton, New York.

Menczer, L. F. (1965). Radioactive ceramic glazes. *Radiol. Health Data* **6,** 656.

Menzel, R. G. (1960). Radioisotopes in soils: Effects of amendments on availability. *In* "Radioisotopes in the Biosphere" (R. S. Caldecott and L. A. Snyder, eds.), pp. 37–46. Univ. of Minnesota Press, Minneapolis.

Menzel, R. G. (1964). Competitive uptake by plants of potassium, rubidium. cesium, calcium, strontium and barium from soils. *Soil Sci.* **77,** 419.

Merriam, G. R., and Focht, E. F. (1957). A clinical study of radiation cataracts and the relationship to dose. *Am. J. Roentgenol., Radium Ther. Nucl. Med.* **77,** 759.

Meyer, H. R., Till, J. E., Bonner, E. S., Bond, W. D., Morse, L. E., Tennery, V. S., and Yalcintas, M. G. (1979). Radiological impact of thorium mining and milling. *Nucl. Saf.* **20,** 319–330.

Miettinen, J. K. (1969). Enrichment of radioactivity by arctic ecosystems in Finnish Lapland. *Proc. Natl. Symp. Radioecol., 2nd* CONF-670503, pp. 23–31. USAEC, Washington, D.C.

Miller, C. W. (1984a). Atmospheric dispersion and deposition. *In* "Models and Parameters for Environmental Radiological Assessments" (C. W. Miller, ed.), DOE Critical Review Series, DE81027154 (DOE/TIC-11468), pp. 11–19. NTIS, Springfield, Virginia.

Miller, C. W., ed. (1984b). "Models and Parameters for Environmental Radiological Assessments," DOE Critical Review Series, DE81027154 (DOE/TIC-11468). NTIS, Springfield, Virginia.

Miller, C. W., and Hoffman, F. O. (1982). An analysis of reported values of the environmental half-time for radionuclides deposited on the surfaces of vegetation. *Environ. Migr. Long-Lived Radionuclides, Proc. Int. Symp., Knoxville, Tenn., 1981.*

Miller, E. C. (1966). "The Integrity of Reactor Pressure Vessels," Rep. No. 15. Nucl. Safety Inf., Oak Ridge Nat. Lab., Oak Ridge, Tennessee.

Minogue, R. B. (1978). NRC's role in regulating consumer products. *In* "Radioactivity in Consumer Products" (A. A. Moghissi, P. Paras, M. W. Carter, and R. F. Barker, eds.), Rep. NUREG/CP-0001, pp. 11–17. USNRC, Washington, D.C.

Mistry, K. B., Bharathan, K. G., and Gopal-Ayengar, A. R. (1970). Radioactivity in the diet of population of the Kerala coast including monazite bearing high radiation areas. *Health Phys.* **19,** 353–542.

Miyake, Y., and Saruhashi, K. (1960). Vertical and horizontal mixing rates of radioactive material in the ocean. *Disposal Radioact. Wastes, Proc. Sci. Conf., Monaco, 1959.*

Mochizuki, Y., Mowafy, R., and Pasternack, B. (1963). Weights of human thyroids in New York city. *Health Phys.* **9,** 1299–1301.

Moeller, D. W., and Fujimoto, K. (1984). Cost evaluation and control measures for indoor radon progeny. *Health Phys.* **46,** 1181–1193.

Moghissi, A. A., ed. (1983). Radioactive Waste Containing Transuranic Elements. (Special Issue.) *Nucl. Chem. Waste Manage.* **4**(1).

Moghissi, A. A., and Carter, M. W., eds. (1973). "Tritium." Messenger Graphics, Phoenix, Arizona.

Moghissi, A. A., and Lieberman, R. (1970). Tritium body burden of children, 1967–1968. *Radiol. Health Data Rep.* **11**, 227–231.

Moghissi, A. A., Godbee, H. W., Ozker, M. S., and Carter, M. W. (1978a). "Nuclear Power Waste Technology." Am. Soc. Mech. Eng., New York.

Moghissi, A. A., Paras, P., Carter, M. W. and Barker, R. F. (1978b). "Radioactivity in Consumer Products," Rep. NUREG/CP-0001. USNRC, Washington, D.C.

Mohammadi, H., and Mehdizadeh, S. (1983). Re-identification of ^{232}Th content and relative radioactivity measurements in a number of imported gas mantles. *Health Phys.* **44**, 649–653.

Monson, R. R., and MacMahon, B. (1984). Prenatal X-ray exposure and cancer in children. *In* "Radiation Carcinogenesis: Epidemiology and Biological Significance" (J. D. Boice and J. F. Fraumeni, eds.), pp. 97–105. Raven, New York.

Montgomery, D. M., Kolde, H. E., and Blanchard, R. L. (1977). "Radiological Measurements at the Maxey Flats Radioactive Waste Burial site—1974 to 1975," Rep. EPA-520/5-76/020. Off. Radiat. Programs, USEPA, Cincinnati, Ohio.

Morse, J. G. (1963). Energy for remote areas. *Science* **139**, 1175–1180.

Morse, R. S., and Welford, G. A. (1971). Dietary intake of ^{210}Pb. *Health Phys.* **21**, 53–55.

Muller, H. J. (1927). Artificial transmutation of the gene. *Science* **66**, 84–87.

Muller, R. M., Sprugel, D. G., and Kohn, B. (1978). Erosional transport and deposition of plutonium and cesium in two small midwestern watersheds. *J. Environ. Qual.* **7**, 171–174.

Musgrave, G. W. (1947). The quantitative evaluation of factors in water erosion—A first approximation. *J. Soil Water Conserv.* **2**, 133–138.

Muth, H., Schraub, A., Aurand, K., and Hantke, H. H. (1957). Measurements of normal radium burdens. *Br. J. Radiol., Suppl.* **7**.

Myrloi, M. G., and Wilson, J. G. (1951). On the proton component of the vertical cosmic-ray beam at sea level. *Proc. Phys. Soc., London, Sect. A* **64**, 404.

National Academy of Engineering (1972). "Engineering for the Resolution of the Energy-Environment Dilemma." Comm. Power Plant Siting, Natl. Acad. Sci., Washington, D.C.

National Academy of Sciences–National Research Council (1956). "Pathologic Effects of Atomic Radiation," Publ. 452. Washington, D.C.

National Academy of Sciences–National Research Council (1957a). "Disposal of Radioactive Wastes on Land," Publ. 519. Washington, D.C.

National Academy of Sciences–National Research Council (1957b). "The Effects of Atomic Radiation on Oceanography and Fisheries," Publ. 551. Washington, D.C.

National Academy of Sciences–National Research Council (1959). "Radioactive Waste Disposal into Atlantic and Gulf Coastal Waters," Publ. 655. Washington, D.C.

National Academy of Sciences–National Research Council (1961a). "Effects of Inhaled Radioactive Particles," Publ. 848. Washington, D.C.

National Academy of Sciences–National Research Council (1961b). "Long-Term Effects of Ionizing Radiations from External Sources," Publ. 849. Washington, D.C.

National Academy of Sciences–National Research Council (1962). "Disposal of Low-Level Radioactive Waste into Pacific Coastal Waters," Publ. 985. Washington. D.C.

National Academy of Sciences–National Research Council (1963). "Damage to Livestock from Radioactive Fallout in Event of Nuclear War," Publ. 1078. Washington, D.C.

National Academy of Sciences–National Research Council (1964). "Civil Defense," Project Harbor Summary Report, Publ. 1237. Washington, D.C.

National Academy of Sciences–National Research Council (1969). "Civil Defense," Little Harbor Report. A Report to the Atomic Energy Commission by a Committee of the National Academy of Sciences, Washington, D.C.

National Academy of Sciences–National Research Council (1970). "Disposal of Solid Radioactive Wastes in Bedded Salt Deposits." Washington, D.C.

National Academy of Sciences–National Research Council (1978). "Radioactive Wastes at the Hanford Reservation: A Technical Review." Natl. Acad. Press, Washington, D.C.

National Academy of Sciences–National Research Council (Committee on the Biological Effects of Ionizing Radiation) (1980). "The Effects on Populations of Exposure to Low Levels of Ionizing Radiation: 1980." Natl. Acad. Press, Washington, D.C.

National Academy of Sciences–National Research Council (1981). "Radioactive Waste Management at the Savannah River Plant: A Technical Review." Natl. Acad. Press, Washington, D.C.

National Academy of Sciences–National Research Council (1982). "Outlook for Science and Technology: The Next Five Years." Freeman, San Francisco, California.

National Academy of Sciences–National Research Council (Board on Radioactive Waste Management) (1983a). "A Study of the Isolation System for Geologic Disposal of Radioactive Wastes." Natl. Acad. Press, Washington, D.C.

National Academy of Sciences–National Research Council (1983b). "Changing Climate," Report of the Carbon Dioxide Assessment Committee. Washington, D.C.

National Academy of Sciences–National Research Council (1984). "Assigned Share for Radiation as a Cause of Cancer: Review of Radioepidemiologic Tables Assigning Probabilities of Causation," Final Report, Board of Radiation Effects Research, Committee on Radioepidemiologic Tables. Natl. Acad. Press, Washington, D.C.

National Academy of Sciences–National Research Council (1985a). "The Management of Radioactive Waste at the Oak Ridge National Laboratory: A Technical Review." Natl. Acad. Press, Washington, D.C.

National Academy of Sciences–National Research Council (1985b). "Review of the Methods Used to Assign Radiation Doses to Service Personnel at Nuclear Weapons Tests." Natl. Acad. Press, Washington, D.C.

National Advisory Committee on Oceans and Atmosphere (1984). "Nuclear Waste Management and the Use of the Sea," Special Report to the President and the Congress. Washington, D.C.

National Committee on Radiation Protection and Measurements (1941). Safe handling of radioactive luminous compounds (NCRP Rep. No. 5). *NBS Handb. (U.S.)* No. 27.

National Committee on Radiation Protection and Measurements (1954). Permissible dose from external sources of ionizing radiation (NCRP Rep. No. 17). *NBS Handb. (U.S.)* No. 59.

National Council on Radiation Protection and Measurements (1971). "Basic Radiation Criteria," NCRP Rep. No. 39. Bethesda, Maryland.

National Council on Radiation Protection and Measurements (1975a). "Review of the Current State of Radiation Protection Philosophy," NCRP Rep. No. 43. Bethesda, Maryland.

National Council on Radiation Protection and Measurements (1975b). "Krypton-85 in the Atmosphere—Accumulation, Biological Significance, and Control Technology," NCRP Rep. No. 44. Bethesda, Maryland.

National Council on Radiation Protection and Measurements (1975c). "Natural Background Radiation in the United States," NCRP Rep. No. 45. Bethesda, Maryland.

National Council on Radiation Protection and Measurements (1975d). "Alpha-Emitting Particles in Lungs," NCRP Rep. No. 46. Bethesda, Maryland.

National Council on Radiation Protection and Measurements (1977a). "Cesium-137 from the Environment to Man: Metabolism and Dose," NCRP Rep. No. 52. Bethesda, Maryland.

National Council on Radiation Protection and Measurements (1977b). "Review of NCRP Radiation Dose Limit for Embryo and Fetus in Occupationally-Exposed Women." NCRP Rep. No. 53. Bethesda, Maryland.

National Council on Radiation Protection and Measurements (1977c). "Radiation Exposure from Consumer Products and Miscellaneous Sources." NCRP Rep. No. 56. Bethesda, Maryland.

National Council on Radiation Protection and Measurements (1977d). "Protection of the Thyroid Gland in the Event of Releases of Radioiodine," NCRP Rep. No. 55. Bethesda, Maryland.

National Council on Radiation Protection and Measurements (1979). "Tritium in the Environment," NCRP Rep. No. 62. Bethesda, Maryland.

National Council on Radiation Protection and Measurements (1980a). "Influence of Dose and Its Distribution in Time on Dose-Response Relationships for Low-LET Radiations," NCRP Rep. No. 65. Bethesda, Maryland.

National Council on Radiation Protection and Measurements (1980b). "Krypton-85 in the Atmosphere—With Specific Reference to the Public Health Significance of the Proposed Controlled Release at Three Mile Island." Bethesda, Maryland.

National Council on Radiation Protection and Measurements (1982). "Preliminary Evaluation of Criteria for the Disposal of Transuranic Contaminated Waste," Report Task Group on Criteria for the Disposal of Transuranic (TRU) Contaminated Waste. Bethesda, Maryland.

National Council on Radiation Protection and Measurements (1983). "Iodine-129: Evaluation of Releases from Nuclear Power Generation," NCRP Rep. No. 75. Bethesda, Maryland.

National Council on Radiation Protection and Measurements (1984a). "Radiological Assessment: Predicting the Transport, Bioaccumulation, and Uptake by Man of Radionuclides Released to the Environment," NCRP Rep. No. 76. Bethesda, Maryland.

National Council on Radiation Protection and Measurements (1984b). "Exposures from the Uranium Series with Emphasis on Radon and its Daughters," NCRP Rep. 77. Bethesda, Maryland.

National Council on Radiation Protection and Measurements (1984c). "Evaluation of Occupational and Environmental Exposures to Radon and Radon Daughters in the United States," NCRP Rep. No. 78. Bethesda, Maryland.

National Council on Radiation Protection and Measurements (1985a). "Induction of Thyroid Cancer by Ionizing Radiation," NCRP Rep. No. 80. Bethesda, Maryland.

National Council on Radiation Protection and Measurements (1985b). "Carbon-14 in the Environment," NCRP Rep. No. 81. Bethesda, Maryland.

National Council on Radiation Protection and Measurements (1986). "Neptunium: Radiation Protection Considerations." In preparation.

National Oceanic and Atmospheric Administration (1979). "Assimilative Capacity of U.S. Coastal Waters for Pollutants" (E. D. Goldberg, ed.). U.S. Dep. Commer., Washington, D.C.

National Radiological Protection Board (1984). "Assessment of Radiation Exposure to Members of the Public in West Cumbria as a Result of the Discharges from BNFL, Sellafield" (G. S. Linsley, J. Dionian, J. R. Simmonds, and J. Burges), Rep. R170. NRPB, London.

National Safety Council (1984). "Accident Facts." Chicago, Illinois.

Nefzger, M. D., Miller, R. J., and Fujino, T. (1968). Eye findings in atomic bomb survivors of Hiroshima and Nagasaki: 1963–1964. *Am. J. Epidemiol.* **89**, 129.

Neill, R. H. (1978). The role of the Bureau of Radiological Health in controlling radioactivity in consumer products. *In* "Radioactivity in Consumer Products" (A. A. Moghissi. P. Paras, M. W. Carter, and R. F. Barker, eds.), Rep. NUREG/CP-0001, pp. 38–39. USNRC, Washington, D.C.

Nero, A. V., Jr. (1985). What we know about indoor radon. Testimony prepared for hearings on "Radon Contamination: Risk Assessment and Mitigation Research." held by the Subcommittee on Natural Resources, Agricultural Research and Environmental Committee on Science and Technology, U.S. House of Representatives (Oct. 10, 1985). USGPO, Washington, D.C.

Newell, R. W. (1971). The global circulation of atmospheric pollutants. *Sci. Am.* **224,** 32–47.

New York Academy of Medicine (1983). "Resolution Concerning the Disposal in New York City of Biomedical Wastes Containing *De Minimis* Levels of Radioactivity." NYAM Comm. Public Health, New York.

New York Academy of Sciences (1981). "The Three Mile Island Nuclear Accident: Lessons and Implications" (T. H. Moss and D. L. Sills, eds.), *Ann. N.Y. Acad. Sci.* **365.**

New York State (1959). "Protection from Radioactive Fallout," Special Task Force Report to Gov. Nelson A. Rockefeller. Albany, New York.

New York State Low-Level Waste Group (1983). "New York State Must Act to Assure Uninterrupted Disposal Capability for its Low-Level Radioactive Waste," Position Paper.

Ng, Y. C. (1982). A review of transfer factors for assessing the dose from radionuclides in agriculture products. *Nucl. Saf.* **23,** 57–71.

Nishita, H., Romney, E. M., and Larson, K. H. (1961). Uptake of radioactive fission products by crop plants. *J. Agric. Food Chem.* **9,** 101.

Nuclear News (1985). World list of nuclear power plants. Feb., pp. 73–92.

Nydal, R. (1968). Further investigation in the transfer of radiocarbon in nature. *J. Geophys. Res.* **73,** 3617–3635.

Oakley, D. T. (1972). "Natural Radiation Exposure in the U.S.," Rep. ORD/SID 7201. USEPA, Washington, D.C.

Oak Ridge National Laboratory (1970). "Siting of Fuel Reprocessing Plants and Waste Management Facilities," Rep. ORNL-4451. USAEC, Washington, D.C.

Oak Ridge National Laboratory (1980). "Spent Fuel and Waste Inventories and Projections," Rep. ORO-778 prepared for U.S. Dep. Energy, Off. Nucl. Waste Manage. NTIS, Springfield, Virginia.

O'Donnell, F. R. (1978). Assessment of radiation doses from radioactive materials in consumer products: Methods, problems, and results. *In* "Radioactivity in Consumer Products" (A. A. Moghissi, P. Paras, M. W. Carter, and R. F. Barker, eds.). Rep. NUREG/CP-0001, pp. 241–252. USNRC, Washington, D.C.

Office of Technology Assessment (1979). "The Effects of Nuclear War." Washington, D.C.

Okrent, D. (1965). A look at fast reactor safety. *Nucl. Saf.* **6,** 317–342.

Okrent, D. (1981). "Nuclear Reactor Safety." Univ. of Wisconsin Press, Madison.

Onishi, Y., Seine, R. J., Arnold, E. M., Cowan, C. E., and Thompson, F. L. (1981). "Critical Review: Radionuclide Transport, Sediment Transport, and Water Quality Mathematical Monitoring and Radionuclide Adsorption/Desorption Mechanisms," Rep. NUREG/-CR-1322. Pac. Northwest Lab., Richland, Washington.

Ophel, I. L., and Judd, J. M. (1966). Effects of internally deposited radionuclides on the thermal tolerance of fish. *Disposal Radioact. Wastes Seas, Oceans Surf. Waters. Proc. Symp., Vienna* pp. 825—833.

Optical Manufacturers Association (1975). "Ophthalmic Glass Radiological Standards." Arlington, Virginia.

Organization for Economic Cooperation and Development (1980). "Radiological Significance and Management of ^{3}H, ^{14}C, ^{85}Kr, and ^{129}I Arising from the Nuclear Fuel Cycle." OECD, Paris.

Otway, H., and Thomas, K. (1982). Reflection on risk perception and policy. *Risk Anal.* **2,** 69–82.

Palmer, H. E., and Beasley, T. M. (1965). Iron-55 in humans and their foods. *Science* **149,** 431–432.

Parker, F. L., Schmidt, G. D., Cottrell, W. B., and Mann, L. A. (1961). Dispersion of radiocontaminants in an estuary. *Health Phys.* **6,** 66–85.

Parker, F. L., Churchill, M. A., Andrew, R. W., Frederick, B. J., Carrigan, P. H., Jr., Cragwall, J. S., Jr., Jones, S. L., Struxness, E. G., and Morton, R. J. (1966). Dilution, dispersion and mass transport of radionuclides in the Clinch and Tennessee Rivers. *Disposal Radioact. Wastes Seas, Oceans Surf. Waters, Proc. Symp., Vienna* pp. 33–55.

Parker, H. M., and Healy, J. W. (1956). Effects of an explosion of a nuclear reactor. *Proc. Int. Conf. Peaceful Uses At. Energy, 1st, Geneva, 1955* p. 482.

Parker, L., Belsky, J. L., Yamamoto, T., Kawamoto, S., and Keehn, R. J. (1974). Thyroid carcinoma after exposure to atomic radiation. *Ann. Intern. Med.* **80,** 600–604.

Parsly, L. F. (1971). "Removal of Elemental Iodine from Reactor Containment Atmospheres by Spraying," Rep. ORNL-4623. Oak Ridge Nat. Lab., Oak Ridge, Tennessee.

Pearce, D. W., Lindroth, C. E., Nelson, J. L., and Ames, L. L. (1960). A review of radioactive waste disposal to the ground at Hanford. *Disposal Radioact. Wastes, Proc. Sci. Conf., Monaco, 1959.*

Pelletieri, M. W., and Welles, B. W. (1985). "History of Nuclear Materials Transportation and Packaging Research and Development Sponsored by the U.S. Federal Government," Rep. SAND85-7153. Sandia Nat. Lab., Albuquerque, New Mexico.

Pendleton, R. C., and Hanson, W. C. (1958). Absorption of cesium-137 by components of an aquatic community. *Proc. U.N. Int. Conf. Peaceful Uses At. Energy, 2nd, Geneva* **18,** 419–422.

Pendleton, R. C., Lloyd, R. D., and Mays, C. W. (1963). Iodine-131 in Utah during July and August, 1962. *Science* **141,** 640–642.

Pendleton, R. C., Mays, C. W., Lloyd, R. D., and Brooks, A. L. (1964). Differential accumulation of ^{131}I from local fallout in people and milk. *Proc. Hanford Symp. Biol. Radioiodine* p. 72. Pergamon, Oxford.

Penna Franca, E., Fiszman, M., Lobão, N., Costa Ribeiro, C., Trindade, H., Dos Santos, P. L., and Batista, D. (1968). Radioactivity of Brazil nuts. *Health Phys.* **14,** 95–99.

Penna Franca, E., Fiszman, M., Lobão, N., Trindade, H., Costa Ribeiro, C., and Santos, P. L. (1970). Radioactivity in the diet in high background areas on Brazil. *Health Phys.* **19,** 657–662.

Penna Franca, E., Campos, E., Lobão, N., Trindade, H., Sachett, I., and Eisenbud, M. (1986). Radium mobilization and transport at a large thorium ore deposit in Brazil. *Braz. J. Environ. Radiat.* (in press).

Perkins, R. W., and Nielsen, J. M. (1965). Cosmic-ray produced radionuclides in the environment. *Health Phys.* **11,** 1297–1304.

Persson, B. R. (1972). Radiolead (^{210}Pb), polonium (^{210}Po) and stable lead in the lichen, reindeer, and man. *In* "The Natural Radiation Environment II" (J. A. S. Adams, W. M. Lowder, and T. F. Gesell, eds.), USDOE CONF-720805-P2, pp. 347–367. NTIS, Springfield, Virginia.

Pertsov, L. A. (1964). "The Natural Radioactivity of the Biosphere." Atomizdat, Moscow. (Translated by Israel Program for Scientific Translations, Jerusalem, 1967).

Peterson, H. T., Jr. (1983). Terrestrial and aquatic food chain pathways. *In* "Radiological Assessment" (J. E. Till and H. R. Meyer, eds.). Rep. NUREG/CR-3332, Chap. 5. USNRC, Washington, D.C.

Peterson, H. T., Jr. (1984). Regulatory implications of radiation dose–effect relationships. *Health Phys.* **47,** 345–359.

Peterson, K. R. (1970). An empirical model for estimating world-wide deposition from atmospheric nuclear detonations. *Health Phys.* **18,** 357–378.

Petterssen, S. (1958). "Introduction to Meteorology," 2nd Ed. McGraw-Hill, New York.

Petterssen, S. (1968). "Introduction to Meteorology," 3rd Ed. McGraw-Hill, New York.

Petukhova, E. V., and Knizhnikov, V. A. (1969). "Dietary Intake of Sr-90 and Cs-137," Publ. A/AC.82/G/L.1245. United Nations Sci. Comm., Sales Sect., New York.

Philip, P. C., Jayaraman, S., and Pfister, J. (1984). Environmental impact of incineration of low-level radioactive wastes generated by a large teaching medical institution. *Health Phys.* **46,** 1123–1126.

Pickering, R. J., Carrigan, P. H., Jr., Tamura, T., Abee, H. H., Beverage, J. W., and Andrew, R. W., Jr. (1966). Radioactivity in bottom sediment of the Clinch and Tennessee Rivers. *Disposal Radioact. Wastes Seas, Oceans Surf. Waters, Proc. Symp., Vienna* pp. 57–88.

Pimpl, M., and Schuttelkopf, H. (1981). Transport of plutonium, americium, and curium from soils into plants by root uptake. *Nucl. Saf.* **22,** 214–225.

Pohl-Ruling, J., and Fischer, P. (1979). The dose-effect relationship of chromosome aberrations to alpha and gamma irradiation in a population subjected to an increased burden of natural radioactivity. *Radiat. Res.* **80,** 61–81.

Pohl-Ruling, J., and Scheminzky, F. (1954). Das Konzentrationsverhaltnis Blut/Luft bei der Radon-inhalation und die Radon-aufnahme in den Menschliehen Korper im Radioaktiven Thermalstollen von Badgastein/ Bockstein. *Strahlentherapie* **95,** 267.

Porter, C. R., Phillips, C. R., Carter, M. W., and Kahn, B. (1967). The cause of relatively high Cs-137 concentrations in Tampa, Florida, milk. *Proc. Int. Symp. Radioecol. Concentration Processes, 1966* pp. 95–101. Pergamon, Oxford.

Prantl, F. A., Tracy, B. L., and Quinn, M. J. (1980). Health significance of the radioactive contamination of soils and plants in Port Hope, Ontario. *Annu. Conf. Can. Radiat. Prot. Assoc., 1st, Montreal.*

Preston, A., and Jeffries, D. F. (1969). The I.C.R.P. critical group concept in relation to the Windscale discharges. *Health Phys.* **16,** 33–46.

Price, K. R., Carlile, J. M. V., Dirkes, R. L., Jaquish, R. E., Travathan, M. S., and Woodruff, R. K. (1985). "Environmental Monitoring at Hanford for 1984," Rep. PNL-5407 (UC-41 & 11). Pac. Northwest Lab., Richland, Washington.

Prichard, H. M., and Gesell, T. F. (1983). Radon-222 in municipal water supplies in the Central United States. *Health Phys.* **45,** 991–993.

Pritchard, D. W. (1960). The application of existing oceanographic knowledge to the problem of radioactive waste disposal into the sea. *Disposal Radioact. Wastes, Proc. Sci. Conf., Monaco.*

Pritchard, D. W. (1967). What is an estuary: physical viewpoint. *In* "Estuaries," Publ. No. 83. Am. Assoc. Adv. Sci., Washington, D.C.

Pritchard, D. W., Reid, R. O., Okubo, A., and Carter, H. H. (1971). Physical processes of water movement and mixing. *In* "Radioactivity in the Marine Environment," pp. 90–136. Natl. Acad. Sci., Washington, D.C.

Raabe, O. G. (1984). Comparison of the carcinogenicity of radium and bone-seeking actinides. *Health Phys.* **46,** 1241–1258.

Radford, E. P., Jr., and Hunt, V. R. (1964). Polonium-210: A volatile radioelement in cigarettes. *Science* **143,** 247–249.

Rahola, T., and Miettinen, J. K. (1973). Accumulation of ^{137}Cs in Finnish Lapps. *Arch. Environ. Health* **26,** 67–69.

Rajewsky, B., and Stahlhofen, W. (1966). [210]Po activity in the lungs of cigarette smokers. *Nature (London)* **209,** 1312–1313.

Ramsey, R. W., Jr. (1981). "Hearings on Uranium Ore Residues: Potential Hazards and Disposition, June 24–25, 1981," pp. 310–335. USGPO, Washington, D.C.

Rand Corporation (1953). "Worldwide Effects of Atomic Weapons," Rep. R-251-AEC. Santa Monica, California.

Randerson, D., ed. (1984). "Atmospheric Science and Power Production," Rep. DOE/-TIC-27601. USDOE, Washington, D.C.

Rankama, K. (1954). "Isotope Geology." McGraw-Hill, New York.

Rankama, K., and Sahama, T. G. (1950). "Geochemistry." Univ. of Chicago Press. Chicago, Illinois.

Ray. S. S., and Parker, F. G. (1977). "Characterization of Fly Ash from Coal-Fired Power Plants," Rep. EPA-600-7-77-010. Off. Energy, Miner. Ind., USEPA, Washington, D.C.

Reid, D. G., Sackett, W. M., and Spaulding, R. F. (1977). Uranium and radium in livestock feed supplements. *Health Phvs.* **32,** 535–540.

Reid, G. K., and Wood, R. D. (1976). "Ecology of Inland Waters and Estuaries," 2nd Ed. Van Nostrand, New York.

Reid, G. W., Lassovszky, P., and Hathaway, S. (1985). Treatment, waste management and cost for removal of radioactivity from drinking water. *Health Phys.* **48,** 671–694.

Reiter, E. R. (1974). "The Role of the General Circulation of the Atmosphere in Radioactive Debris Transport," Rep. COO-1340-38. USAEC, Washington, D.C.

Revelle, R., and Schaefer. M. B. (1957). "General Considerations Concerning the Ocean as a Receptacle for Artificially Radioactive Materials," Publ. No. 551. Natl. Acad. Sci., Washington, D.C.

Revelle, R., Folsom, T. R., Goldberg, E. D., and Isaacs, J. D. (1956). Nuclear science and oceanography. *Proc. Int. Conf. Peaceful Uses At. Energy, 1st, Geneva 1955* p. 177.

Rice, P. D., Sjoblom, G. L., Steele, J. M., and Harvey, B. F. (1982). "Environmental Monitoring and Disposal of Radioactive Wastes from U.S. Naval Nuclear-Powered Ships and their Support Facilities, 1981," Naval Nuclear Propulsion Program, Rep. NT-82-1. Dep. Navy, Washington, D.C.

Ritchie, J. C., and McHenry, J. R. (1973). Vertical distribution of fallout ^{137}Cs in cultivated soils. *Radiat. Data Rep.* **14,** 727–728.

Ritchie, J. C., Clebsch, E. E. C., and Rudolph, W. K. (1970). Distribution of fallout and natural gamma radionuclides in litter, humus and surface mineral soil layers under natural vegetation in the Great Smoky Mountains, North Carolina–Tennessee. *Health Phys.* **18,** 479–489.

Roberts, H., Jr., and Menzel, R. G. (1961). Availability of exchangeable and nonexchangeable strontium-90 to plants. *J. Agric. Food. Chem.* **9,** 95.

Robinson, A. R., and Marietta, M. G. (1985). "Research, Progress and the Mark A Box Model for Physical, Biological and Chemical Transports," Rep. SAND84-0646, UC-70. Sandia Natl. Lab., Albuquerque, New Mexico.

Robinson, E. W. (1968). "The Use of Radium in Consumer Products," Rep. MORP 68-5. U.S. Public Health Serv., Washington, D.C.

Roche-Farmer, L. (1980). "Study of Alternative Methods for the Management of Liquid Scintillation Counting Wastes," Rep. NUREG-0656. Div. Tech. Inf., USNRC, Washington, D.C.

Rodden, D. J. (1948). Letter from Rodden, U.S. Dep. of Commerce, National Bureau of Standards, Washington. D.C. to F. M. Belmore, U.S. Atomic Energy Commission. Office of New York Directed Operations, New York, dated Sept. 20, 1948.

Roessler, C. E., Smith, Z. A., Bolch, W. E., and Prince, J. R. (1979). Uranium and radium-226 in Florida phosphate materials. *Health Phys.* **37**, 269–277.

Roessler, C. E., Kautz, R., Bolch, W. E., Jr., and Wethington, J. A., Jr. (1980). The effect of mining and land reclamation on the radiological characteristics of the terrestrial environment of Florida's phosphate regions. *In* "Natural Radiation Environment III" (T. F. Gesell and W. M. Lowder, eds.), CONF-780422, Vol. 2, pp. 1476–1493. USDOE, Washington, D.C.

Roessler, C. E., Roessler, G. S., and Bolch, W. E. (1983). Indoor radon progeny exposure in the phosphate mining region: A review. *Health Phys.* **45**, 389–396.

Rogovin, M., and Frampton, G. J. (1980). A sequence of physical events. *In* "Three Mile Island," Vol. II, Part 2, pp. 309–340. Report to the Commissioners and the Public. Spec. Inquiry Group, Nucl. Regul. Comm., Washington. D.C.

Romney, E. M., Wallace, A., Schulz, R. K., and Dunaway, P. (1982). Plant root uptake of ^{239,240}Pu and ^{241}Am from soils containing aged fallout materials. *In* "Environmental Migration of Long-Lived Radionuclides," IAEA-SM-257/83. IAEA, Vienna.

Roser, F. X., and Cullen, T. L. (1964). External radiation measurements in high background regions of Brazil. *In* "The Natural Radiation Environment I" (J. A. S. Adams and W. M. Lowder. eds.), pp. 825–836. Univ. of Chicago Press, Chicago, Illinois.

Roser, F. X., Kegel, G., and Cullen, T. L. (1964). Radiogeology of some high-background areas of Brazil. *In* "The Natural Radiation Environment I" (J. A. S. Adams and W. M. Lowder, eds.). pp. 855–872. Univ. of Chicago Press, Chicago, Illinois.

Rothman, S., and Lichter, S. R. (1985). Elites in conflict: Nuclear energy, ideology, and the perception of risk. *J. Contemp. Stud.* **8**, 23–44.

Rowland, R. F., Stehney, A. F., and Lucas, H. F. (1983). Dose response relationships for radium induced bone sarcomas. *Health Phys.* **44**, Suppl. 1, 15–31.

Rundo, J., Failla, P., and Schlenker, R. A. (1983). Foreword. *Health Phys.* **44**, Suppl. 1, 1–3.

Russell, R. S. (1965). Interception and retention of airborne material on plants. *Health Phys.* **11**, 1305–1315.

Russell, R. S. (1966). "Radioactivity and Human Diet." Pergamon, Oxford.

Russell, R. S., and Bruce, R. S. (1969). Environmental contamination with fallout from nuclear weapons. *Environ. Contam. Radioact. Mater., Proc. Semin., Vienna.*

Russell, W. L. (1968). Recent studies on the genetic effects of radiation in mice. *Pediatrics* **41**, 223–230.

Ryan, M. T. (1981). Radiological impacts of uranium recovery in the phosphate industry. *Nucl. Saf.* **22**, 70–76.

Saenger, E. L., Thoma, G. E., and Tompkins, E. (1968). Incidence of leukemia following treatement of hyperthyroidism. *J. Am. Med. Assoc.* **205**, 147–154.

Schaefer, H. J. (1971). Radiation exposure in air travel. *Science* **173**, 780–783.

Scheminzky, F. (1961). 25 Jahre Baderforschung in Gastein mit Verzeichnis der Wissenschaftlichen Veroffenlichungen bis 1960. *Sonderabdruck Badgasteiner Bladebl.* Nos. 35 and 36.

Schiager, K. J. (1974). Analysis of radiation exposures on or near uranium mill tailings piles. *Radiat. Data Rep.* **15**, 411–425.

Schiager, K. J. (1986). Disposal of uranium mill tailings. *Proc. Annu. Meet. Natl. Counc. Radiat. Prot. Meas. 21st, 1985,* pp. 149–162.

Schlenker, R. A. (1982). Risk estimates for bone. *In* "Critical Issues in Setting Radiation Dose Limits," Proc. 17th Annu. Meet. NCRP, pp. 153–163. Natl. Councl. Radiat. Prot. Meas. Bethesda, Maryland.

Schneider, K. J., Bradshaw, R. L., Blasewitz, A. G., Blomeke, J. O., and McClain W. C. (1971). Status of solidification and disposal of highly radioactive liquid wastes from nuclear power in the U.S.A. *Proc. Environ. Aspects. Nucl. Power Stn.* p. 369.

Schreckhise, R. G., and Cline, J. F. (1980a). Uptake and distribution of ^{232}U in peas and barley. *Health Phys.* **38,** 341–343.

Schreckhise, R. G., and Cline, J. F. (1980b). Comparative uptake and distribution of Pu, Am, Cm and Np in four plant species. *Health Phys.* **38,** 817–824.

Schulz, R. K. (1965). Soil chemistry of radionuclides. *Health Phys.* **11,** 1317–1324.

Scott, N. S. (1897). X-ray injuries. *Am. X-Ray J.* **1,** 57–65.

Scott, R. L., Jr. (1971). Fuel-melting incident at the Fermi reactor on Oct. 5, 1966. *Nucl. Saf.* **12,** 123–134.

Seaborg, G. T. (1958). "The Transuranic Elements." Addison-Wesley, Reading, Massachusetts.

Seaborg, G. T. (1981). "Kennedy, Krushchev, and the Test Ban." Univ. of California Press, Berkeley.

Seaborg, G. T., and Bloom, J. (1970). Fast breeder reactors. *Sci. Am.* **223**(5), 13–21.

Seelentag, W., and Schmier, H. (1963). Radiation exposure from luminous watch dials. *Radiol. Health Data* **4,** 209–213.

Sehmel, G. A. (1984). Deposition and resuspension. *In* "Atmospheric Science and Power Production" (R. Randerson, ed.), Rep. DOE/TIC-27601, pp. 533–583. USDOE, Washington, D.C.

Shafer, C. K. (1959). "Testimony before Joint Committee on Atomic Energy," Hearings on Biological and Environmental Effects of Nuclear War. USGPO, Washington, D.C.

Shapley, D. (1971). Rocky Flats: Credibility gap widens on plutonium plant safety. *Science* **174,** 569–571.

Shearer, S. D., Jr., and Lee, G. F. (1964). Leachability of radium-226 from uranium mill solids and river sediments. *Health Phys.* **10,** 217–227.

Shearer, S. D., Jr., and Sill, C. W. (1969). Evaluation of atmospheric radon in the vicinity of uranium mill tailings. *Health Phys.* **17,** 77–88.

Shelton, F. H. (1959). Statement. *In* "Fallout from Nuclear Weapons Tests," Vol. I, pp. 772–778. Hearings before the Joint Committee on Atomic Energy, Special Subcommittee on Radiation. USGPO, Washington, D.C.

Sheppard, M. I. (1980). "The Environmental Behavior of Radium," Rep. AECL-6796. At. Energy Can., Whiteshell.

Shipman, T. L., ed. (1961). Acute radiation death resulting from an accidental nuclear critical excursion. *J. Occup. Med.* **3,** No. 3, Spec. Suppl.

Shleien, B., Glavin, T. P., and Friend, A. G. (1965). Particle size fractionation of airborne gamma emitting radionuclides by graded filters. *Science (Washington, D.C. 1883–)* **147,** 290–292.

Shleien, B., Cochran, J. A., and Magno, P. J. (1970). Strontium-90, strontium-89, plutonium-239, and plutonium-238 concentrations in ground-level air, 1964–1969. *Environ. Sci. Technol.* **4,** 598–602.

Shleien, B., Tucker, T., and Johnson, D. W. (1977). "The Mean Active Bone Marrow Dose to the Adult Population of the United States from Diagnostic Radiology," Publ. (FDA) 77-8013. U.S. Dep. Health, Educ. Welfare, Food Drug Adm., Washington, D.C.

Shore, R. E., Woodward, E. D., and Hempelmann, L. H. (1984). Radiation-induced thyroid cancer. *In* "Radiation Carcinogenesis: Epidemiology and Biological Significance" (J. D. Boice and J. F. Fraumeni, eds.), pp. 131–138. Raven, New York.

Simpson, H. J., Linsalata, P., Olsen, C. R., Cohen, N., and Trier, R. M. (1986). Transport of fallout and reactor radionuclides in the drainage basin of the Hudson River estuary. *In* "Environmental Research for Actinide Elements" (J. E. Pinder III, ed.). NTIS, Springfield, Virginia. In press.

Simpson, R. E., and Shuman, F. G. D. (1978). The use of uranium in ceramic tableware. *In* "Radioactivity in Consumer Products" (A. A. Moghissi, P. Paras, M. W. Carter, and R. F. Barker, eds.), Rep. NUREG/CP-0001, pp. 470–474. USNRC, Washington, D.C.

Sinclair, W. K. (1984). Letter from President, NCRP to W. D. Ruckelshaus, N. J. Palladino, and D. P. Hodel, dated October 15, 1984. Nat. Counc. Radiat. Prot. Meas., Bethesda, Maryland.

Slinn, W. G. (1984). Precipitation scavenging. In "Atmospheric Science and Power Production" (D. Randerson, ed.), Rep. DOE/TIC-27601. USDOE, Washington, D.C.

Slovic, P., Fischhoff, B., and Lichtenstein, S. (1980). Facts and fears: Understanding perceived risk. In "Societal Risk Assessment" (R. C. Schwing and W. A. Albers, eds.). Plenum, New York.

Smith, B. M., Grune. W. N., Higgins, F. B., Jr., and Terrill, J. G., Jr. (1961). Natural radioactivity in ground water supplies in Maine and New Hampshire. J. Am. Water Works Assoc. **53,** 75.

Smith, F. A., and Dzuiba. S. P. (1949). Preliminary observations of the uranium content of photographic materials. Unpubl. memo. Univ. of Rochester, Rochester, New York.

Smith, J. M., Fowler, T. W., and Goldin, A. S. (1982). "Environmental Pathway Models for Estimating Population Risks from Disposal of High-Level Radioactive Waste in Geologic Repositories, Draft Report," EPA 520/5-80-002. USEPA, Washington, D.C.

Smith, P. G., and Doll, R. (1982). Mortality of patients with ankylosing spondylitis after a simple treatment course with X rays. Br. Med. J. No. 284, 449–460.

Soldat, J. K. (1963). The relationship between I-131 concentrations in various environmental samples. Health Phys. **9,** 1167–1171.

Soldat, J. K. (1965). Environmental evaluation of an acute release of ^{131}I to the atmosphere. Health Phys. **11,** 1009–1015.

Solon, L. R., et al. (1958). External environmental radiation measurements in the United States. Proc. U.N. Int. Conf. Peaceful Uses At. Energy, 2nd, Geneva p. 740.

Spalding, R. F., and Sackett, W. (1972). Uranium in runoff from the Gulf of Mexico distributive province: Anomalous concentrations. Science **175,** 629–631.

Spiers, F. W. (1966). Dose to bone from strontium-90: Implications for the setting of maximum permissible body burden. Radiat. Res. **28,** 624–642.

Spiers. F. W. (1968). "Radioisotopes in the Human Body: Physical and Biological Aspects." Academic Press, New York.

Spiess, H., and Mays, C. W. (1970). Bone cancers induced by ^{224}Ra (Th X) in children and adults. Health Phys. **19,** 713–729.

Spitsyn, V. I., and Balukova, V. D. (1979). The scientific basis for, and experience with, underground storage of liquid radioactive wastes in the U.S.S.R. In "Scientific Basis for Nuclear Waste Management" (G. J. McCarthy, ed.), p. 237. Plenum, New York.

Spitsyn, V. I., et al. (1958). A study of the migration of radioelements in soils. Proc. U.N. Int. Conf. Peaceful Uses At. Energy, 2nd, Geneva p. 2207.

Spitsyn, V. I., et al. (1960). Sorption regularities in behavior of fission product elements during filtration of their solutions through ground. Disposal Radioact. Wastes, Proc. Sci. Conf., Monaco, 1959.

Starr, C. (1969). Social benefits versus technological risk. Science **165,** 1232–1238.

Stebbins, A. K. (1961). "Second Special Report on High Altitude Sampling Program," Rep. No. 539B. Def. At. Support Agency, Washington, D.C.

Stebbins, A. K., and Minx, R. F. (1962). "The High Altitude Sampling Program," Testimony before Joint Committee on Atomic Energy, Hearings on Radioactive Fallout. USGPO, Washington, D.C.

Steinberg, E., and Glendenin. L. (1956). Proc. Int. Conf. Peaceful Uses At. Energy, 1st, Geneva, 1955 p. 614.

Steinhausler, F. A. (1975). Long-term measurements of Rn-222, Rn-220, Pb-214 and Pb-212 concentrations in the air of private and public buildings and their dependence on meteorological parameters. Health Phys. **29,** 705–713.

Stevens, D. L., Jr. (1963). "A Brief History of Radium." Bur. Radiol. Health, U.S. Public Health Serv.. Washington, D.C.

Stewart, A., and Kneale, G. W. (1970). Radiation dose effects in relation to obstetric x-rays and childhood cancers. *Lancet*, 1185–1188.

Stewart, N. G., *et al.* (1957). "World-Wide Deposition of Long-Lived Fission Products from Nuclear Test Explosions," Rep. MP/R2354. U.K. At. Energy Auth., Res. Group.

Stigall, G. E., Fowler, T. W., and Krieger, H. L. (1971). ^{131}I discharges from an operating boiling water reactor nuclear power station. *Health Phys.* **20**, 593–599.

Stoller, S. M., and Richards, R. B. (1961). "Reactor Handbook," Vol. II. Wiley (Interscience), New York.

Stonier, T. (1964). "Nuclear Disaster." World Publ., Cleveland, Ohio.

Strand, J. A., Fujihara, M. P., Burdett, R. D., and Poston, T. M. (1977). Suppression of the primary immune response in rainbow trout, *Salmo gairdneri*, sublethally exposed to tritiated water during embryogenesis. *J. Fish. Res. Board Can.* **34**, 1293–1304.

Suess, H. E. (1958). The radioactivity of the atmosphere and hydrosphere. *Annu. Rev. Nucl. Sci.* **8**, 243.

Summerlin, J., and Prichard, H. M. (1985). Radiological health implications of lead-210 and polonium-210 accumulations in LPG refineries. *Am. Ind. Hyg. Assoc.* **46**, 202–205.

Sun, S., Meng, X., Yuan, L., You, Z., Liu, S., Yang, X., Yang, L., and Chen, H. (1981). Etiological analysis of lung cancers among tin miners at Gejiu, Yunnan Province. *Fushe Fanghu* **2**, 1–8.

Sundaram, K. (1977). Down's syndrome in Kerala. *Nature (London)* **267**, 728.

Sunta, C. M., David, M., Abani, M. C., Basu, A. S., and Nambi, K. S. V. (1982). Analysis of dosimetry data of high natural radioactivity areas of southwest coast of India. *In* "The Natural Radiation Environment IV" (K. G. Vohra, U. S. Mishra, K. C. Pillai, and S. Sadasivan, eds.). Wiley Eastern, Bombay/New Delhi.

Sutton, O. G. (1953). "Micrometeorology." McGraw-Hill, New York.

Sverdup, H. V., Johnson, M. W., and Fleming, R. H. (1963). "The Oceans: Their Physics, Chemistry and General Biology." Prentice-Hall, Englewood Cliffs, New Jersey.

Tabor, W. H. (1963). Operating experience of the Oak Ridge research reactor through 1962. *Nucl. Saf.* **5**, 116–123.

Tait, J. H. (1983). Uranium enrichment. *In* "Nuclear Power Technology" (W. Marshall, ed.), Vol. 2. Oxford Univ. Press (Clarendon), London and New York.

Tajima, E., and Doke, T. (1956). Airborne radioactivity. *Science* **123**, 211–214.

Tamplin, A. R., and Fisher, H. L. (1966). "Estimation of Dosage to Thyroids of Children in the U.S. from Nuclear Tests Conducted in Nevada during 1952 through 1955," Rep. UCRL-147-7. Lawrence Radiat. Lab., Univ. of California, Berkeley.

Tanner, A. B. (1964). Radon migration in the ground: A review. *In* "The Natural Radiation Environment I" (J. A. S. Adams and W. M. Lowder, eds.), pp. 161–190. Univ. of Chicago Press, Chicago, Illinois.

Tanner, A. B. (1980). Radon migration in the ground: A supplementary review. *In* "Natural Radiation Environment III" (T. F. Gesell and W. M. Lowder, eds.), CONF-780422, Vol. 1, pp. 5–56. USDOE, Washington, D.C.

Taylor, L. S. (1971). "Radiation Protection Standards." Chem. Rubber Publ. Co., Cleveland, Ohio.

Taylor, L. S. (1981). The development of radiation protection standards (1925–1940). *Health Phys.* **41**, 227–232.

Templeton, W. L., Nakatani, R. E., and Held, E. E. (1971). Radiation effects. *In* "Radioactivity in the Marine Environment," pp. 223–239. Natl. Acad. Sci., Washington, D.C.

Thompson, D. L. (1978). *In* "Radioactivity in Consumer Products" (A. A. Moghissi, P. Paras, M. W. Carter, and R. F. Barker, eds.), Rep. NUREG/CP-0001, pp. 475–478. USNRC, Washington, D.C.

Thompson, R. C. (1982). Neptunium—the neglected actinide: A review of the environmental and biological literature. *Radiat. Res.* **90,** 1–32.

Thompson, T. J., and Beckerley, J. G. (1964). "The Technology of Nuclear Reactor Safety," Vol. 1. MIT Press, Cambridge, Massachusetts.

Tichler, J., and Benkovitz, C. (1984). "Radioactive Materials Released from Nuclear Power Plants," Rep. NUREG/CR-1906, BNL-NUREG-51581, Vol. 2. USNRC, Washington, D.C.

Till, J. E., and Meyer, H. R., eds. (1983). "Radiological Assessment." USNRC, Washington, D.C.

Tobias, C. A., and Todd, P. (1974). "Space Radiation Biology and Related Topics." Academic Press, New York.

Totter, J. R., and MacPherson, H. G. (1981). Do childhood cancers result from prenatal x-rays? *Health Phys.* **40,** 511–524.

Totter, J. R., Zelle, M. R., and Hollister, H. (1958). Hazard to man of carbon-14. *Science* **128,** 1490–1495.

Trabalka, J. R., and Garten, C. T., Jr. (1983). Behavior of the long-lived synthetic elements and their natural analogs in food chains. *Adv. Radiat. Biol.* **10,** 39–104.

Trabalka, J. R., Eyman, L. D., and Auerbach, S. I. (1980). Analysis of the 1957–1958 Soviet nuclear accident. *Science* **209,** 345–353.

Tracy, B. L., Prantl, F. A., and Quinn, J. M. (1984). Health impact of radioactive debris from the satellite Cosmos 954. *Health Phys.* **47,** 225–233.

Triffet, T. (1959). Basic properties and effects of fallout. *In* "Biological and Environmental Effects of Nuclear War," Hearings before the Joint Committee on Atomic Energy. USGPO, Washington, D.C.

Trunk, A. D., and Trunk, E. V. (1981). Three Mile Island: A resident's perspective. *In* "The Three Mile Island Nuclear Accident: Lessons and Implications" (T. H. Moss and D. L. Sills, eds.), *Ann. N.Y. Acad. Sci.* **365,** 175–185.

Tsivoglou, E. C. (1959). Radioactive waste disposal to surface waters. *In* "Industrial Radioactive Waste Disposal," Hearings before the Joint Committee on Atomic Energy. USGPO, Washington, D.C.

Tsivoglou, E. C., *et al.* (1960a). Estimating human radiation exposure on the Animas River. *J. Am. Water Works Assoc.* **52,** 1271.

Tsivoglou, E. C., Stein, M., and Towne, W. S. (1960b). Control of radioactive pollution of the Animas River. *J. Water Pollut. Control Fed.* **32,** 262.

Tsuzuki, M. (1955). The experience concerning radioactive damage of Japanese fishermen by Bikini fallout. *Muench. Med. Wochenschr.* **97,** 988.

Turco, R. P., Toon, O. B., Ackerman, T. P., Pollack, J. B., and Sagan, C. (1983). Nuclear winter: global consequences of multiple nuclear explosions. *Science* **222,** 1283–1929.

Turner, D. B. (1970). "Workbook of Atmospheric Dispersion Estimates." USEPA, Washington, D.C.

Unger. W. E., Browder, F. N., and Mann, S. (1971). Nuclear safety in American radiochemical processing plants. *Nucl. Saf.* **23,** 234.

United Kingdom Agricultural Research Council (1961). "Strontium-90 in Milk and Agricultural Materials in the United Kingdom 1959–1960," Rep. No. 4.

United Kingdom Atomic Energy Office (1957). "Accident at Windscale No. 1 Pile on October 10, 1957." HM Stationery Off., London.

United Kingdom Atomic Energy Office (1958). "Final Report on the Windscale Accident." HM Stationery Off.. London.

United Nations (1956). *Proc. Int. Conf. Peaceful Uses At. Energy, 1st, Geneva, 1955* 17 vols.

United Nations (1981). "Report of the Working Group on the Use of Nuclear Power in Outer Space on the Work of its Third Session." U.N. Document A/AC.105/287. New York.

United Nations Scientific Committee on the Effects of Atomic Radiation (1958). 13th Annual Session, United Nations, New York.

United Nations Scientific Committee on the Effects of Atomic Radiation (1962). 17th Annual Session, Suppl. No. 16 (A/5216). United Nations, New York.

United Nations Scientific Committee on the Effects of Atomic Radiation (1964). 19th Session, Suppl. No. 14 (A/5814). United Nations, New York.

United Nations Scientific Committee on the Effects of Atomic Radiation (1966). 21st Session, Suppl. No. 14 (A/6314). United Nations, New York.

United Nations Scientific Committee on the Effects of Atomic Radiation (1969). 24th Session, Suppl. No. 13 (A/7613). United Nations, New York.

United Nations Scientific Committee on the Effects of Atomic Radiation (1977). 32nd Session, Suppl. No. 40 (A/32/40). United Nations, New York.

United Nations Scientific Committee on the Effects of Atomic Radiation (1982). 37th Session, Suppl. No. 45 (A/37/45). United Nations, New York.

U.S. Atomic Energy Commission (1957a). "Theoretical Possibilities and Consequences of Major Accidents in Large Nuclear Power Plants," Rep. WASH-740. Washington, D.C.

U.S. Atomic Energy Commission (1957b). "Atomic Energy Facts." Washington, D.C.

U.S. Atomic Energy Commission (1960a). "Hazards Summary Report on Consolidated Edison Thorium Reactor," Docket No. 50-3, Exhibit K-5 (Revision 1). Washington, D.C.

U.S. Atomic Energy Commission (1960b). "Summary of Available Data on the Strontium-90 Content of Foods and of Total Diets in the United States," Rep. HASL-90. New York.

U.S. Atomic Energy Commission (1961). "Investigation Board Report on the SL-1 Accident." Washington, D.C.

U.S. Atomic Energy Commission (1969). "Report on the May 11, 1969 Fire at the Rocky Flats Plant near Boulder, Colorado," USAEC Press Release No. M-257. Washington, D.C.

U.S. Atomic Energy Commission (1970a). "The Nuclear Industry-1970." Washington, D.C.

U.S. Atomic Energy Commission (1970b). "Survey Report on Structural Design of Piping Systems and Components." TID-25553. Washington, D.C.

U.S. Atomic Energy Commission (1972a). "Biomedical Implications of Radiostrontium Exposure." *AEC Symp. Ser.*

U.S. Atomic Energy Commission (1972b). "Grand Junction Remedial Action Criteria," 10CFR12. Federal Register 12/6/72, pp. 25918–25919. Washington, D.C.

U.S. Atomic Energy Commission (1973). "Report on the Investigation of the 106 T Tank Leak at the Hanford Reservation, Richland, Washington." Richland Oper. Off., Richland, Washington.

U.S. Atomic Energy Commission (1974). "High-Level Radioactive Waste Management Alternatives," WASH-1297. Washington, D.C.

U.S. Atomic Energy Commission (1975). "Operational Accidents and Radiation Exposure Experience Within the U.S. Atomic Energy Commission 1943–1975," Rep. WASH 1192. Washington, D.C.

U.S. Department of Agriculture (1957). "Soils." U.S. Dep. Agric., Washington, D.C.

U.S. Department of Agriculture (1967). "Food Consumption Households in the United States, Spring, 1965," A preliminary report, Agricultural Research Service, Consumer and Food Economics Research Div., ARS 62-16 (Aug., 1967). Washington, D.C.

U.S. Department of Defense (1961). "Fallout Protection." U.S. Dep. Defense, Washington, D.C.

U.S. Department of Energy. Environmental Measurements Laboratory (formerly U.S. AEC HASL). Data from ^{90}Sr analyses of foods obtained at market in New York and San Francisco have been published at intervals for many years. New York.

U.S. Department of Energy (1978). "Long-Term Management of Defense High-Level Radioactive Wastes," Rep. DOE/EIS-0023-D. Washington, D.C.

U.S. Department of Energy (1979a). "Draft Environmental Impact Statement, Waste Isolation Pilot Plant" (April, 1979), pp. 9–55. Washington, D.C.

U.S. Department of Energy (1979b). "Final Environmental Impact Statement: Management of Commercially Generated Radioactive Waste," DOE/EIS-0046G (3 vols.), p. 126. Washington, D.C.

U.S. Department of Energy (1980a). "Management of Commercially Generated Radioactive Waste," DOE/EIS-0046F, Vol. 1. Final Environmental Impact Statement. NTIS, Springfield, Virginia.

U.S. Department of Energy (1980b). "Grand Junction Remedial Action Program." DOE/-EV/01621-T1. Washington, D.C.

U.S. Department of Energy (1981). "Background Report for the Uranium Mill Tailings Remedial Action Program," Rep. DOE/EP-001. Washington, D.C.

U.S. Department of Energy (1983). "Spent Fuel and Radioactive Waste Inventories, Projections, and Characteristics," Rep. DOE/NE-0017/2. NTIS, Springfield, Virginia.

U.S. Department of Energy (1984a). "Commercial Power 1984: Prospects for the United States and the World." Energy Information Administration Rep. DOE/EIA 0438(84). NTIS, Springfield, Virginia.

U.S. Department of Energy (1984b). 10CFR Part 960; Nuclear Waste Policy Act of 1982: General Guidelines for the Recommendation of Sites for the Nuclear Waste Repositories. Fed. Regist. 49, 44714–47770.

U.S. Department of Energy (1984c). "Safety and Health Highlights, Fiscal Year 1984." Washington, D.C.

U.S. Department of Energy (1984d). "U.S. Uranium Mining and Milling Industry: A Comprehensive Review." Rep. DOE/S-0028. Washington, D.C.

U.S. Department of Energy (1985a). "Domestic Uranium Mining and Milling Industry—1984 Viability Assessment," Rep. DOE/EIA-0477. Washington, D.C.

U.S. Department of Energy (1985b). "Energy Systems Acquisition Project Plan: Formerly Used MED/AEC Sites Remedial Action Program (FUSRAP)," Oak Ridge Operations Office (Apr. 1985). Oak Ridge, Tennessee.

U.S. Department of Energy (1985c). Office of Civilian Radioactive Waste Management "Annual Report to Congress," Rep. DOE/RW-0004/1. NTIS, Springfield, Virginia.

U.S. Department of Energy (1985d). "Announced U.S. Nuclear Tests," Rep. NVO 209 (Jan. 1985). Washington, D.C.

U.S. Department of Health and Human Services (1985). "Report of the National Institutes of Health Ad Hoc Working Group to Develop Radioepidemiological Tables," NIH Publ. No. 85-2748. Washington, D.C.

U.S. Department of State (1946). "International Control of Atomic Energy," Publ. No. 2702. Washington, D.C.

U.S. Environmental Protection Agency (1972). "Estimates of Ionizing Radiation Doses in the United States 1960–2000," Report of Special Studies Group, Division of Criteria and Standards, Office of Radiation Programs. Washington, D.C.

U.S. Environmental Protection Agency (1979). "Radiological Impact Caused by Emissions of Radionuclides into Air in the United States," Rep. EPA 520/7-79-006, Washington, D.C.

U.S. Environmental Protection Agency (1982). "Final Environmental Impact Statement for Remedial Action Standards for Inactive Uranium Processing Sites," Rep. EPA 520/4-82-013-1. Washington, D.C.

U.S. Environmental Protection Agency (1983). Environmental Standards for Uranium and Thorium Mill Tailings at Licensed Commercial Processing Sites; Final Rule. *Fed. Regist.* **48**.

U.S. Environmental Protection Agency (1984). "Report on the Scientific Basis of EPA's Proposed National Emission Standards for Hazardous Air Pollutants for Radionuclides," Report of Subcommittee of the Science Advisory Board. Washington, D.C.

U.S. Environmental Protection Agency (1985). "Environmental Standards for the Management and Disposal of Spent Nuclear Fuel, High Level and Transuranic Radioactive Wastes," 40 CFR Part 191. Washington, D.C.

U.S. Government Printing Office (1959). "Report of Fallout Prediction Panel," Hearings before Joint Committee on Atomic Energy on Fallout from Nuclear Weapons Tests. Washington, D.C.

U.S. Nuclear Regulatory Commission (1975). "Reactor Safety Study: An Assessment of Accident Risks in U.S. Commercial Nuclear Power Plants," Executive Summary. WASH-1400 (NUREG 75/014). Washington, D.C.

U.S. Nuclear Regulatory Commission (1976). "Regulatory Guide 1.113, Estimating Aquatic Dispersion of Effluents from Accidental and Routine Reactor Releases for the Purpose of Implementing Appendix I." Off. Stand. Dev., Washington, D.C.

U.S. Nuclear Regulatory Commission (1977a). "Regulatory Guide 1.111: Methods for Estimating Atmospheric Transport and dispersion of Gaseous Effluents in Routine Releases from Light-Water-Cooled Reactors, Revision 1." Off. Stand. Dev., Washington, D.C.

U.S. Nuclear Regulatory Commission (1977b). "Regulatory Guide 1.109 Revision 1: Calculation of Annual Doses to Man from Routine Releases of Reactor Effluents for the Purpose of Evaluating Compliances with 10 CFR Part 50, Appendix I." Off. Stand. Dev., Washington, D.C.

U.S. Nuclear Regulatory Commission (1980a). "Final Generic Environmental Impact Statement on Uranium Milling," Rep. NUREG 0706. Washington, D.C.

U.S. Nuclear Regulatory Commission (Advisory Committee on Reactor Safeguards) (1980b). "An Approach to Quantitative Safety Goals for Nuclear Power Plants," Rep. NUREG 0739. Washington, D.C.

U.S. Nuclear Regulatory Commission (1981). Nuclear Regulatory Commission Standards for Protection Against Radiation, 10CFR 20.306, Disposal of Specific Wastes. *Fed. Regist.* **46,** 16230.

U.S. Nuclear Regulatory Commission (1983). "NRC Occupational Radiation Exposure at Commercial Nuclear Power Reactors, 1982," Rep. NUREG-0703. Washington, D.C.

U.S. Office of the Federal Register (a). "Code of Federal Regulations, Title 10, Energy, Parts 0-199." USGPO, Washington, D.C. The most recent code should be consulted.

U.S. Office of the Federal Register (b). "Code of Federal Regulations, Title 49, Chapter 1, Research and Special Programs Administration, Subchapter C, Hazardous Materials Regulations, Parts 171–179." USGPO, Washington, D.C. The most recent code should be consulted.

U.S. Public Health Service (1971). "State and Federal Control of Health Hazards from Radioactive Materials Other Than Materials Regulated Under the Atomic Energy Act of 1954," Rep. BRH/DMRE 71-4. Washington, D.C.

U.S. Weather Bureau (1955). "Meterology and Atomic Energy." USGPO, Washington, D.C.

U.S.S.R. State Committee on the Utilization of Atomic Energy (1986). The Accident at the Chernobyl Power Plant and Its Consequences. Information compiled for the IAEA Experts' Meeting, August 25–29, 1986, Vienna.

Upton, A. C. (1977). Radiobiological effects of low doses: Implications for radiological protection. *Radiat. Res.* **71,** 51–74.

Upton, A. C., Chase, H. B., Hekhuis, G. L., *et al.* (1966). Radiobiological Aspects of the Supersonic Transport: A Report of the ICRP Task Group on the Biological Effects of High-Energy Radiations. *Health Phys.* **12**, 209–226.

Uzinov, I., Steinhausler, F., and Pohl, E. (1981). Carcinogenic risk of exposure to radon daughters. *Health Phys.* **41**, 807–813.

Van Middlesworth, L. (1954). Radioactivity in animal thyroids from various areas. *Nucleonics* **12**, 56.

Vennart, J. (1981). Limits for Intakes of Radionuclides by Workers, ICRP Publ. 30. *Health Phys.* **40**, 477–484.

Villforth, J. C. (1964). Problems in Radium control. *Public Health Rep.* **79**, 337–342.

Villforth, J. C., Robinson, E. W., and Wold, G. J. (1969). A review of radium incidents in the U.S.A. *Handl. Radiat. Accid., Proc. Symp., Vienna.*

Volchok, H. L. (1970). "Worldwide Deposition of ^{90}Sr through 1969," Rep. HASL-227. USAEC, New York.

Volchok, H. L., and Kleinman, M. T. (1971). "Worldwide Deposition of Sr-90 through 1970," Rep. HASL-243. USAEC, New York.

Volchok, H. L., Knuth, R., and Kleinman, M. T. (1972). "Plutonium in the Neighborhood of Rocky Flats, Colorado: Airborne Respirable particles," Rep. HASL-246. USAEC, New York.

von Hevesey, G. (1966). Radioactive tracers and their application. *Isot. Radiat. Technol.* **4**, 9–12.

Wadleigh, C. H. (1957). Growth of plants. *In* "Soils," pp. 38–48. U.S. Dep. Agric., Washington, D.C.

Wald, N. (1980). The Three Mile Island incident in 1979: the state response. *In* "The Medical Basis for Radiation Accident Preparedness" (K. F. Hubner and S. A. Fry, eds.), pp. 491–500. Elsevier/North-Holland, New York.

Wald, N., Thoma, G. E., Jr., and Broun, G., Jr. (1962). Hematologic manifestations of radiation exposure in man. *Prog. Hematol.* **3**.

Wallace, R. (1973). "Comparison of Calculated Cosmic-Ray Doses to a Person Flying in Subsonic and Supersonic Aircraft," Rep. LBL-1505, UC-41 Health and Safety, p. 21. Lawrence Berkeley Lab., Univ. of California, Berkeley. (Also U.S. At. Energy Rep. TID-4500 R 61.)

Walton, G. N. (1961). Fission and fission products. *In* "Atomic Energy Waste, Its Nature, Use, and Disposal" (E. Glueckauf, ed.), Chap. 1. Wiley (Interscience), New York.

Watters, R. L., and Hansen, W. R. (1970). The hazards implication of the transfer of unsupported ^{210}Po from alkaline soil to plants. *Health Phys.* **18**, 409–413.

Watters, R. L., Edington, D. N., Hakonson, T. E., Hanson, W. C., Smith, M. H., Whicker, F. W., and Wildung, R. E. (1980). Synthesis of the research literature. *In* "Transuranic Elements in the Environment" (W. C. Hanson, ed.), Rep. DOE/TIC-22800. Tech. Inf. Cent./U.S. Dep. Energy, Washington, D.C.

Webb, J. H. (1949). The fogging of photographic film by radioactive contaminants in cardboard packaging materials. *Phys. Rev.* **76**, 375–380.

Webb, J. W., and Voorhees, L. D. (1984). Revegetation of uranium mill tailings sites. *Nucl. Saf.* **25**, 668–675.

Weems, S. J., Lyman, W. G., and Haga, P. B. (1970). The ice-condenser reactor containment system. *Nucl. Saf.* **11**, 215–222.

Wegst, A. V., Pelletier, C. A., and Whipple, G. H. (1964). Detection and quantitation of fallout particles in a human lung. *Science* **143**, 957–959.

Wei, L., ed. (1980). Health survey in high background radiation areas in China. Report by High Background Radiation Research Group, China. *Science* **209**, 877–880.

Wei, L., *et al.* (1985). Report of Third Stage (1982–1984) Health Survey in high background area in Yangjiang, China. *Chin. J. Radiat. Med. Prot.* **5**, 144–153.

Weinberg, A. M., and Wigner, E. P. (1958). "The Physical Theory of Neutron Chain Reactors," Univ. of Chicago Press, Chicago, Illinois.

Weiss, E. S., Rallison, M. L., London, W. T., and Thompson, G. D. C. (1971). Thyroid nodularity in southwestern Utah school children exposed to fallout radiation. *Am. J. Public Health* **61**, 241–249.

Welford, G. A., and Sutton, D. (1957). "Determination of the Uranium Content of the National Bureau of Standards Iron and Steel Chemical Standards," Rep. NYOO-4755. USAEC, Washington, D.C.

Wenslawski, F. A. and North, H. S. Jr. (1979). Response to a widespread, unauthorized dispersal of radioactive waste in the public domain, P/53. *Proc. Health Phys. Soc. Midyear Top. Symp., 12th., Low-Level Radioact. Waste Manage.* Rep. EPA 520/3-79-002. Off. Radiat. Programs, USEPA, Washington, D.C.

Whicker, F. W. (1980). Ecological effects of transuranics in the terrestrial environment. *In* "Transuranic Elements in the Environment" (W. C. Hanson, ed.), Rep. DOE/TIC-22800, pp. 701–713. Tech. Inf. Cent./U.S. Dep. Energy, Washington, D.C.

Whitehead, D. C. (1984). The distribution and transformations of iodine in the environment. *Environ. Int.* **10**, 321–339.

Wick, R. R., and Gossner, W. (1983). Late effects of ^{224}Ra treated ankylosing spondylitis patients. *Health Phys.* **44**, Suppl. 1, 187–192.

Wilkening, M. H. (1952). Natural radioactivity as a tracer in the sorting of aerosols according to mobility. *Rev. Sci. Instrum.* **23**, 13.

Wilkening, M. H. (1964). Radon-daughter ions in the atmosphere. *In* "The Natural Radiation Environment I" (J. A. S. Adams and W. M. Lowder, eds.), p. 359–368. Univ. of Chicago Press, Chicago, Illinois.

Wilkening, M. H.(1982). Radon in atmospheric studies: A review. *In* "The Natural Radiation Environment IV" (K. G. Vohra, U. C. Mishra, K. C. Pillai, and S. Sadasivan, eds.), pp. 565–574. Wiley Eastern, Bombay/New Delhi.

Williams, A. R. (1982). Biological uptake and transfer of ^{226}Ra: A review. *Environ. Migr. Long-Lived Radionuclides, Proc. Int. Symp., Knoxville, Tenn., 1981* IAEA-SM-257/92.

Wilson, D. W., Ward, G. M., and Johnson, J. E. (1969). A quantitative model of the transport of Cs-137 from fallout to milk. *Environ. Contam. Radioact. Mater., Proc. Semin., Vienna.*

Wilson, J. W. (1981). Solar radiation monitoring for high altitude aircraft. *Health Phys.* **41**, 607–617.

Wischmeier, W. H., and Smith, D. O. (1978). "Predicting Rainfall Erosion Losses—A Guide to Conservation Planning," USDA Handbook No. 537. USGPO, Washington, D.C.

Wittels, M. C. (1966). Stored energy in graphite and other reactor materials. *Nucl Saf.* **8**, 134.

Wolff, T. A. (1984). "Transportation of Nuclear Materials," Rep. SAND 84-0062. Sandia Natl. Lab., Albuquerque, New Mexico.

Wollenberg, H. A., and Smith, A. R. (1966a). Radioactivity of cement raw materials. U.S. Atomic Energy Commission, Report UCRL-16878.

Wollenberg, H. A., and Smith, A. R. (1966b). A concrete low background counting enclosure. *Health Phys.* **12**, 53–60.

World Health Organization (1986). Working Group on Assessment of Radiation Dose Committment in Europe due to the Chernobyl Accident. ICP/COR 129(S) 5134V, July 22, 1986.

Wrenn, M. E. (1968). The dosimetry of ^{55}Fe. *Proc. Congr. Radiat. Prot., 1st* pp. 843–850. Pergamon, Oxford.

Wrenn, M. E., and Cohen, N. (1967). Iron-55 from nuclear fallout in the blood of adults:

Dosimetric implications and development of a model to predict levels in blood. *Health Phys.* **13,** 1075–1082.

Wrenn, M. E., and Cohen, N. (1979). Dosimetric and risk/benefit implications of americium-241 in smoke detectors disposed of in normal wastes. *Proc. Health Phys. Soc. Midyear Top. Symp., 12th, Low-Level Radioact. Waste Manage.* Rep. EPA 520/3-79-002. Off. Radiat. Programs, USEPA, Washington, D.C.

Wrenn, M. E., Mowafy, R., and Laurer, G. R. (1964). ^{95}Zr-^{95}Nb in human lungs from fallout. *Health Phys.* **10,** 1051–1058.

Wrenn, M. E., Lentsch, J. W., Eisenbud, M., Lauer, J., and Howells, G. P. (1972). Radiocesium distribution in water, sediment, and biota in the Hudson River estuary from 1964 through 1970. *Proc. Natl. Symp. Radioecol., 3rd* CONF-710501-P2, pp. 752–768. USAEC, Washington, D.C.

Wrenn, M. E., Singh, N. P., Cohen, N., Ibrahim, S. A., and Saccomanno, G. (1981). "Thorium in Human Tissues," Rep. NUREG/CR-1227. NTIS, Springfield, Virginia.

Wrenn, M. E., Durbin, P. W., Howard, B., Lipszstein, J., Rundo, J., Still, E. T., and Willis, D. L. (1985). Metabolism of ingested U and Ra. *Health Phys.* **48,** 601–633.

Wyckoff, H. O. (1980). "From 'Quantity of Radiation' and 'Dose' to 'Exposure' and 'Absorbed Dose'—An Historical Review," NCRP Taylor Lecture No. 4. Natl. Counc. Radiat. Prot. Meas., Bethesda, Maryland. (Also in *Proc. Annu. Meet., Quant. Risk Stand. Setting. 16th* Natl. Counc. Radiat. Prot. Meas., Bethesda, Maryland.)

Wyerman, T. A., Farnsworth, R. K., and Stewart, G. L. (1970). Tritium in streams in the United States, 1961–1968. *Radiol. Health Data Rep.* **11,** 421–439.

Yamagata, N., and Yamagata, T. (1960). The concentration of cesium-137 in human tissues and organs. *Natl. Inst. Public Health (Jpn.)* **9,** 72.

Zach, R., and Mayoh, K. R. (1984). Soil ingestion by cattle: A neglected pathway. *Health Phys.* **46,** 426–431.

Zapp, F. C. (1969). Testing of containment systems used with light-water-cooled power reactors. *Nucl. Saf.* **10,** 308–315.

Zelle, M. R. (1960). Radioisotopes and the genetic mechanism: Mutagenic aspects. *In* "Radioisotopes in the Biosphere," pp. 160–180. Univ. of Minnesota Press, Minneapolis.

Index